HYPERMEDIA SEDUCTION FOR TERRORIST RECRUITING

NATO Science for Peace and Security Series

This Series presents the results of scientific meetings supported under the NATO Programme: Science for Peace and Security (SPS).

The NATO SPS Programme supports meetings in the following Key Priority areas: (1) Defence Against Terrorism; (2) Countering other Threats to Security and (3) NATO, Partner and Mediterranean Dialogue Country Priorities. The types of meeting supported are generally "Advanced Study Institutes" and "Advanced Research Workshops". The NATO SPS Series collects together the results of these meetings. The meetings are co-organized by scientists from NATO countries and scientists from NATO's "Partner" or "Mediterranean Dialogue" countries. The observations and recommendations made at the meetings, as well as the contents of the volumes in the Series, reflect those of participants and contributors only; they should not necessarily be regarded as reflecting NATO views or policy.

Advanced Study Institutes (ASI) are high-level tutorial courses to convey the latest developments in a subject to an advanced-level audience.

Advanced Research Workshops (ARW) are expert meetings where an intense but informal exchange of views at the frontiers of a subject aims at identifying directions for future action.

Following a transformation of the programme in 2006 the Series has been re-named and re-organised. Recent volumes on topics not related to security, which result from meetings supported under the programme earlier, may be found in the NATO Science Series.

The Series is published by IOS Press, Amsterdam, and Springer Science and Business Media, Dordrecht, in conjunction with the NATO Public Diplomacy Division.

Sub-Series

A.	Chemistry and Biology	Springer Science and Business Media
B.	Physics and Biophysics	Springer Science and Business Media
C.	Environmental Security	Springer Science and Business Media
D.	Information and Communication Security	IOS Press
E.	Human and Societal Dynamics	IOS Press

http://www.nato.int/science
http://www.springer.com
http://www.iospress.nl

Sub-Series E: Human and Societal Dynamics – Vol. 25 ISSN 1874-6276

Hypermedia Seduction for Terrorist Recruiting

Edited by

Boaz Ganor

Founder and Executive Director of the International Institute for Counter-Terrorism (ICT) and deputy dean of the Lauder school of Government and Diplomacy at the Interdisciplinary Center Herzliya, Israel

Katharina Von Knop

Adjunct Professor, Program on Terrorism and Security Studies George C. Marshall Center, Germany

and

Carlos Duarte

Associated Professor, IADE – Design School and researcher member of UNIDCOM/IADE, Lisbon, Portugal

IOS Press

Amsterdam • Berlin • Oxford • Tokyo • Washington, DC

Published in cooperation with NATO Public Diplomacy Division

Proceedings of the NATO Advanced Research Workshop on Hypermedia Seduction for Terrorist Recruiting
Eilat, Israel
17–21 September 2006

ISBN 978-1-58603-761-1
Library of Congress Control Number: 2007932354

Publisher
IOS Press
Nieuwe Hemweg 6B
1013 BG Amsterdam
Netherlands
fax: +31 20 687 0019
e-mail: order@iospress.nl

Distributor in the UK and Ireland
Gazelle Books Services Ltd.
White Cross Mills
Hightown
Lancaster LA1 4XS
United Kingdom
fax: +44 1524 63232
e-mail: sales@gazellebooks.co.uk

Distributor in the USA and Canada
IOS Press, Inc.
4502 Rachael Manor Drive
Fairfax, VA 22032
USA
fax: +1 703 323 3668
e-mail: iosbooks@iospress.com

PRINTED IN THE NETHERLANDS

Hypermedia Seduction for Terrorist Recruiting
B. Ganor et al. (Eds.)
IOS Press, 2007

Preface

This book includes the main lectures given in the framework of the NATO-sponsored Advanced Research Workshop on Hypermedia Seduction for Terrorist Recruiting, held 17–21 September 2006 in Eilat, Israel. The workshop brought together researchers from diverse disciplines for five days of intensive discussions. Participants included experts on security, Islamic studies, design and marketing, and psychology from different countries including Israel, USA, England, Russia, Germany, Portugal, the Netherlands, Denmark, Italy, India and Ukraine.

The workshop's primary objective was to understand the efficacy of terrorists' use of the Internet and the scope of this activity, and to analyze the content, graphics and source code. Topics covered included the use of the Internet for psychological warfare in general and for terrorist "narrow-casting" to specific audiences. The discussion focused in particular on the Global Jihad. Terrorist websites were analyzed in terms of common graphical and linguistic motifs. Different methodologies for targeting different audiences were discussed, including the increasing outreach on the part of extremist groups to young children in both Muslim countries and the West. Participants learned how online communities evolve, and how membership in an online community can promote isolation and radicalization, particularly among immigrant societies. The psychological side of these communities was also dealt with, including extremists' manipulation of the younger members of immigrant communities.

The ultimate goal of this workshop was not only to evaluate the effectiveness of terrorist propaganda on the web and its influence on the target audience, but to recommend practical steps that can be taken to counter this effectiveness. This included the exploitation of vulnerabilities resulting from jihadists' behavior online, and the analysis of jihadist forums and websites for intelligence gathering.

The meeting concluded with an outline of a working plan for countering online recruitment by terrorist organizations. Participants agreed that in order to do this, it is important to target the not-yet-persuaded potential recruits—in particular the younger generation, who are the primary targets of extremist online recruitment. There are phases in the radicalization process when people are most susceptible to manipulation; we need to aim our own counter-radicalization message at those who are at this phase. The psychology of this is very important: extremists exploit the need for belonging, the need for roots and authenticity, the "angry young man" syndrome.

We learned that there is a whole "science of seduction" that we can use to our advantage to counter the extremists' messages. To put this to good use, we need to understand our audience and learn to identify with them enough to know what messages are effective. Branding is an effective way of marketing alternative viewpoints, but first we need to arrive at a common lexicon and a common sense of our own identity, in order to know what it is we're selling.

At the same time, the extremists themselves, in their discussions in the public space of the Internet, give us some hints of their own vulnerabilities. There are plenty of things that we can do in cyberspace to hamper their operations. Some of these possibilities are discussed in the lectures that follow.

To counter the extremists' message, we will have to understand that we are competing for the same audience that they are. We need to learn to operate in the same spheres in which the extremists themselves operate. At the same time, this must be carried out in a sensitive and well thought-out manner, in order to avoid alienating the very audience that we would like to affect. For this and other reasons, it was decided that the best agents for countering extremists on the web are not governments but private individuals, in particular, those in the communities most at risk of extremist propaganda.

Overall, the workshop provided a lively and stimulating venue for discussing one of today's most pressing issues. Moreover, the discussions helped to lay the foundation for future work in this important area.

The Editors

Contents

The Seductive Web: Technology as a Tool for Persuasion

The Seductive Web: Terrorism, Media and Psychological Effects

Understanding Symbolism in Islamic Fundamentalism and the Imagery of Jihad

Hypermedia Seduction for Terrorist Recruiting
B. Ganor et al. (Eds.)
IOS Press, 2007

Terrorism: Perception, Illusion and the Art of Political Warfare

Peter S. PROBST
Terrorism Consultant

Abstract. Terrorism is effective because we don't understand it and, too often, treat it as either a criminal offense or a form of conventional warfare. We do not address it as a form of asymmetric conflict and psychological warfare, but instead try to combat it with traditional military tactics. Too often, these merely play into terrorist hands. By adopting the language and mindset of conventional warfare, we constrain our thinking, and, by extension, our options. As my father used to say: "When the only tool you have is a hammer, everything looks like a nail." To be successful, we must expand our operational tool kit, with heavy emphasis on PsyOps and Covert Action. These are critical if we are to be successful in the protracted conflicted in which we are currently engaged.

Keywords. Psychological warfare, political warfare, asymmetric conflict, perception management, *Protocols*, propaganda, conspiracy theories, media, nation-building,

1. Misperception and Mindset

Reality, like beauty, is in the eye of the beholder. The shaping of that reality to your own advantage is, in essence, the goal of psychological warfare. Reality, whether we like to admit it or not, is subjective and malleable. It is filtered through a variety of lenses that include culture, history, personal experience, mythology, and fears, as well as a host of other factors which may be primal and specific to the individual (or to his or her community and culture).

Terrorism is effective because we don't understand it, and we all too often treat it either as a criminal offense or as a form of conventional warfare. We don't address it as we should, as a form of asymmetric conflict and psychological warfare. Instead, we try to combat it with traditional military tactics, which too often redound to the benefit of the terrorists. By adopting the language and mindset of conventional warfare, we reinforce that mindset, and interpret events through the lens of conventional military thinking. Most often, this is counterproductive when dealing with asymmetric forms of conflict, such as insurgency and terrorism. The rigidity of this approach discourages innovation and non-traditional thinking. As my father used to say, "When the only tool you have is a hammer, everything looks like a nail."

Some years ago, there was a very fine Japanese movie called *Rashomon*, in which four characters witnessed a rape and murder, and faithfully reported what they believed they had seen. No two stories matched. In fact, they were at great variance. Seasoned investigators know that eye-witness testimony is highly subjective and often unreliable. *Rashomon* adroitly demonstrated this fact. The mind is not a camera. It does not record and "freeze" objective reality. Memories are constantly being resorted, modified and influenced, by any number of mechanisms and events that are too complex to address in this forum.

Perception, too, is highly malleable. Perception management is the stock and trade of the political operative, whether he works for a legitimate political party, an intelligence service, or a terrorist group. In some instances, perception management can be used for good, such as defusing mass panic in the face of a natural or other disaster. In such instances, the information provided must be factual, because if it is exposed as a lie, public confidence in governmental authority drops like a stone. Thereafter, the next government pronouncement is likely to be met with doubt and derision.

2. Old Messages, New Faces

Increasingly, governments, political parties, special interest groups and militant extremists all depend on the services of "spinmeisters". Spinmeisters can take an event or fact that well may be true, but present the public with an interpretation that plays to their political advantage while simultaneously destroying the credibility of an opponent. To be successful, such a narrative must resonate and play to deeply rooted fears, myths and beliefs. At its most effective, it strikes a primal chord.

The interpretative narrative is tailored to each particular target audience, and reinforces a human predisposition to believe in things. It dovetails and reinforces preconceived notions, suspicions, fears, and worldviews, together with the target group's particular interpretation of history, their view of the adversary, and their particular cultural and historical mythologies. The story is fine-tuned to resonate with the collective consciousness.

One of the most effective psychological operations ever is based on a proven forgery. It was initiated in the late 1890s or early 1900s by the Tsarist secret police, the Okhranka [1]. The Okhranka widely circulated the *Protocols of the Learned Elders of Zion,* a document that purported to be the Jewish plan for world conquest. It found a ready audience not only in Russia but also in the Europe of that day.

The *Protocols* have subsequently been successfully used by the Tsars, the Communists, and the Fascists. Today it is republished and circulated by Arab governments and Muslim extremists. It is even a bestseller in Japan, which has only a miniscule Jewish population—in other words, the phenomenon of anti-Semitism without Jews.

The *Protocols* have been so effective and so long-lived precisely because they reinforce, via purported documentary "evidence," deeply held beliefs and fears. The document provides a ready explanation for events that are complex, difficult to understand, and often at variance with the target population's traditional belief system and political mythology. The *Protocols* play on the fears and prejudices which many people absorbed in their earliest childhoods. The document has tremendous impact because it seems to provide independent confirmation and simple, apparently logical explanations of events that are incredibly complex and frightening, and over which the individual has little or no control. The use of the *Protocols* as a propaganda weapon continues today.

The Charter of Hamas explicitly refers to the *Protocols,* and promotes it as the secret plan of the Jews to achieve world domination. Article 32 of the Hamas Charter states: "The Zionist plan is limitless. After Palestine, the Zionists aspire to expand from the Nile to the Euphrates. When they will have digested the region they overtook, they will aspire to further expansion, and so on. Their plan is embodied in the *Protocols of the Elders of Zion*, and their present conduct is the best proof of what we are saying." [2]

The Palestinian National Authority (PNA) controlled media frequently refers to the *Protocols,* and it is taught in many of the madrassas. The 25 January 2001 edition of *Al-Hayat al-Jadida,* the official PNA daily newspaper, referred to the *Protocols* on its "Political National Education" page as follows: "Disinformation has been one of the bases of morale [sic] and psychological manipulation among the Israelis... The *Protocols of the Elders of Zion* did not ignore the importance of using propaganda to promote the Zionist goals."

Later that year, the same newspaper published the following statement: "The purpose of the military policy is to impose this situation on the residents and force them to leave their homes, and this is done in the framework of the *Protocols of Zion...*" [3].

The Grand Mufti of Jerusalem, Shayk Ekrima Sa'id Sabri, appeared on the Saudi satellite channel *Al-Majd* on 20 February 2005. Commenting on the assassination of former Lebanese Prime Minister, Rafik Hariri, he stated: "Anyone who studies *The Protocols of the Elders of Zion* and specifically the Talmud will discover that one of the goals of these *Protocols* is to cause confusion in the world and to undermine security throughout the world."

On 19 May 2005, the *New York Times* reported that the Palestinian National Authority's Minister of Information, Nabil Shaath, had removed an Arabic translation of the *Protocols* from the ministry's website.

The bottom line is that an effective propaganda product, even when repeatedly discredited, can retain its power for centuries. It may continue to do immense damage, even when used by people and governments with diametrically opposed ideologies and interests.

3. The Nature of Psychological Warfare

Psychological warfare during times of conflict may be carried out on both the tactical and strategic levels. The more sophisticated operations will be run as a campaign, the broad objectives of which are supported by complementary and mutually supportive tactical operations. Such operations are tailored to influence specific target audiences, and they are crafted to produce a particular set of behaviors. The tactician seeks to tap into primal psychological needs and deeply held belief systems, which to a large extent are molded by the culture, history, fears and expectations of the targeted groups.

Terrorism is one facet of what the British call political warfare, which is essentially asymmetric warfare on the strategic level, and in which psychological operations play a major role. Other tactics that are used in political warfare include political action, disinformation, media operations, agents of influence, black operations, and direct action.

A terrorist group will often oscillate between emphasizing political operations and emphasizing terrorist operations. To be most effective, "political action" activities and acts of terrorism must be complementary and mutually supporting, and the emphasis will depend on the reality on the ground, as interpreted by the terrorist group.

In our struggle against terrorism, we focus almost exclusively on the carnage, the inherent drama and the horror of unspeakable acts. But violence is only one arrow in the terrorist's operational quiver. Of equal, and at times indeed very great, importance are the complementary political and psychological operations. These make use of conventional media, the Internet, leaflets, rumor and wall graffiti as means to disseminate the message.

The primary aim of terrorism is to immobilize the target groups and foster panic, despair and, ultimately, psychological surrender. The aim of the concomitant political and psychological operations is to convince target groups that the terrorists' acts are justified, and that to resist them is to court death. Media serves as the transmission belt for the message. This includes not only print media, radio and television, but also the Internet, with its panoply of sites, chat room, bulletin boards and blogs.

Some of the most potent forms of information comprise rumor and conspiracy theory. When people are frightened and doubt the veracity of conventional sources of information, rumors and conspiracy theories fill the void. This is particularly true in countries where the press is government-controlled or heavily censored, with few outlets for dissenting opinion or opposing views. The more outlandish and convoluted the rumor, the more likely it is to be believed. Effective rumors play on primal fears, and all too often those fears overpower reason.

The Arab Street is often viewed as the quintessential rumor mill, but Europe and the United States do not lag far behind. In America, stories still circulate about mysterious black helicopters and shadowy "men-in-black". This is no movie, but is a "reality" for many who have bought into the belief that a malignant United Nations is geared to invade the country and impose a one-world government.

In the United States, "talk radio" is of particular importance. These are radio programs that feature verbal commentary; they are run by radio hosts who frequently have a flamboyant image and a crass or sarcastic manner. The hosts flaunt language and views that are often anything but politically correct, and the programs themselves are openly partisan and often inflammatory. Each one advocates a particular brand of politics and a particular political and social agenda. These programs tend to play to the darker emotions and beliefs, and seek to inflame and legitimize these.

Today, in many parts of the Gulf Coast that were ravished by Hurricane Katrina, people believe that the dikes protecting New Orleans were purposely breached or bombed by the white establishment. The purported motive was the destruction of historically black neighborhoods, to make room for white developers. Another story that circulates in our black communities is that AIDS was developed by the American military to destroy America's black population.

In Mexico and Central America, particularly in the impoverished rural areas and urban slums, many believe that Americans are adopting Latin American babies so as to sell their body parts to American laboratories, and to wealthy people who need organ transplants. This theme was originally spread by the Russian Intelligence Service, and over the years has developed a life of its own.

When you stop and think about it, do such stories sound any less believable than if a friend had told you—prior to the apprehension of Richard Reid—that somewhere in London there was a large black man with a bomb in his shoe, who was planning to blow up an airliner? One person's reality is another person's paranoia.

Wall graffiti are another extremely effective medium for spreading the message. For those who live in neighborhoods where graffiti sprout like mushrooms after the rain, such messages are, to the pedestrian, almost like an RSS news-feed scrolling at the bottom of your computer screen. After a while, you don't even read them, but they influence you on an almost subliminal level.

In some countries and neighborhoods, such fear-arousing slogans appear on almost every block, conveying the message that a terrorist group is omnipresent—and, perhaps, omnipotent. I remember walking the streets of Belfast and marveling at the graffiti posted by both the Provisional Irish Republican Army (PIRA) and the rival Protestant Paramilitaries. These were not simple slogans but entire wall murals, works of art depicting important operations and events, as well as portraits of martyrs. The effect is extremely powerful and emotionally evocative, even to an outsider.

4. The Message and the Medium

A seasoned psychological warrior will choose the medium that will be most accepted and trusted by his target. If he is targeting intellectuals or the political elite, he will seek to plant his disinformation in respected newspapers and magazines. If he lives in the United States, he will also aim for the respected Sunday morning news and interview programs. If his target audience is "blue collar" and somewhat less educated, he might choose talk radio. And increasingly, for the younger audience, he will choose the Internet, which has almost a counter-culture appeal.

More conventional outlets are the church or mosque. Pronouncements by the priest or imam carry the weight of religious dictum, and are often viewed as coming from the mouth of God. This is why in Soviet Russia, the secret police kept such a tight reign on the Russian Orthodox Church. Its priests and bishops were co-opted by the state, and served not only as active intelligence sources but also as disseminators of state propaganda and disinformation. The radical mosque today similarly serves to indoctrinate and motivate. But it has an additional function, in that it provides a venue for identifying, assessing and recruiting potential jihadists. As in the Soviet example, it

5. The Power of Language

Terrorists are particularly adroit in their use of language, and have been successful in persuading the media, much of the public and even some government officials to adopt their lexicon, frames of reference and justifications for their actions. Terrorists describe the savage murders of social workers, news reporters, business people, teachers, civil servants and even religious leaders as "executions". Thus they convey the impression that the killings are somehow legitimate and follow due process, and adhere to the customary laws of war—or are at least mandated by a higher religious precept or authority.

They similarly describe their own captured hit men as "political prisoners" or "prisoners-of-war", in an effort to legitimize their crimes. Gunmen who terrorize peaceful citizens with killing, kidnapping or torture are not called murderers or criminals, but are lionized as "freedom fighters" or "martyrs".

Too often, out of sheer disgust, and, perhaps, intellectual laziness, we dismiss terrorists as "thugs and murderers". This tendency is understandable, but dangerous. It sells the terrorists short, and leaves the public lacking even a basic understanding of the essentially political nature of the conflict. This failure to educate target audiences leaves them vulnerable to terrorist propaganda and the sophistries of terrorist supporters.

There is no better example of how such sophistry sells than the widely embraced platitude "One man's terrorist is another man's freedom fighter." Terrorists are not freedom fighters, nor do they constitute a legitimate guerrilla force. Terrorists deliberately target non-combatants to shock and intimidate. They target the innocent: the school teacher, the editor, the political leader, the mother and child. Freedom fighters or guerrillas, in contrast, operate against legitimate military targets, and in the same way as conventional military forces they are bound by the Geneva Conventions and the Customary Laws of War. For freedom fighters, violations of these standards are deplorable anomalies. For terrorists, such violations are acts to be celebrated, and the perpetrators thereof lionized.

In order to successfully fight the terrorists, it is imperative to promote a clearer understanding of how they operate, so that target audiences can better understand their tactics and strategies and learn to resist their psychological blandishments and rampant intimidation. It is particularly important to aggressively counter pernicious efforts to establish a moral equivalence between operations undertaken by democratic states and those launched by terrorists or terrorist groups. Part of this effort must include developing a better understanding of the power of the word, and the role of nuance.

6. Media Co-opted

Unfortunately, Western media is often complicit in adopting the vocabulary of the terrorists, and thereby promoting their view of the struggle and their justifications for their actions. Words have power. Words can inspire people to kill or even to turn themselves into human bombs. Even the most well-meaning individual can be seduced into adopting the terrorists' frame of reference and their ostensible agenda.

Once you begin using your adversary's vocabulary, you have tacitly adopted his frame of reference, and are on the road to legitimizing his agenda. Terrorists are characterized in much of the media as "fighters," such as in "Hezbollah fighters" or "Hamas fighters". And terrorist bands and movements are called insurgents, a name which implicitly provides them with inherent legitimacy codified in international law.

Reporters who are covering a conflict often adopt the terrorist's lexicon so as to curry favor, in the hopes that they will not be targeted, as well as to maintain access to terrorist sources and sympathizers. But words have power. There is a world of difference between describing an individual as a "Hezbollah fighter" or, in contrast, as a "Hezbollah terrorist". Too often, reporters effectively become an important extension of the terrorists' propaganda operation. Such reporters try to justify their approach as an effort to provide "balanced coverage." It simply does not wash.

When terrorists hide their rocket launchers or anti-aircraft missiles on the grounds of hospitals or schools, or in civilian neighborhoods, they are committing a gross violation of the Geneva Convention. In effect, they are holding the civilian population hostage. But this point is rarely noted by Western media.

A photograph of civilian casualties and the dead is essentially neutral. It is how the photograph is captioned, and the accompanying narrative, that are important. If the photograph shows a picture of a grieving mother holding her dead infant and attributes their deaths to an American bomb or rocket, the reader's response is readily predictable. If, however, the mother and child are described as victims of the terrorist practice of launching their rocket and artillery attacks from within densely populated civilian areas—home to many women and children—the reader's reaction will be very different.

The Western media all too often focuses on the deaths of innocent civilians—complete with gruesome pictures—and report those deaths as the result of American bombs or Israeli rockets. They ignore the fact that the civilian population was deliberately put in jeopardy so their deaths could provide grist for the terrorists' propaganda mill.

Thus, Western governments are twice victimized. Firstly, by the terrorist rocket attack itself, and secondly by the adverse publicity that the necessary counterattack generates. We ran into this problem in Vietnam. Israel is faced with the same set of issues in Lebanon. The appropriate response to such an issue is a generous dose of truth and education.

An interesting example of the power of captioning is illustrated by an Associated Press photograph that appeared in the 30 September 2000 edition of *The New York Times.* It was reprinted in many other publications. It showed a snarling Israeli policeman standing over a bloodied teenager, and was captioned "An Israeli Policeman and a Palestinian on the Temple Mount." The picture was an accurate reflection of reality. The caption was not. The youth was not Palestinian, but was instead a Jewish student from Chicago. And the Israeli policeman had not attacked him, as was the implied by the caption; quite the opposite. The policeman was protecting the student from a Palestinian mob that had pulled him from a cab and beaten and stabbed him.

When it comes to news photographs or video footage, everything depends on the caption and the accompanying narrative. The picture elicits powerful emotions, and the caption focuses and directs those emotions. Captions identify the hero, the villain and the victim. The reader takes it from there.

7. Exploitation of Media

Intelligence services that engage in Media Operations will often pay reporters to slant their stories, or to insert a particular fact or fabrication that the service wants disseminated. Some intelligence services will pay a reporter to write a particular headline and lead paragraph, and then have the reporter graft this press placement onto a completely unrelated story that the reporter may be preparing. I have seen instances of very incendiary headlines and lead paragraphs, but with the body of the story being devoted to a completely different topic, such as advances in medicine or the legacy of the pre-Columbian Olmec civilization.

This kind of operation is based on the premise that most people read only the headline and lead paragraph of a story, and then page to the sports section or their horoscope. It is a cynical approach, but apparently it works. It is also cost effective, which is important for a service with a tight budget.

A key issue is how best to get your message out, particularly if the traditional media and wire services tend to be more sympathetic to your adversary. There are any number of lessons from the Cold War that are transferable—with a bit of tailoring—to the problems we face today. The Soviets relied heavily on human nature. Reporters, like most of us, tend to be lazy. So to make their job easier, the story was prepared and tailored to generally conform to the language, slant and writing style of the individual reporter and the particular news outlet. Interwoven in the text, however, was the message or theme that the Soviets wanted to promote.

In most cases a cash payment was made for such services, and similar payments or gifts might be made up through the newspaper hierarchy, to ensure that the piece was published, as well as to foster continued access. Having the article appear in a known communist publication was acceptable, but in essence it meant preaching to the converted. How much better to tailor the message so that it could be published in a socialist, liberal, or even mainstream publication.

Mainstream and anti-communist media were something of the gold standard, as they provided credibility and access to an audience that media sympathetic to the Soviets could never provide. Monthly payments or, perhaps, special favors for the newspaper's publisher and editor would often open the door. These could take the form of trips to conferences in Paris or Moscow, and the bestowing of honorific titles and awards, complete with media coverage and accommodation in luxury resorts. In the world of psychological operations, it is important to remember that money isn't the only currency. Access and power are often more important. If you have these, the money will follow.

If the above tactics did not work, the Soviets might buy an interest in the newspaper or radio station through their front men. If this also failed, they could always approach a competing outlet that had a similar political orientation. If that did not work, they could, as a last resort, establish a "competing" news organ, which, although fully controlled in reality, would attempt to portray itself as independent but progressive.

They might also seek to silence critics through blackmail. Some prominent reporters and at least one well known publisher were compromised in this fashion, having been photographed engaging in extramarital or homosexual affairs. The more prominent individuals were not asked to reverse their well-known political positions. They did, however, significantly curtail their hard-edged criticism, and provided what their Soviet contacts called more "balanced" coverage.

8. The Rhythm of Terrorism

People have often commented on the ebb and flow of terrorist operations, but often do not understand the reason behind this. Sometimes lessening in ops tempo may be due to increased government success, but at other times it may be part of a clever and productive political and psychological strategy.

It is a bit like the interrogation technique of "good cop/bad cop", which is the staple fare of American police dramas. This is a common interrogation technique, in which one officer presents a sympathetic and supportive facade when interrogating a prisoner, whereas his partner, in contrast, is openly contemptuous and brutal. The aim is to lower the psychological resistance of the suspect, in anticipation that one or the other officer will obtain a confession.

In a somewhat similar vein, a group may carry out heinous acts of terrorism, and, when they feel they have gained all they can or that the target is vulnerable, they will then propose a truce and negotiations. Once the gains from the negotiations have been exhausted, the group may then revert to terrorism. In this manner they whip-saw the targeted government and the civilian population between periods of optimism and periods of dread. The ultimate result is the psychological exhaustion of their enemies, and the loss of the will to resist. Often, only when the targeted government and citizens believe that psychological surrender will result in their annihilation do they find the strength to soldier on.

9. Counter-Operations

Terrorist groups, too, are particularly vulnerable to psychological operations. For example, one weakness that is characteristic of virtually all clandestine organizations is virulent paranoia. Through a mix of psychological and black operations, this paranoia can be exacerbated to create an atmosphere so poisoned by suspicion and distrust that the group turns inward on itself, and essentially self-destructs.

A less dramatic outcome may be that the organization spends much time and effort seeking to uncover non-existent spies and informers, and therefore has less time and fewer resources to spend terrorizing the targeted government and civilian population.

Another approach is to promote suspicion and distrust among the terrorist leadership. For example, if the second-in-command of a clandestine organization is particularly able and charismatic, it may be possible to develop and nurture a rivalry between him and the leader where none existed before. An article may, for example, be planted in a Third World newspaper or on the Internet, stating that the leader of the group is basically a figurehead, and that the real brains of the operation is his loyal subordinate, who essentially keeps him on track and provides both the operational and ideological inspiration that has made the organization a success.

Such an article might just be enough to plant concerns in the mind of the leader, who may worry that such praise will fuel his subordinate's ambitions. Another completely separate media outlet might, in a follow-on complementary story, speculate that the subordinate resents the leader's rise to prominence. The reason cited for this supposed resentment might be that the leader's success rests on the subordinate's hard work and talent, and the subordinate believes he has not received sufficient recognition or reward. A third placement might suggest that the leader is jealous of his subordinate's abilities, and plans to marginalize him or even remove him from the scene.

Once such placements are published, and especially if they contain some juicy (and preferably true) tidbits, it is likely the story will be picked up by mainstream media. Then it will be circulated within the region, or perhaps even on a wider basis. At each iteration, more detail and variations on the theme may be added, some of which may be true and others total fabrication. Placement in multiple outlets provides seemingly independent confirmation that further enhances the credibility of the story. This is basically a variation on Shakespeare's "Othello", in which both Othello and Desdemona are destroyed through a psychological operation carried out by the villain, Iago.

An interesting thing about such operations is that they can be based on minimal intelligence. Their credibility rests primarily on inference, the inherent fear of betrayal that permeates most clandestine groups, and perhaps a few nuggets of factual information—gleaned by the intelligence service. All of these can be used to enhance credulity in the minds of the prime targets.

In the above hypothetical example, even if both of the individual terrorist targets believe the stories to be the work of a foreign intelligence service, suspicion lingers. This suspicion is often fed by the concerns of loyal supporters. In such a scenario, there are also the unanswered questions: Why do our adversaries believe my compatriot could be successfully manipulated? What flaws of character or intellect do they see in me?

Thus, even if both targets know they are being manipulated, they may fear their ostensible rival might not be so astute and may thus be prone to acting on the disinformation. This further feeds the paranoia, the end result being that there is significant impetus to play it safe and strike first. The Soviets were masters of the game, and of this type of operation.

A variation on the theme might be to publicly promote a total incompetent, thus assisting him in reaching a position of power within the terrorist organization. The aim here is twofold. Firstly, far more capable individuals are kept out of the limelight and are not equally promoted to power. Secondly, such operations are carried out in the hope that the person whose career is being promoted will so severely mismanage his responsibilities that his bungling will damage the organization. This would promote internal dissension, further diminishing the group's effectiveness.

10. Nation Building as Political Warfare

Political warfare carried out by a government may include operations that range from the most benign nation building activities, designed to win popular support, to some extremely hard-edged paramilitary operations. Soft operations such as nation building most often constitute an integral component of political warfare. Hezbollah is master of this art form, namely benign operations designed to win the loyalty of target groups—or at least to coerce their neutrality.

There is, however, a much less benign complement, which may extend across the spectrum of violence. This component of terrorism is designed for maximum psychological impact: to cow the opposition, to intimidate the more moderate, and to serve as a constant reminder to supporters and the opposition alike that the price of defection or opposition is death. If you cooperate, you and your family will benefit on any number of different levels. If you resist, you will die.

Government-sponsored nation building efforts and reconstruction programs are very vulnerable to disruption. Unless the government can first provide security, nation building will not work. Security is key. The terrorists have a much easier job than the government. All they need to do is to assassinate aid workers, randomly bomb newly constructed civic works, and threaten or kill those who participate in the rebuilding effort.

When government programs are curtailed or non-existent, the civilian population begins to looks to the terrorists to provide the services that would normally be the province of the central government. This is the Stockholm Syndrome on a communal level. In essence, the terrorists out-govern the government, and thereby acquire legitimacy and accrue political power. Such activities also bolster the terrorist argument that they are the only organization that truly cares for the people. They are on the ground working daily with the populace, whereas the central government rarely maintains an effective presence in the rural areas or the urban slums.

The terrorists begin to be viewed as a shadow or alternative government, and often begin to pursue a complementary political strategy by forming a parallel political movement or party. They portray the government as weak and corrupt which, all too often, is in fact the case. The terrorists' political and civic activities teach by example. So do the activities of the government, but often theirs is the wrong lesson.

Only authoritarian governments can successfully use terror tactics over the long term against their adversaries. Established democracies and constitutional governments are much more restricted in their choice of options, as terror tactics violate the traditions and beliefs of their constituents, and the ideological underpinnings that provide the government its legitimacy. Democratic governments are held to a much higher standard than their insurgent or terrorist adversaries, by both the international community and their own citizens. Therefore, they must be more clever and cannot rely on brute force alone. Their operations cannot violate the collective conscience, but must be much more nuanced and surgical in nature.

11. Conclusion

What we are engaged in is not a war in a conventional sense, but rather a protracted conflict. Our adversaries view this protracted conflict as a continuation of their campaign against the Crusaders—now led by the United States and Israel. They take the long-range view, which is a perspective we are only now beginning to adopt.

This is not a conflict that can be won with bombs, guns and technology. It is a most dangerous form of conflict, because our adversaries are motivated by unquestioned religious fervor, which has the demonstrated ability to motivate people to carry out suicide attacks and other unspeakable cruelties. Religious wars are usually the most brutal, the most cruel and the most unforgiving. And now, these same people have access to the requisite knowledge and skills for building Improvised Weapons of Mass Destruction.

Fighting terrorism is not a game for impatient people. One must carefully think through second and third order consequences. It is a bit like three-dimensional chess. Good intelligence, tenacity, an engaged public, uncommon courage and disciplined restraint are required. The way we choose to carry out this endeavor will largely determine the winners, the losers and the price paid by each.

References

[1] Wikipedia, "The Protocols of the Elders of Zion". Accessed February 2007.
[2] Hamas Charter, available at http://www.yale.edu/lawweb/avalon/mideast/hamas.htm. Accessed February 2007.
[3] *Al-Hayat al-Jadida*, 18 November 2001.

Hypermedia Seduction for Terrorist Recruiting
B. Ganor et al. (Eds.)
IOS Press, 2007

Islam & Political symbolism – Hezbollah as a case study

Dr. Eitan AZANI
Deputy Executive Director at the International Institute for Counter-Terrorism (ICT) at the Interdisciplinary Center in Herzliya

Abstract The Shiites make extensive use of shaping new symbols and adding new meaning to old traditions as a meaningful means for social mobilization, broadening their support base, recruitment of new activists and the encouragement of proactive actions in order to achieve the organization's interests. The internet is one of the most significant platforms for Shiite terrorist groups, especially Hezbollah, to spread their messages and shape public opinion by using existing Shiite symbols and creating new symbols for the promotion of Shiite interests. The article will address this phenomenon by analyzing and discussing concrete examples from Shiite websites.

Background

Hezbollah is a fundamentalist Shiite terrorist group that was founded in the heyday of the Lebanese political system crisis as an expression of regional and intra-Lebanese political and social processes. In the summer of 1982, a group of young religious clerics, who were alumni of the religious colleges in Najef, together with Shiite combatants, founded Hezbollah. They did this with Iranian support and blessing and based on the infrastructure of pre-movement groups, Islamic organizations and institutions that were previously formed amongst the Shiite group and on the fringes of the Amal group during the seventies and the early eighties.

One of the central issues addressed by Hezbollah leaders since the movement's inception has been the way in which the movement should operate in order to broaden its organizational infrastructure and solidify its position as a broad popular movement with roots in the Shiite sect, Lebanon and the regional arena. The strategy they chose is conducting activities in the following four spheres:

A. Religious ideology sphere – the activities in this sphere are intended to enlist the society and instill in them motifs such as religious activism, determination and willingness to make personal sacrifices for the betterment of the nation by using rhetoric and symbolism from the Shiite heritage and world.
B. Social sphere – the goal was to "capture the hearts" of the Shiite population by investing in social projects and taking over the educational system through the generous aid money they received from Iran.
C. Military sphere – the movement's military operations were intended to expel all the foreigners from Lebanon. Its ability to bring about the Israeli

withdrawal from Lebanon in May of 2000 gave it broad support amongst the Shiite population, but also among other ethnic groups, and it became a model of emulation for other terrorist organizations in the Middle-East.

D. Media/awareness sphere – this sphere is characterized by the running of an indoctrination and propaganda campaign in the intra-Lebanese arena alongside psychological warfare against its opponents in the regional and international arenas. The movement uses its media outlets in order to achieve its goals, especially the Al-Mannar television station, Radio Nur, the written press and the movement's websites. At the same time, the movement utilizes the exposure it receives in other media networks in order to disseminate its messages. The movement's leadership understands the importance of the media in the fight over public opinion. It uses the media widely and wisely and invests efforts in developing this component, especially the movement's television network and official websites.

The movement shaped the political Islamic discourse in Lebanon. It created an organizational history laden with figures worthy of emulation, mythology, formulated an ideology and goals based on Khomeini's teachings, defined its objectives in the Lebanese and pan-Islamic arenas, and pledged its allegiance to the Iranian leader Khomeini and his successor khamenei.

The content, subject, and goals of Shiite Islamic discourse are intended to serve radical Islam's vision common to all Shiite and Sunni fundamentalist movements. This vision is based on the idea of the "Umma" (Islamic nation) – establishing the Islamic community of the believers that would unify all Muslims in the world and return Islam to its leading position in the world. The tools for implementing the vision are Jihad (holly war) and Da'awa. Both concepts are taken from classical Islam and express a yearning for the glory days of Islam. The recruitment of the believers for missions of Da'awa and Jihad is done by the religious clerics by exploiting Islamic heritage and the various means of congregation (Friday prayers at the mosque, religious ceremonies, holidays, religious processions and funerals) in order to increase the motivation for enlistment into the organization and the support of its ideas. These capabilities increased exponentially in the internet and electronic communication age, which enable the movement to reach millions of viewers around the world.

From a Shiite Islamic theological point of view, the world is divided into three spheres:

- Dar Al Iman – The realm of belief – the sphere controlled by Shiite belief
- Dar Al Islam – The realm of Islam – the sphere controlled by the Sunni
- Dar Al Harb – The realm of war – the sphere controlled by the infidels and has yet to be subjected to Islam.

Da'awa and Jihad do not contradict each other, rather they support and complete each other. However, while Da'awa is based on non-violent means that are intended to fix the Islamic society through education, preaching and social solidarity, Jihad, on the other hand, aims to achieve its goals through military means.

Shiite rhetoriticians make use of the Shiite heritage based on Khomeini's teachings in order to recruit believers into their ranks. This heritage, which fosters in the believer the feeling of ongoing discrimination against the Shiites by the Sunnis, made a transition from political passivity to political activism in the 20th century. This transition laid the groundwork for the Iranian Islamic revolution in 1979 and the awakening of Shiite activism that was based on Khomeini's teachings.

The use of symbolism is very common in Shiism. The Shiite clerics' special status as the ones who can determine religious law (Ijtihad) and the fact that they are role models for emulation (Marja Taqlid) enabled them to employ the Shiite heritage in order to promote political goals and ideas. The fundamental principles/"pillars" of modern Shiite thought are based on Khomeini's teachings, on which the modern Shiite political array of symbols in Iran and Shiite communities around the world is based. Khomeini's heritage includes the following central components and principles:

A. Legitimization of Shiite activism –The Shiites, who until the time of Khomeini were passive and oppressed due to being a minority in the Islamic world, underwent a change under Khomeini's leadership. They became an active sect that worked to achieve political goals, including through the use of violence. Khomeini called for action by the oppressed (the Shiites) against the oppressors.
B. Delegitimization of corrupt Islamic regimes – this allows for carrying out measures to topple these regimes through all means necessary, including violence.
C. Definition of the enemies – the "Great Satan" is the US and the "Little Satan" is Israel.
D. The principle of individual self-sacrifice – the willingness to give one's life in Jihad against the enemies of Islam in order to further the Islamic community's goals. By setting this principle, Khomeini established the basis on which the phenomenon of suicide bombing developed as a strategic means used by terrorism.
E. The principle of "the rule of the religious cleric" – this principle is unique to radical Shiite thought and is closely followed in Iran and Hezbollah.

These principles are regularly used in the day-to-day rhetoric used by the Shiite religious clerics. They are especially emphasized during times of struggle, in religious ceremonies, and events initiated by the Shiite leadership around the world. The Iranian and Hezbollah leadership work to instill these ideas amongst target audiences in Lebanon and the Shiite diaspora by any means possible. Following are the events, holidays and ceremonies used in order to promote these ideas:

A. The month of Ramadan is used to bring people closer to Islam and increase the support for the struggle against Shiite discrimination.
B. The Ashura procession reenacts the murder of the Shiite Imam Hussein by the hands of the Sunnis. In the past few decades, this procession has been used to encourage Shiite activism and active resistance.
C. World Jerusalem Day – the ceremony was initiated by Khomeini in order to promote the Sunni and Shiite common goal of liberating Jerusalem.
D. It should be noted that, alongside these events, Hezbollah's leadership organizes conferences and conventions in various locations and cultivates an organizational ethos based heroic stories and current models of emulation such as Sheikh Ragheb Harb and Sheikh Abbas Musawi, who are enshrined in the organizational pantheon and their life stories are studied as a means for the encouragement of enlistment into the organization. Moreover, Hezbollah's calendar of events includes organizational and religious memorial dates that are also used in order to recruit potential target audiences into the organization.

The Ashura procession as a means for shaping Shiite activism

The Ashura, which is the tenth day in the month of 'Muharam', marks the day Hussein son of Ali Bin Abu-Taleb, who is considered by Shiites to be Muhammad's successor, was killed. Every year, the Shiites mark the event with mourning processions in which they reenact the battle. Some of the marchers even flog themselves until they bleed as part of the Shiite culture and desire to experience the pain Ali felt during the battle. To the Shiites, the battle symbolizes the essence of the rivalry between Sunnis and Shiites and the fact that the Shiites are deprived because the prophet's heritage was stolen from them.[1] Beginning in the second half of the twentieth century, and especially after the Islamic revolution, the Ashura processions became an important means for shaping a current Shiite activist consciousness. Symbols and events from distant Shiite history were redefined and adapted to the Shiite group's current situation and experience.[2]

The processions were used, and are still used, in order to enthuse the believers and recruit them for the fight against the enemy, who is accused of oppressing the Shiites. Traditionally, these ceremonies tend to be volatile and cause the development of incidents in Lebanon, mass mobilization and the propagation of hatred towards the West in general and the US and Israel particularly (the great Satan and the little Satan). As the years past and as the conflict between the West and Islam intensified, the Ashura procession spread and became more violent. The Ashura procession in 2006 was among the most violent in the past decades. It took place against the backdrop of the ongoing conflicts, the continued American presence in Iraq and the intensifying conflict between radical Sunnis and Shiites in this arena, and the wave of protest in the Muslim world following the cartoons published in the West, which they felt dishonoured the prophet Muhammad and Islam.[3]

---

[1] http://news.bbc.co.uk/1/hi/world/middle_east/1890541.stm

[2] Shaping Shiite activist awareness in Ashura processions was evident with Khomeini both before and after the revolution. The Shiite cleric imam Mussa Sadr used the Ashura processions already in the seventies of the previous century in order to unite the Shiites in Lebanon

[3] http://www.jpost.com/servlet/Satellite?pagename=JPost/JPArticle/ShowFull&cid=1139395375738

Examples of children's education in the 2006 Ashura procession

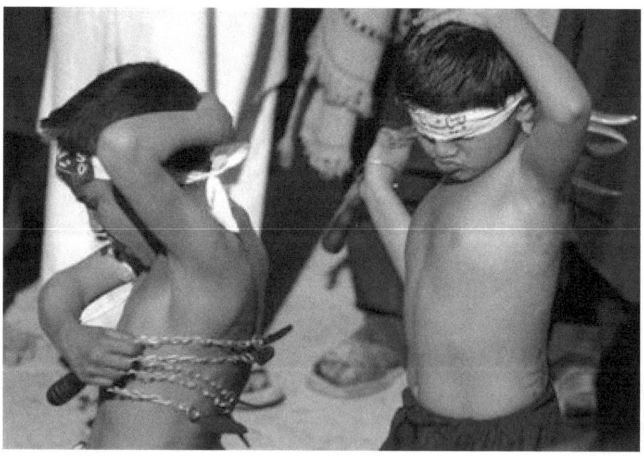

Participation in the 2006 Ashura procession

The reenactment of Hussein Bin Ali's death in the Battle of Karbala during the 2006 Ashura procession in Turkey

The World Jerusalem Day as an impetus for Shiite-Sunni cooperation in the struggle against Israel

The World Jerusalem Day is celebrated every year on the last Friday in the month of Ramadan in Iran, amongst Shiite communities around the world, amongst Sunni communities in the Palestinian territories and in other places as well. The World Jerusalem Day was established by Khomeini in 1979 in order to create a common denominator between Shiite and Sunni organizations with regard to the struggle against Israel. The custom of Jerusalem Day ceremonies spread well beyond Iran's borders and the Shiite community and it is observed by both Shiite and Sunni communities around the world. As the Israeli – Palestinian struggle intensified, World Jerusalem Day became more and more a day of Islamic solidarity with the Palestinian conflict and a stage for voicing belligerent slogans intended to recruit Muslims for the struggle to liberate Jerusalem and destroy the state of Israel. Inciting speeches, calls of death to Israel and America and the threats to harm those states that support Israel that are heard in Iran and amongst Shiite communities around the world are disseminated by the Iranian and Hezbollah television networks and websites. The use of the World

Jerusalem Day processions to promote Jihad against Israel is also common amongst Palestinian communities in the territories and abroad.[4]

http://www.qudsday.com website

[4] See summary of the World Jerusalem Day events of October 2006 – http://www.terrorism-info.org.il/malam_multimedia/English/eng_n/pdf/jerusalem_d_e.pdf. See also summary of the World Jerusalem Day events of October 2005 – http://www.terrorism-info.org.il/malam_multimedia/html/final/eng/eng_n/iran_je_e.htm and the World Jerusalem Day events of October 2003 – http://www.terrorisminfo.org.il/malam_multimedia//ENGLISH/IRAN/PDF/IRAN_1_04.PDF

Hezbollah's "Jerusalem Day Parade", Beirut, October 2006

2003 “Jerusalem Day Parade” in Berlin

سبحان الذي أسرى بعبده ليلا من المسجد الحرام الى المسجد الأقصى

محمد رسول الله

لا إله إلا الله

DEMONSTRATION

ZUM

QUDS-TAG

Samstag, 22.11.2003 – Um 10.00 Uhr
Berlin - Adenauerplatz

In einer Zeit, in der fast alle Macht habenden Politiker, insbesondere der arabischen aber auch der übrigen muslimischen Welt, wetteifern, das Wohlgefallen der zionistischen Besatzungsmacht sowie des amerikanischen Imperialismus zu gewinnen, und sich demütig und ergeben zu den Füßen der neuen, ungerechten Weltordnung werfen, wollen die verantwortungsbewussten Massen der Umma am Quds-Tag (Tag von Jerusalem) zum Ausdruck bringen, dass die Muslime trotz aller Verschwörungen und trotz des Verrats der meisten Regierungen niemals aufgeben werden, ihrer heiligen Stadt, die allen drei abrahamitischen Religionen heilig ist, eines Tages wieder Freiheit und Würde zurück zu geben.

Wir richten daher zum Anlass des Quds-Tages unseren Aufruf an alle muslimischen Geschwister in Deutschland, an der Quds-Demonstration teilzunehmen, um unsere Verantwortung gegenüber Jerusalem, Palästina und allen aktuellen gerechten Sachen unserer islamischen Umma, die Einheit aller Rechtsschulen, politischen Strömungen, Nationen und Völkergruppen und den unaufgebbaren Willen der Umma zu präsentieren.

Organisationskomitee „Demonstration zum Jerusalem-Tag(Quds-Tag)“

In 21st century, the World Jerusalem Day became an international Islamic event. It has turned into a common theme in both Shiite and Sunni Islamic websites. For example, pictures of Palestinian flags and symbols appeared in Iranian, Hezbollah, and Sunni radical websites.[5]

The following picture was taken from an Iranian website and it includes several motifs:

- An Iranian flag on the bottom right part of the picture
- A Palestinian flag signifying the connection between Iran and the Palestinian struggle
- A Hezbollah flag
- Significance: a connection between the three entities
- Additional aspects: a drawing of the moon, which has special meaning in the Shiite world, and also appears on the Iranian flag
- Sources of light, such as the sun, are part of the "light of Islam" theme
- The Omar Mosque signifies the ties to Jerusalem
- The poster reads - "Imam Khamenei: Jerusalem day Islam day"

[5] See http://www.hamasna.com and http://www.qudsday.com

Stories of shaheeds as a means of promoting the idea of self-sacrifice

Terrorist organizations, including Hezbollah, work to maintain and strengthen ties between the activists and the organization and its goals. One of the central tools for shaping this connection is the creation and use of myths, models for emulation from the distant past, but mostly from the organization's recent history. The role models are chosen very carefully, their life and death stories are recounted in the organization's media outlets and passed on to the next generations through the organization's education and Da'awa network. For example, the myth created around two of the movement's senior leaders and founders - Sheikh Ragheb Harb and Sheikh Abbas Musawi – serves the purpose of promoting the idea of self-sacrifice. Both of the leaders were senior religious clerics physically involved in the organization's military operations and both died as shaheeds. Sheikh Ragheb Harb was the Imam of Jibshit, participated in attacks against IDF forces in Lebanon and was killed in 1984. Sheikh Abbas Musawi was the second chairman of Hezbollah and the founder of Hezbollah's operational branch, which he headed for several years. He was killed by the IDF in 1992 on the anniversary of Harb's death. Every year there are memorial ceremonies and rallies in their memory attended by Hezbollah's senior leaders who use the event to call on their followers to continue in the path of Ragheb Harb and Abbas Musawi. Alongside the 'worship' of these two senior shaheeds, there is a glorification of the rest of the shaheeds as well, which usually entails the publication of their pictures on billboards, the internet and the organizational media outlets and their memorialization in the organization's memorial site.

Sheikh Abbas Musawi

Abbas Musawi's gravesite is used in order to promote the heritage of suicide

Sheikh Ragheb Harb – a combination of a religious cleric and fighter and a role model for the Shiite Islamic fighter

Pictures of shaheeds on bulletin boards in Lebanon as a means of promoting the concept of self-sacrifice. There is also a clear attempt to connect the Palestinian arena and Hezbollah

The use of the suicide motif as a means of deterrence

Alongside the use of the shaheeds as a model for self-sacrifice, the Shiites use this means in order to create deterrence on the one hand and a sense of empowerment amongst the target population on the other hand. Groups of shaheeds march in Hezbollah processions in Lebanon on World Jerusalem Day and on other occasions. Nasrallah stated several times that they would use suicide bombings if Israel would enter Lebanon. In addition, the Iranians announced the establishment of shaheed units in order to carry out suicide bombings against Western targets around the world as deterrence against the international system so as to abstain from acting against the Iranian nuclear project.

Following is a photograph of the graduates of a "Martyrs Course" in Iran. Similar photos appeared on various websites and news programs following developments in the crisis surrounding the Iranian nuclear program[6]

[6] http://www.iranfocus.com/modules/news/article.php?storyid=2088

The use of symbols as a means of showing support and promoting Palestinian terrorism

Iran and Hezbollah promote Palestinian terrorism as a means to thwart peace agreements and a calming between the Israelis and Palestinians. They provide financing, training and media support. The internet serves as a significant platform for disseminating these messages by using symbols from the Palestinian struggle in Hezbollah's website and by publishing pictures on Shiite websites or billboards that contain Shiite elements and symbols alongside symbols that express Shiite support of Palestinian terrorist activities.[7]

The Symbol of the Iranian Revolutionary Guard

The Symbol of Hezbollah

The Symbol of the Palestinian Popular Revolutionary Committees

[7] http://www.oppression.org/cgi-bin/viewnews.cgi?category=11&id=977638791

Support of the Palestinian struggle

A Palestinian flag alongside Hezbollah and Lebanon flags during the 2006 "Jerusalem Day Parade" in Beirut

Nassrallah's ties to the Palestinian struggle through pictures on websites and bulletin boards

Symbols and cartoons as a means of propagating the message that the US is the "Great Satan"

The concept of the "Great Satan" and its association with the US was invented by Khomeini. His successors continued breeding the hatred towards the US. American activity in Iraq, the ongoing crisis surrounding the Iranian nuclear program and the increase in the power of the extremists in Iran since 2005 radicalized the expressions used and the entire handling of this issue. The Second Lebanon War was perceived as another step in the American conspiracy to turn the Middle-East into a new and democratic Middle-East, which is quite contrary to the Iranian inclination to turn the Middle-East into an Islamic Middle-East.

The US is the "Great Satan" that is "lighting up" the Middle-East

The poster reads: "The New Middle-East"

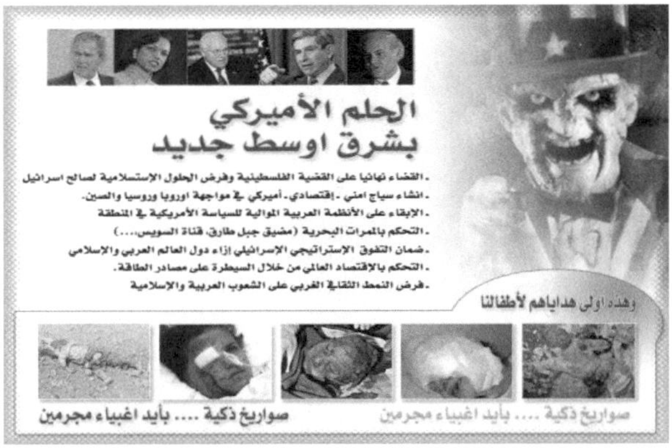

The idea of the American 'conspiracy' regarding the "New Middle-East" as expressed in cartoons –

The US is trying to drag Lebanon towards the "New Middle-East" by using Israel

The Israeli-American conspiracy against Lebanon

The US as the "Great Satan" and Israel as the "Little Satan" in the context of the Second Lebanon War

In summary, the Shiites make extensive use out of shaping new symbols and adding new meaning to old traditions as a meaningful means for social mobilization, broadening their support base, recruitment of new activists and the encouragement of proactive actions in order to achieve the organization's interests. The internet is one of the most significant platforms for Shiite terrorist groups, especially Hezbollah, to spread their messages and shape public opinion by using existing Shiite symbols and creating new symbols for the promotion of Shiite interests.

Hypermedia Seduction for Terrorist Recruiting
B. Ganor et al. (Eds.)
IOS Press, 2007

Radical Islamic Internet Propaganda: Concepts, Idioms and Visual Motifs

Jonathan FIGHEL
The International Institute for Counter-Terrorism (ICT), Senior Researcher

Abstract. Al Qaeda and radical Islamic terrorism rely on future recruits and sympathizers more than on any other factor. Terrorists do not use the Internet for direct operational recruitment, but rather to shape a committed virtual radical Islamic community from which individuals will be identified as potential candidates for recruitment. Radical Islamic websites extensively use text, imagery, audio and special idioms and phrases in their websites in order to enhance its radicalization capability. The article addresses the methods and means of internet sedition.

The Radicalization Process

A considerably large number of radical Islamic and Jihadi websites and forums use Islamic motifs as their web addresses. Calligraphy, visual imagery and semantics are used to attract new potential target audiences and individual visitors. Islamic visual motifs and key words are used to "trap" sympathizers and potential recruits, as they offer the needed seduction through 'virtual' on line Da'awa (Call, propagation proselytizing,) using marketing techniques similar to those employed by commercial enterprises. The Internet is used for initial attraction, ideological recruitment, social support through virtual community bonding and sense of belonging, rather than for direct operational recruitment. The homegrown radicalized youth is expected to easily identify texts and subtexts posted on the message boards.

The first stage of exposure in the on line radicalization process is to use well known and common knowledge Islamic idioms and key words ("buzz" words) that appear on the web's address and link to "information hubs" where new radical Islamist texts, declarations, claims of responsibility, video clips and recordings are posted. These elements will indicate the nature of the site and raise expectations regarding its potential content. Although some of the idioms may seem to be legitimate and innocent, they can have a sub-textual message and a double meaning according to the interpretation of radical Islamic schools of thought.

The second stage in the exposure process is a rapid visual screening process where the eye scans for additional motifs on the homepage. Some of the important elements are radical Islamic visual motifs and calligraphy shapes that can seduce and suggest that this is the website the radical youth was looking for. Some websites use audio recordings of radical Islamic songs or Qur'an versus to enhance the seduction process.

An additional element employed to authenticate the web site as a "genuine" Islamic web site is the use of the Islamic Hijra calendar and dates. "Real" Islamic forums and Jihadi websites usually do not post the Gregorian calendar dates, as it is identified with Western culture and values. Carefully selected Islamic texts, idioms and wording, including imagery, provide interactive ways for Jihadi sympathizers to connect with the ideology. The notion of resonance, the ways in which a message meshes with an audience's existing perceptions, is the outcome of this dialectical process.

Radical Islamic websites have special sections in their homepages for posting future attack threats or intents, claims of responsibility after attacks, attack video clips, video and audio recordings of statements made by terrorist leaders, prisoners and "Shaheeds" videotaped wills, glorification posters, etc.'. The first 30 seconds from entry to the site are the most critical for the viewer to decide whether the site meets his expectations. In order to seduce the viewer, a series of Islamic idioms and key words as visual motifs (in addition to the address name) are used in order to send a signal regarding the internal content. Some well known Jihadi web sites used, and still use, Islamic idioms embedded in their internet address, such as: Alneda.com (The Call), which was known to be the first Al Qaeda website on the internet, Al Sahat (The Fields), Al Qal'ah (The Fortress), Muslm.net, Jihad-algeria.com, Dawah.ws, Salafi.net, Dawahmemo.com, Al Islah (Reform), Mojahedun.com (Jihad warriors), Al Ansar (Supporters or warriors of Islam) and many others.

It should be noted that right after the 9/11 attacks, a considerable number of websites with the word "Jihad" embedded in their address changed their names or addresses by posting new "innocent" Islamic names in different variations as their new address in order to avoid being shut down, hacked or cracked by law enforcement, local authorities and others. A proper scholarly and professional understanding of Islamic culture and religion can show how the explicit texts and key words on the websites can be used to convey a hidden agenda. The Islamic idioms listed below are partial examples of website addresses that are still used for messaging, articles, scholarly dialogue, claims of responsibility after attacks or warnings of future ones.

The use of Islamic terminology and key words:[1] 'Al-Walaa wal-baraa' (absolute allegiance to the community of Muslims and total rejection of non-Muslims), Ansar, Al Neda (The Calling for the true and genuine believers to join "true" Islam), Bilad al-Haramain (The Land of the Two Shrines – referring to Mecca and Medina), Da'awah (proselytizing), Jihad, Janna (Paradise) Majd, Mujahadin, Ribat/Ard al-Ribat (The Land of the 'Frontier Post'), Shaheed, Umma, etc.) are all used as recognized and legitimate historical Islamic symbols and as a basis for creating the new radicalized interpretation by radical Islam.

Visual Islamic motifs and idioms are used as a sophisticated tool to accomplish several objectives for the Jihad propagandists. They create a mental concept and perception of reality for their target audiences. The use of Islamic symbols, images and phrases evokes emotional or historical collective Islamic memories, thus generating a subconscious emotional response. These motifs generate the feelings of deep beliefs or suppressed hidden concepts within the target audience as a tool of communicating ideas.[2]

---

[1] http://answering-islam.org/Books/Hughes/t.htm; http://www.answers.com/topic/mujahideen; http://encycl.opentopia.com/term/Jihad.

[2] http://www.ctc.usma.edu/imagery_introduction.asp.

The motifs employed in the websites are based on mainstream Sunni Islamic tradition, history and cultural references of the Islamic Ummah. A certain timeline related to the successful spread of Islam between the 7th and 10th centuries can be identified within these motifs. These motifs, idioms and calligraphy reflect the desire to portray the nature of "true" original Islam in an extreme fashion of interpretation by fundamental radical thought as a logical improvement of the traditional Islamic theology. The Islamic motifs, in a sophisticated manipulation, are embedded within propaganda; they can be easily recognized as legitimate Islamic historical symbols as a basis for creating new motifs that support a radicalized interpretation. The new radical motifs can easily gain legitimacy, as they appear to be authentic when they are used in proximity to widely accepted mainstream Islamic symbols and cultural references. An example of this technique is the frequent use of a photograph of Osama bin Laden on a horse, which connotes his pseudo proximity or affiliation with the companions of the Prophet Mohammad ("Sahaba") and the successful spread of Islam by his successors, while building Bin Laden's legitimacy to restore the past glory of Islam. One can observe a similar element in the use of the "Islamic Sword" as a symbol for the spread of Islam and its connotation to the radical Islamic notion of "Din Muhammad Bil- Saif" – The spread of Prophets' Muhammad religion by the sword (Jihad).

An additional element that can be identified in most radical Islamic and Jihadi websites is the shape of the Arabic calligraphy that is used. Most Arabic calligraphy used in radical Islamic websites is shaped in old design fashions as a hidden symbol of return to the "true origins" and a rejection of modernity. Banners are often designed in old Arabic scripts and style in contrast to modern regular Arabic fonts. The most popular used scripts are: Thuluth, Eastern Kufi, Old Kufi, Muhaqqaq, Naskhi, Nastaliq and Riq'a.[3] The common denominator of all these Islamic scripts is rooted back in the Islamic "golden age" to which radical Islam in general and Al Qaeda in particular wishes to return in order to restore its glory and reestablish the "Islamic Caliphate".

Al Qaeda and radical Islamic terrorism rely on future recruits and sympathizers more than on any other factor. Terrorists do not use the Internet for direct operational recruitment, but rather to shape a committed virtual radical Islamic community from which individuals will be identified as potential candidates for recruitment. The use and exploitation of the Internet for radical Islamic Dawa and violent ideas is a cause of great concern and a challenge for mainstream moderate Islam, Muslim and Arab governments, Western law enforcement and the intelligence community around the world. The fact that the Internet has become a popular apparatus for early stage mobilization and recruitment poses a huge challenge. Up until now, no real moderate mainstream Islamic alternative was developed or introduced on the Internet. Moreover, no "Counter Sahwa" (radical Islamic awakening) or "Counter Da'awa" websites were developed, designed or posted in order to offer an attractive alternative to influence the "hearts and minds" of potential target audiences as a real countermeasure to brutal radical Islamic dominance on the net. Any future steps or measures to fight the terrorists' use of the Internet should involve a careful interdisciplinary professional examination of new alternatives, contents and motifs that can be adopted and revitalized from the vast monumental and respectable heritage of Islam.

---

[3] http://www.sfusd.k12.ca.us/schwww/sch618/Calligraphy/Calligraphystyles.html
http://www.islamicart.com/main/calligraphy/styles/riqa.html.

Examples of the Arabic font

Anti-US and anti Western sentiment in the Muslim world and in some Arab states is on the rise and fueling radical Islamic ideology. This could prompt some groups to adopt terrorist methods to attack US and Western interests. The radicalization process is occurring quickly, widely, and more anonymously in the Internet age, raising the likelihood of surprise attacks by unknown groups or self radicalized young frustrated individuals whose members and supporters may be difficult to identify in advance. It seems that groups from the marginalized social status will increasingly use the Internet to communicate with their new "virtual community" with whom they identify with, propagandize, recruit and support.[4]

[4] http://ikhwanweb.com/Home.asp?zPage=Systems&System=PressR&Press=Show&Lang=E&ID=5373.

Hypermedia Seduction for Terrorist Recruiting
B. Ganor et al. (Eds.)
IOS Press, 2007

Aspects of Communication and Message Design in the Arab Culture

Mr. Yoav MIMRAN
Associate, Arab Culture and the International Institute for Counter-Terrorism (ICT), Israel

Abstract. Understanding the culture-specific attributes of cognitive style and communication is essential for successful interaction with those members of a culture who use the Internet and other means of communication. The cognitive style of members of the Arab culture is integrative and holistic. Arab culture is described as "high context" and "indirect"; the path of communication is nonlinear and associative, rather than organized. Arab culture is also an oral culture, in which language serves to create an emotional experience rather than merely to transfer information. As an outward manifestation of certain principles of communication, a message in Arab culture is characterized by extensive use of metaphors and symbols, vagueness, repetition and exaggeration. All of these characteristics are perceived as comprising an elevated form of language.

Keywords. Culture, communication, cognitive style, nonlinear, high context, oral, indirect, lawful magic, message design.

Whoever learns the language of a nation secures himself from its evil. —*Arab proverb*

Introduction

Cultures differ from each other, among other things, in the cognitive style of their members, and in their means and principles of communication. These differences are manifestations of important elements of cultures, namely their assumptions about the world, and their values, norms, standard operating procedures, and other factors. Competent communication with individuals or communities depends on knowing their socio-cultural codes, their pragmatic codes (both universal and culture-specific codes), and their linguistic system—the language and pragmatic codes of using it. Communicational competence has behavioral, linguistic and cognitive aspects, and is rooted in historical, social, political and religious contexts.

The Internet, as a platform of communication, is used extensively by terror organizations for various purposes. These purposes include disseminating information and professional knowledge, spreading propaganda to their constituents and non-constituents, fund-raising, and command and control of terror cells.

Jihadi terrorist organizations use the Internet efficiently by combining, on the one hand, ultramodern technologies—which are used to build a "cyber ground" for the "gathering" of communities—and, on the other hand, traditional means of communication. The latter is based on intimate acquaintanceship with Arab culture regarding the act of "gathering the tribe" around a common ideology and aim. In this context, terrorist websites are the display windows of terror organizations, and those organizations strive to realize their vision through such sites.

Thus, terrorist organizations' websites are the best source of information if one wishes to learn about those groups' ideology and agendas, strategies, and means of realization, or to acquire a better understanding of the nature of the dynamics between the organizations and their various target audiences. One may also identify their target audiences' motivations and sensitivities.

In addition, the Internet can and should be used to offer alternative political, social and religious ideologies. Such alternatives would help to convince the constituents of terror organizations to rethink the way of terror, and would also help to prevent new constituents from joining terror organizations. Ultimately, such alternative ideologies would help in building strong and proud communities that publicly oppose those terrorist organizations which claim to be "the genuine Muslims". Alternatives should be offered using many different kinds of websites, including the current terror organizations' own sites, existing non-terrorist websites, or new sites which would be established for this purpose.

Having good culture-specific communicational competency is necessary in order to achieve these goals. Developing such competency is a real challenge, and some would even claim that full communicational competency in a foreign culture is impossible. However, while the question of the extent to which it is possible to understand a foreign culture is essentially theoretical, cross-cultural communication is a practical need in the real world. Effort must be expended to improve cross-cultural communication as far as possible.

This chapter presents the cognitive and cultural factors that influence the nature of communication in the Arab culture, and explores their realization in message design.

1. Cognitive Aspects of Communication and Message Design

Cognition, according to the viewpoint of cultural psychology, is not universal but culture-specific. One's style of thinking reflects one's culture, and is influenced by it; therefore, different cultures are characterized by different cognitive styles. "Cognitive style" is defined as the way in which individuals organize information as well as their own experiences [1]. The term thus refers to the characteristic ways in which people conceptually organize their environment, and spontaneously filter and process stimuli so that their environment assumes psychological meaning [2].

Cognitive style is reflected in patterns of communication—in messages designed by individuals, as well as in paths of communicational interaction (e.g. conversation). As a platform of communication that combines visual and verbal means, Web design reflects the cognitive style of designers, and influences a website's efficiency in conveying messages.

An online experiment that exposed American and Chinese people to sites created by American or Chinese designers found that cognitive differences at the design level exist in the form of cultural style, and that these stylistic differences were perceptible to the users [3]. The results of this research strongly suggest that an awareness of cultural cognitive styles is necessary for the improvement of online communication. The efficacy of Web communication depends, among other things, on "responding to implicit thinking patterns of cross cultural users" [3].

The cognitive style of Arabic culture is *nonlinear* and *holistic*. Nonlinearity relates to the way of organizing and processing information, and it is reflected in patterns of communication which are characterized by multiple themes and the simultaneous bombardment and processing of a variety of stimuli. This pattern of communication characterizes nonlinear cultures. The relation between the different themes is associative rather than causative, the path of speaking is meandrous rather than straight, and so is the act of communication. Holistic style refers to perception that is integrative, rather than analytical and differentiative.

Zebian and Denny [4] found a preference for integrative thinking and an avoidance of differentiative thinking in Middle Eastern groups, compared with Euro-Canadian groups, which preferred differentiative thinking to integrative thinking. The Middle Eastern groups included immigrants who supposedly had been influenced by a Western environment, yet had preserved elements of their native cultural cognitive style.

2. Behavioral and Social Aspects of Communication in the Arab Culture

The behavioral and social aspects of communication include the manner in which individuals and groups within Arabic society build and maintain social relationships; the context in which these relationships occur is emphasized. Arabic culture is a *high context culture.*

The terms “high context” or “low context” refer to the relationship between the content and the context of a message, or how much of the meaning of the message is found in its context versus in its code (i.e. language). High context cultures are characterized by extensive use of non-verbal codes, such as body language or visual cues. Interpretation of the message depends on the environment in which the interaction occurs, and relies on values, norms and social codes. The burden of extracting the exact meaning is carried by the listener, while only part of the information is transmitted in a clear-cut manner.

Accordingly, the communication style of Arab culture is *indirect* and *ambiguous*; that is, the meaning of messages is implicit rather than direct, vague and circular rather than straight and clear.

Furthermore, as another behavioral feature, members of indirect cultures prefer to create emotional empathy even though they might have to compromise on precision. These characteristics are strongly connected with collectivist societies, where individuals are dependent on each other and have incentives to maintain good relationships amongst themselves.

3. Socio-linguistic Aspects of Communication in the Arab Culture: the Role and Function of Language

No people in the world manifest such enthusiastic admiration for literary expression, and are so moved by the word, whether spoken or written, as the Arabs. Modern audiences in Damascus, Baghdad and Cairo can be stirred to the highest degree by the recital of poems—which may be only vaguely comprehended, and by the delivery of orations in the classical tongue—though these too may be only partially understood.

The rhythm, the rhyme, and the music produce in them an effect they call “lawful magic” [5]. Arabic language has a unique status for Arabs, for religious, aesthetic and nationalist reasons. As such, the Qura’an Arabic language is considered to be a “miracle of God”—“*iAjaaz - al qura'an*”—a language which cannot be created by any human being. As an art form, Arabic poetry has been admired by the Arabs since the ancient times.

Arabic is perceived as a common denominator unifying the Arab nation. Arabic culture is described as an *oral culture*, in which the role of language is not only to transfer information, but to create emotional and social experiences. Such experiences are mediated by the heavy use of symbolism, extensive relating of historical events, and quotations of qura'anic verses. Communication within oral cultures is also characterized by extensive use of aural ornaments.

4. Principles of Message Design in the Arab Culture

Principles of message design are the concrete manifestation of the communicational characteristics of a culture. They are important for both conveying and interpreting information. In the following section, these principles—as found in Arab culture—are presented, and are linked to the relevant aspects of communication described above.

4.1. Repetition

Repetition is defined as "reiteration, or repeating the same word, or the same sense in different words, for the purpose of making a deeper impression on the audience" [6]. Repetition is seen as having an aesthetic value in Arabic culture, and is highly appreciated as a demonstration of one's own vocabulary and the beauty and richness of the language itself. In Western cultures, in contrast, repetition might be seen as redundant, or as "repeating unnecessary words" [7]. It is regarded mostly as a legitimate strategy for emphasizing the main topics in a message.

Repetition is strongly connected to the oral tradition of Arab culture, as it helps to keep a high degree of mental concentration amongst listeners. From an audience-decoder perspective, where an integrative and holistic cognitive style is being used, repetition is not perceived as an interruption of the flow of information.

4.2. Metaphors, Images and Symbols

The intensive use of metaphors, images and symbols in the communication style of Arab culture relates to the fact that this is an oral cultural tradition, with a holistic and integrative cognitive style rather than a differentiative and analytical one. A picturesque example or illustration can convey a message better than an analytical explanation supported by facts.

Metaphors, images and symbols create strong emotional effects in listeners, and are therefore powerful means of conveying a message. Metaphors that might sound or look strange to Westerners are highly effective for Arab listeners. Well-known examples that demonstrate these principles are the "knight of believers", bin Laden's epithet; and the "Mother of all wars", Saddam Hussein's promise to Iraqi troops and the Arab world. I would suggest that from the perspective of Arab culture, the phrase "Mother of all wars" is considered to be strong, influential and threatening.

"Innocence like the innocence of the wolf of Josef's blood" is another example. Here, a family uses the well-known qura'anic version of the story of Josef and his brothers to assert its blamelessness whilst in conflict with another family [8].

Understanding the historical, social and religious contexts, and knowing the symbolic world of Arab culture, is mandatory if one is to create the desired impact when conveying a message, or to gain hearts, or to provoke. Such knowledge is also essential if one is to avoid making unnecessary mistakes and creating an undesired impact. One example of provoking bad feelings might be President Bush's usage of the term "crusade" to describe the military operation in Iraq. This phrase has had a negative impact on the Arab world, and has been used by Islamic jihad terror organizations in their propaganda.

Other examples relating to the social and cultural contexts would be the names given by America to radio station broadcasts in the Middle East, such as "sawa" and the television station "Hurriyya". "Sawa" in Arabic means "together", a name that emphasizes the strong collectivist aspects of the Arab culture and society. However, this name has no political meaning; in particular, it does not relate to the loaded relationship between the West and the Arab world. "Hurriyya", the name of the television station, means "freedom", a term that has strong internal and external political connotations, and which might create antagonism.

4.3. Proverbs and Maxims

Arabs' extensive use of proverbs and maxims is deeply rooted in the Arab tradition; the use of proverbs creates a strong emotional resonance, as it connects Arabs to their glorious past, and emphasizes their ancient common cultural background.

Proverbs have strong persuasive power, as they combine the authenticity of the speaker and the reliability of the message with aesthetic value. In another light, proverbs are well used as a strategy for delivering messages implicitly, reflecting the indirect nature of the Arab culture.

4.4. Exaggeration

Exaggeration is favored as a technique for stressing a point, and for creating the desired atmosphere and a strong impression. This is another technique which is considered to have aesthetic value in Arabic culture. Interestingly, both the words "exaggeration"—"*mubaalagha*" and "eloquence"—"*balaagha*" stem from the same root in Arabic, so they share a semantic core.

A good example of exaggeration and eloquence is found in a citation of Abu al Bakr Naji, one of Al-Qaida's ideologists. He states, "Oh people! The viciousness of the Russian soldier is *twice* that of the American soldier. If the Americans suffer one *tenth* of the casualties the Russian suffered in Afghanistan and Chechnya, they will flee and never look back." [9] The author uses "twice" and "tenth" to demonstrate his idea efficiently, and to create an emotional resonance within his audience or community.

4.5. Vagueness

Vagueness, as a principle of message design, is mostly relevant to interpersonal communication. It relates to the high context characteristic of the Arab culture, where messages can be designed in a vague way since the context will support them. In interaction between individuals and groups, vague messages help to create social harmony and emotional experiences. In general, members of Arabic culture are tolerant of inaccurate and vague messages.

5. Summary

Cross-cultural communication is always challenging, due to the fact that all sides involved have to overcome latent and overt obstacles related to the patterns, contents and contexts of the process. The extent to which communication in Arab culture relies on patterns and context rather than on content, along with the sensitivity of the historical, political and religious background, makes communication between Western and Arab cultures especially complicated.

Developing communicational competency is essential in order to act efficiently and to achieve the desired results in the Internet arena. The principles presented in this chapter may help one to develop competency when communicating with members of the Arab culture. However, this is not the only knowledge one requires. An understanding of the linguistic system is also necessary for conveying messages, and a familiarity with the social, religious, historical and political contexts is necessary for the content aspect of the message.

The efficacy of websites is influenced by design, and thus is strongly related to culture-specific cognitive styles. The styles of both the designers and the surfers must be taken into consideration, and more research is needed to determine the most favorable designs.

References

[1] Laschinger & Boss, 1984 at: Cognitive Style Versus Learning Style, Peter H. Heinman, www.personality-project.org/perproj/others/heinmen/cog.htm

[2] Goldstein and Blackman, 78, Cognitive Style: Five Approaches and Relevant Research. New York: Wiley.

[3] Faiola & Mtei 2005, Cultural Cognitive Style and Web Design: Beyond behavioral Inquiry into Computer Mediated Communication. Journal of Computer – Meditated Communication, 11(1), article 18. http://jcmc.indiana.edu/vol11/issue1/faiola.html

[4] Zebian S., and Denny , J. P. 2001, Integrative Cognitive Style in Middle Eastern and Western Groups: Multidimensional classification and major and minor property sorting. Journal of Cross Cultural Psychology, 32 (1), 58-75

[5] Philip Hitty, *History of the Arabs, Palgrave Macmillan,10th edition, 2002*

[6] *Hypertext Webster Gateway Dictionary, http://bennetyee.org/http_webster.cgi , repetition(Rhet)*

[7] *Encarta Dictionary, Microsoft, 2006*

[8] Public announcement, *Al-quds*, Palestinian newspaper, 12.6.06

[9] Abu al Bakr Naji, *The management of Savagery,* The Most Critical Stage Through Which the Umma Will Pass, 2004. translation by William McCants, Combating Terrorism Center at West Point, 2006 http://ctc.usma.edu/naji/asp

Further References

Announcement in *Al-quds*, Palestinian newspaper, 22.6.06

Zaharna, R.S. (1998). "Integrating Intercultural Communication and International Public Relations: An 'In-Awareness' Model, http://list.msu.edu/cgi-bin/wa?A2=ind9812b&l=aejmc&T=0&P=4406

Radical Islamist Websites – An Overview

Hypermedia Seduction for Terrorist Recruiting
B. Ganor et al. (Eds.)
IOS Press, 2007

Using the Internet for Terrorist Recruitment and Mobilization

Gabriel WEIMANN[1]

Department of Communication, University of Haifa, Israel
weimann@soc.haifa.ac.il
http://soc.haifa.ac.il/~rsso343/

Abstract. Without recruitment, terrorism cannot prevail, survive and develop. Recruitment provides the killers, the suicide bombers, the kidnappers, the executioners, the engineers, the soldiers, and the armies of future terrorism. The Internet has become a useful instrument for modern terrorists' recruitment. It combines several advantages for the recruiters: it makes information gathering easier for potential recruits by offering more information, more quickly, and in multimedia format; the global reach of the Net allows groups to publicize events to more people; and, by increasing the possibilities for interactive communication, new opportunities for assisting groups are offered, along with more chances for contacting the group directly. Also, through the use of discussion forums and chat rooms, the Internet also allows members or potential members of the group to engage in debate with one another. Online recruitment by terrorist organizations is said to be widespread, though the Internet is used more for initial attraction, ideological recruitment and social support than for direct recruitment. Moreover, the online process is more often activated to reward recruits and suicide terrorists, thus serving as an additional indirect recruitment initiative. Finally, terrorist recruiters may use interactive Internet technology to roam online chat rooms looking for receptive members of the public, particularly young people, using sophisticated profiling procedures.

Keywords. Terrorism, internet, recruitment, mass media.

---

[1] Dr. Gabriel Weimann is a Full Professor of Communication at Haifa University, Israel. This chapter is based on a research project funded by the United States Institute of Peace, Washington, DC, where the author was a Senior Fellow. His project is presented in his recent book, *Terror on the Internet*, published in Washington DC (2006, USIP Press).

> Extremists are more and more making extensive use of the internet. Websites are difficult to monitor and trace; they can be established anywhere and have global reach; they are anonymous, cheap and instantaneous; and it requires no special expertise to set up a website. The internet is widely used for propaganda; training (including in weapons and explosives); to claim responsibility for attacks; and for grooming through chat rooms and elsewhere.
> *—From the report on the London bombings on 7th July 2005* [1]

Introduction

In the summer of 2006, during the violent clashes between Israel and the Hezbollah, an Iranian website posted a draft call online, urging volunteers to join the war against Israel. On July 26 2006, the Iranian news agency Alborz announced that volunteers could sign up online for the "Army of the Fighters of Muhammad." The report explained that the Army of the Fighters of Muhammad was recruiting fighters to be sent to Lebanon, and that those interested in signing up could also call a telephone number in Tehran (88938821). According to the report, the minimum age for enlisting is sixteen. This is one example of the modern techniques of Internet-based recruitment, as explored in this chapter.

1. Recruitment and Terrorism's Endurance

Without recruitment, terrorism can not prevail, survive and develop. Recruitment provides the killers, the suicide bombers, the kidnappers, the executioners, the engineers, the soldiers, and the armies of future terrorism. As RAND's Hoffman argues: "Al-Qaida's resiliency and longevity are predicated not on the total number of Jihadists that it might have trained in the past but on its continued ability to recruit, to mobilize, and to animate both actual and would-be fighters, supporters, and sympathizers." [2]

Marc Sageman studied the background, recruitment and motivations of over 150 members of four specific terrorist networks: the Central Staff around Osama bin Laden, the Core Arabs, the Maghred Arabs, and the Southeast Asians [3]. His book, *Understanding Terror Networks*, highlights the role of social networks in the recruitment of modern terrorists. The study documents the role of kinship and social networking as means of recruiting, as means of providing social and psychological security, and as means of radicalizing and mobilizing supporters. For example, he found that 68 percent of the sample of terrorists had friends in the jihad, or had joined as groups. An additional 20 percent had close relatives who were already members. Sageman argues that recruitment is a bottom-up, self-selected process rather than a top-down "seek out and recruit" process.

Following the American campaign in Iraq, Al-Qaida and other terror organizations have intensified their efforts to recruit young Muslim men, directing the rising anger about the American presence in the Middle East. The war in Iraq has swollen the ranks of Al-Qaida and has "galvanized its will" by increasing radical passions among Muslims, argued the annual report of the London-based International Institute for

Strategic Studies, *The Military Balance*. The Institute believes the network is present in more than sixty countries, has a rump leadership intact, and that there are more than 18,000 potential terrorists at large, with recruitment continuing.

These days, governments throughout Mexico and Central America are on the alert as evidence grows that Al-Qaida members are looking for recruits to carry out attacks in Latin America—the potential last frontier for international terrorism. The territory could be a perfect staging ground for Osama bin Laden's militants, with homegrown rebel groups, drug and people smugglers, and corrupt governments. As Hoffman argues [2, 4], Al-Qaida's resiliency (and longevity) is predicated by its continued ability to recruit and mobilize actual and would-be fighters, supporters, and sympathizers.

Despite the attacks on Al-Qaida, the arrests and personnel losses, and the dismantling of its operational bases and training camps in Afghanistan, Al-Qaida managed to launch an efficient propaganda campaign. A professionally produced Al-Qaida recruitment video, for example, was circulated around various foreign Muslim communities in the hope of attracting new martyrs to bin Laden's cause. The seven-minute videotape openly extolled the virtues of martyrdom, and solicited new recruits to bin Laden's cause. It presented various scenes of jihadists in combat, followed by the images of 27 martyrs shown in rapid succession, with the name of each listed, where they were from, and where they died.

Twelve of the martyrs were then featured in a concluding segment, accompanied by the voiceover: "They rejoice in the bounty provided by Allah: And with regard to those left behind who have not yet joined them in their bliss, the martyrs glory in the fact that on them is no fear, nor have they cause to grieve." Abu Ghaith, the group's chief spokesman, argued: "Those youths that destroyed Americans with their planes, they did a good deed. There are thousands more young followers who look forward to death like Americans look forward to living."

According to US government officials, Al-Qaida now uses chat rooms to recruit Latino Muslims with US passports, in the belief that they will arouse less suspicion as operatives than would Arab-Americans.

Since the outbreak of "Al Aksa Intifada" in 2001, there has been a growing trend in the use of Palestinian children aged eleven to seventeen, who are recruited by terrorist organizations to perpetrate suicide attacks in Israel. During this period, forty female terrorists were recruited to carry out suicide bombing attacks. Salah Shehade, one of the former leaders of Hamas in the Gaza Strip, discussed the use of children in terrorist attacks in an interview on the *Islam On Line* Internet site (May 26 2002). He said that the children must be trained well prior to perpetrating terrorist attacks, and must be recruited into a special branch within the organization's military apparatus, in order to instill the jihad culture and teach them right from wrong.

Expressions such as these represent one of the primary elements in convincing parents to send their children to perpetrate terrorist attacks. In many countries wracked by ethnic, political, or religious violence in the developing world, such as Algeria, Colombia, and Sri Lanka, new members of terrorist organizations are recruited at younger and younger ages. Adolescents and teenagers in these countries are often receptive to terrorist recruitment, because they have witnessed killings first-hand, and thus see violence as the only way to deal with grievances and problems. Militant groups in Jammu and Kashmir are increasingly recruiting children of the Gujjar and Bakerwal tribes in the hills of Udhampur, and using them in operations against security forces. Most of these children are between nine and fifteen years and belong to the higher reaches of the Pir Panjal, and are soft targets for terrorist groups.

2. Online Recruitment

The Internet has become a useful instrument for modern terrorists' recruitment. It combines several advantages for the recruiters: it makes information gathering easier for potential recruits, by offering more information, more quickly, and in multimedia format; the global reach of the Net allows groups to publicize events to more people; and, by increasing the possibilities for interactive communication, new opportunities for assisting groups are offered, along with more chances for contacting the group directly. Finally, through the use of discussion forums and chat rooms, the Net also allows members—or potential members—of the group to engage in debate with one another.

Let us illustrate this use of the Internet by modern terrorists with few examples. These examples are taken from our archive of terrorist websites. The project was hosted and funded by United States Institute of Peace, which summarized eight years (1988 to 2006) of monitoring terrorist presence on the Net [5, 6, 7]. Our scan of the Internet revealed over 4,800 websites serving terrorists and their supporters.

Adam Gadahn was an average American boy, who loved watching television and listening to music; he was raised by parents who farmed goats in Riverside County, California. However, at the age of seventeen, Adam discovered Islam on the Web, and soon started a conversion process that ended with his joining Al-Qaida. In 2003, the US federal authorities labeled him an enemy, saying he was one of seven people suspected of being Al-Qaida operatives who may be working to mount a terrorist attack in the United States.

Gadahn, now 25, has attended Al-Qaida training camps and worked as a translator for the terrorist group, the FBI said. Authorities said he may be using the names Adam Pearlman or Abu Suhayb Al-Amriki. In his own words, Gadahn described how the Internet was the gate for his recruitment:

> The turning point, perhaps, was when I moved in with my grandparents here in Santa Ana, the county seat of Orange, California. My grandmother, a computer whiz, is hooked up to America Online and I have been scooting the information superhighway since January. But when I moved in, with the intent of finding a job (easier said than done), I begin to visit the religion folders on AOL and the Usenet newsgroups, where I found discussions on Islam to be the most intriguing. You see, I discovered that the beliefs and practices of this religion fit my personal theology and intellect as well as basic human logic. [8]

In October 2004, a videotape in which a masked man claiming to be American and warning of coming attacks in the United States drew the scrutiny of federal intelligence agencies. They found out that the person was Adam Gadahn. The speaker identified himself as "Assam the American," and referred to Osama bin Laden and his top deputy as "our leaders." In some respects, Gadhan's case is reminiscent of that of John Walker Lindh, the "American Taliban" from Marin County, who embraced Islam, traveled to Afghanistan, met Osama bin Laden and was captured by US forces in December 2001.

Another modern recruitment operation emerged on the Web. In June 2004 the magazine *Insight Online* obtained a recruitment appeal seeking Iranians to commit suicide in deadly attacks on American and coalition forces in Iraq, on Israel, and on US citizens around the world [9]. It was posted on the Internet by an Iranian recruiting group. Reuter's news agency reported (on June 5 2004) that the group is known as the Committee for the Commemoration of Martyrs of the Global Islamic Campaign. It quoted the group's spokesman, Mohammad Ali Samadi, as saying "some 10,000 people have registered their names to carry out martyrdom operations on our defined targets."

Ali Samadi, according to Reuters, stated that the alleged independent group's targets include "mainly the occupying American and British forces in the holy Iraqi cities, all the Zionists in Palestine, and Salman Rushdie the only nonmilitary target for us because we believe his attack against Islam was much worse than a military assault." The recruitment application was translated from Farsi as follows:

> In the Name of God
> Preliminary Registration for Martyrdom Operations
> I ____________, child of ____________, born 13______ [Islamic calendar], the City of: ________________ proclaim my preparedness for carrying out martyrdom operations:
> ___ against the occupiers of the holy sites [referring to Najaf, Karbala, and other places in Iraq].
> ___ against the occupiers of [Jerusalem].
> ___ for carrying out the death sentence of the infidel Salman Rushdie.
> Also, I would like to become an active member of the Army of Martyrs of the International Islamic Movement. Yes ____, No _____
> Contact telephone:
> Applicant's address:
> Applicant's signature:

In October 2004 *Intelwire* presented a recruiting video of Al-Qaida young boys training in terrorist techniques [10]. The video includes a statement read in Arabic by a young son of Osama bin Laden. The video was hosted on an Arabic-language website connected to Tawhid Wal Jihad, the terror organization led by Abu Musab al-Zarqawi. The video, titled "Hamza Laden", appears to show young boys between the ages of ten and fourteen engaged in commando training, including marching in formation, running a military-style obstacle course, rappelling and reading written statements in Arabic. All these examples show how modern terrorists are using advanced communication technologies to attract and seduce potential followers, recruits, supporters and would-be terrorists.

3. Methods Of Online Recruitment

The Internet can be used to recruit and mobilize supporters to play a more active role in support of terrorist activities or causes. In addition to seeking converts by using the full panoply of website technologies (audio, digital video, etc.) to enhance the presentation of their message, terrorist organizations capture information about the users who browse their websites. Users who seem most interested in the organization's cause or well suited to carrying out its work are then contacted.

Sophisticated methods are used by terrorists to refine or customize recruiting techniques on the Net. Zanini and Edwards state:

> Using some of the same marketing techniques employed by commercial enterprises, terrorist servers could capture information about the users who browse their web sites, and then later contact those who seem most interested. Recruiters may also use more interactive Internet technology to roam online chat rooms and cyber cafes looking for receptive members of the public, particularly young people. The Dutch General Intelligence and Security Service's Annual Report for 2004 notes the importance of the internet, specifically in the radicalization of parts of the Muslim communities in the Netherlands, through 'virtual' *dawa* (online radical sermons) and increasingly through unmonitored chat rooms, where intensive exchange of Islamic ideas is taking place. Electronic bulletin boards and user nets can also serve as vehicles for reaching out to potential recruits. Interested computer users around the world can be engaged in long-term 'cyber relationships' that could lead to friendship and eventual membership. [11]

However, it should be noted that modern terrorists do not recruit directly online; they use the Net only to identify, profile and select potential candidates for recruitment. Afraid of having their groups infiltrated by security agencies and counter-terrorism forces, they will use the Internet only for the early stages of the recruitment process. Once identified, potential recruits are bombarded with religious decrees, propaganda, and training manuals on how to become a part of the "global jihad movement". Those who become ensnared, either by rhetoric or through curiosity, are then guided through an online maze of secret chat rooms, or instructed to download software called PalTalk. This software enables users to speak to each other on the Internet without fear of being monitored, at which point the personal online indoctrination begins.

Thomas Hegghammer, a Norwegian researcher, divides the jihadi Internet community into three categories [12]. First, there are the message boards where one finds political and religious discussions taking place among sympathizers and potential recruits. Among the most important message boards for Al-Qaida sympathizers are Al Qal'ah ("The Fortress"), Al Sahat ("The Fields"), and Al Islah ("Reform"). These message boards, Hegghammer argues, provide links to the second type of Internet community, the "information hubs," where new radical-Islamist texts, declarations, and recordings are posted. These sites are often posted in the "communities" sections of popular Western sites such as Yahoo, Lycos, and others. Among these, the most important one is the "Global Islamic Media." Finally, there are the third type of sites, the "mother sites", which are run by people who get their material directly from the ideologues or operatives.

In addition, some would-be recruits use the Internet to advertise themselves to terrorist organizations. In 1995, as reported by Verton in *Black Ice,* Ziyad Khalil enrolled as a computer science major at Columbia College in Missouri. He also became a Muslim activist on the campus, developing links to several radical groups and operating a website that supported Hamas. Thanks in large part to his Internet activities, he came to the attention of bin Laden and his lieutenants. Khalil became Al-Qaida's procurement officer in the United States, arranging purchases of satellite telephones, computers and other electronic surveillance technologies, and helping bin Laden communicate with his followers and officers.

In February 2004, a US National Guardsman was arrested and accused of trying to provide information to the Al-Qaida terrorist network. Spc. Ryan G Anderson was being held at Fort Lewis, charged of "aiding the enemy by wrongfully attempting to communicate and give intelligence to the Al-Qaida terrorist network." Anderson had converted to Islam five years before his arrest. The National Guardsman was held after allegedly offering military information through an Internet chat room. He thought the people he was contacting were Muslim terrorists, but in fact they were FBI agents running a "sting" operation.

CNN reported that the information the suspect was trying to sell concerned protective equipment for armored vehicles deployed in Iraq. An investigation by the *National Review Online* reporter Michelle Malkin of Usenet discussion forums yielded some interesting clues about Anderson:

> Using the e-mail address wensler@wsunix.wsu.edu, Anderson posted prolifically to a strange and volatile variety of Internet newsgroups, including forums for gun enthusiasts, skinheads, cinema, games, Islam, and Arabic culture. He offered opinions on everything from the movie *Starship Troopers* to sniping weapons and presidential politics. Putting aside the chronic misspellings, the 154 messages are an instructive glimpse into the mind of an immature American youth with a potentially dangerous longing to belong… On September 26, 1997, Anderson — now with a new, adopted Muslim name, Amir, posted to soc.culture.arabic: "Salaam Alaaykum all, I am a Muslim convert studying at Washington State University for an Asian History Major, my focus is on Arabic nations…" In October 1997… Anderson offered a conspiracy theory posted to soc.culture.somalia, soc.culture.palestine, alt.religion.islam and alt.culture.somalia: "Is it possible Mossad replaced the Sheik entirely? With the amount of knowledge they have on people, I wouldn't put it entirely beyond them to be able to substitute an imposter." [13]

In September 2004 Anderson was found guilty on all five counts of trying to help the Al-Qaida terrorist network.

More typically, however, terrorist organizations go looking for recruits rather than waiting for them to present themselves. The SITE Institute, a Washington DC-based terrorism research group that monitors Al-Qaida's Internet communications, has provided chilling details of a high-tech recruitment drive launched in 2003 to recruit fighters to travel to Iraq and attack US and coalition forces there. Potential recruits are bombarded with religious decrees and anti-US propaganda, provided with training manuals on how to be a terrorist, and—as they are led through a maze of secret chat rooms—given specific instructions on how to make the journey to Iraq [14].

Rita Katz, the SITE Institute's director, argues, “Al-Qaida's use of the Internet is amazing. We know from past cases—from captured Al-Qaida fighters who say they joined up through the Internet—that this is one of the principal ways they recruit fighters and suicide bombers.” [15]

On June 6 2003, Abu Thur, a computer programmer in Malaysia, posted a message to the Arabic-language “himmame” message board, hosted by Lycos in the United Kingdom. The message read:

> Dear Brothers,
> I have already succeeded with the grace of Allah and his help, to go to Kurdistan for Jihad through one of the brothers in this forum. Praise be to Allah, I have fought there, by the grace of God and his bounty. But Martyrdom was not granted to me, and therefore I ask Allah to give me more lifetime and to make my deeds good. I ask anyone who has the capacity to organize for me to go to another Jihad front to correspond with me. [16]

Asad Allah, an Egyptian based in Malaysia, posted a similar message, asking for help from forum members: “I told myself that if I am already here [in Malaysia], I might as well fulfill my Jihad, far away from the Egyptian authorities… I failed to contact or get to know someone who might help me with fulfilling Jihad here. Should I go back [to Egypt]?” [16]

In one particularly graphic exchange in a secret Al-Qaida chat room in early September 2003, an unknown Islamic fanatic, with the user name “Redemption Is Close,” wrote, “Brothers, how do I go to Iraq for Jihad? Are there any army camps and is there someone who commands there?” Four days later he had a reply from “Merciless Terrorist”, who wrote: “Dear Brother, the road is wide open for you—there are many groups, go look for someone you trust, join him, he will be the protector of the Iraqi regions and with the help of Allah you will become one of the Mujahidin”.

“Redemption is Close” then pressed for more specific information on how he could wage jihad in Iraq. “Merciless Terrorist” sent him a propaganda video and instructed him to download the software called PalTalk, which enables users to speak to each other on the Internet without fear of being monitored [17].

Many terrorist websites stop short of enlisting recruits for violent action, but they do encourage supporters to show their commitment to the cause in other tangible ways. “How can I help the struggle: A few suggestions,” runs a heading on the *Kahane Lives* website, while “Action alert: What you can do” is a feature on the *Shining Path's* website.

The power of the Internet to mobilize activists is illustrated by the response to the arrest of Abdullah Ocalan, leader of the Kurdish terrorist group the PKK. When Turkish forces arrested Ocalan, tens of thousands of Kurds around the world responded with demonstrations within a matter of hours—thanks to sympathetic websites urging supporters to protest.

In the summer of 2002, Israeli police arrested three Jerusalem Arabs linked to Hamas, who allegedly planned to carry out a mass poisoning of clients in a Jerusalem café. One of the three, Othman Kiania, was employed as a chef at the restaurant, and planned to mix poison in the drinks. His partners were Sufian Abdo and Mussa Nasser, who ran a Hamas Internet forum. According to the *Jerusalem Post*,

> Abdo had contacted various members of Hamas via e-mail early this year, expressing his willingness to take part in terrorist activities. For the next eight months, he and Nasser kept in contact with Hamas officials, by providing them with the name of an Egyptian engineer who could help them develop Kassem rockets and raise funds for Hamas. In July, Abdo told them he wanted to carry out a suicide bombing to avenge the death of Salah Shehadeh [a well-known Hamas leader]. They suggested he carry out a mass poisoning at a large Jerusalem cafe, and he agreed. [18]

They were tracked down and arrested by Israeli security agents just a day before executing the plan.

The use of the Internet by radical Muslim terrorists was formally approved by Sheikh Abdul Aziz Al-Alshaikh, the Grand Mufti of Saudi Arabia and the highest official cleric in the country. In December 2002 the Saudi Information Agency obtained a fatwa issued by the Grand Mufti approving cyberterrorism. This special fatwa appeared in the religious magazine *Al-Dawa*, published in Riyadh. The fatwa, in the form of question-and-answer, read:

> **Question:** If there were websites on the Internet that are hostile to Islam, and broadcasting immoral materials, is it permissible for me to send it viruses to disable and destroy these websites?
> Abdul Aziz Saleh Al-Morashid – Erqa.
> **Answer:** If these websites are hostile to Islam and you could encounter its evilness with goodness; And to respond to it, refute its falsehood, and show its void content; that would be the best option. But if you are unable to respond to it, and you wanted to destroy it and you have the ability to do so, it's ok to destroy it because it is an evil website. [19]

Following the publication of this fatwa, Saudi hackers started to attack FBI and Pentagon websites. In an interview with one of these hackers, published in the *Al-Riyadh* newspaper on September 5 2001, he reported attacking over 1,000 websites in the US, including many government official websites.

4. Targeting Children

The Internet is very popular among children and youth. Terrorists know this, and are using the Internet to target children. One of Hamas's websites, Al-Fateh ("The Conqueror"), is updated every other week and is designed for children, what with its cartoon-style design and colorful children's stories. The website's title promises "pages discussing Jihad, scientific pages, the best stories, not be found elsewhere, and unequalled tales of heroism." The Al-Fateh site has a link to the official Hamas site, www.palestine-info.com.

Among its attractive graphics, drawings, children's songs and stories ("The Thrush", "The Troubles of Fahman the Donkey", texts written by children themselves, etc.), the Al-Fateh website posts messages promoting suicide terrorists. Thus, in

October 2004, the website presented the picture of a decapitated head of a young female, Zaynab Abu Salem, a suicide bomber. On September 22 2004, she detonated an explosive belt at a soldiers' hitchhiking stop at in Jerusalem, killing two Border Patrol policemen and wounding seventeen civilians.

The text accompanying the horrible picture praises the act, arguing that she is now in paradise, a "shaheeda" like her male comrades: "The perpetrator of the suicide bombing attack, Zaynab Abu Salem. Her head was severed from her pure body and her headscarf remained to decorate [her face]. Your place is in heaven in the upper skies, oh, Zaynab, sister [raised to the status of heroic] men" [20].

The same website, among its cartoons and children's stories, posted the last will of a Hamas suicide bomber, who on June 1 2001 perpetrated a suicide bombing attack at the Dolphinarium in Tel Aviv, a club frequented by teenagers. The attack resulted in the deaths of 21 Israeli civilians, most of them teenagers, and 83 wounded. In his online will, the suicide bomber writes to "the Engineer, Yahya Ayyash" (a Hamas terrorist who was an expert in constructing bombs and explosive belts until he was killed by the Israelis): "you taught us… that the true heroes are those who write the history of their people in blood… I will turn my body into shrapnel and bombs, which will chase the children of Zion, will blow them up and will burn what is left of them… There is nothing greater than killing oneself on the land of Palestine, for the sake of Allah." [20]

The Hezbollah, too, targets children and adolescents. In 2003, Hezbollah began online marketing of a computer game simulating terrorist attacks on Israeli targets. The computer game, called *Special Force,* was developed by the Hezbollah Central Internet Bureau, and its development, according to a report in Lebanon's *Daily Star*, took two years [21].

The game places the player in different Hezbollah operations against Israelis; the producers of *Special Force* argue that it is based on actual Hezbollah battles with Israeli forces. *Special Force* can be played in Arabic, English, French, and Farsi, and is available on one of the Hezbollah websites. The violent game also features a training mode, where players can practice their shooting skills on targets such as Israeli Prime Minister Sharon and other Israeli political and military figures. A good performance earns a certificate presented by Hezbollah leader Sheikh Hassan Nasrallah in a "cyberceremony".

The game ends with an exhibit of Hezbollah "martyrs", fighters killed by Israel. On the game's cover, and in the online version, there is a message to users that states: "The designers of Special Force are very proud to provide you with this special product, which embodies objectively the defeat of the Israeli enemy and the heroic actions taken by heroes of the Islamic Resistance in Lebanon. Be a partner in the victory. Fight, resist and destroy your enemy in the game of force and victory."

Mahmoud Rayya from the Hezbollah, in an interview with the *Daily Star*, said that the decision to produce the game was made by leaders of Hezbollah officials. He stated, "In a way, Special Force offers a mental and personal training for those who play it, allowing them to feel that they are in the shoes of the resistance fighters." [21]

5. Conclusion

The future of terrorism relies more on future recruitment than on any other single factor. Although terrorists do not use the Internet for direct recruitment, but only to identify and profile potential candidates for recruitment, this use of the Net is nonetheless alarming. The fact that the Internet has become a more popular apparatus for the early stages of recruitment and mobilization is challenging governments, security agencies and counterterrorism services all over the world.

Moreover, it also challenges the future of the Internet, since any attempt to limit or minimize Internet usage by terrorists implies imposing restrictions on the Internet's free flow of information, and an abuse of privacy. Thus, any step or measure to fight the terrorists' use of the Net should involve a careful examination of the prices in terms of civil liberties. We need to find a "golden way" that will minimize the abuse of the Internet by terrorism without harming civil liberties.

Without recruitment, terrorism cannot prevail, survive and develop. Recruitment provides the killers, the suicide bombers, the kidnappers, the executioners. The Hotel Intercontinental trade-offs between security and civil liberties, when it comes to the Internet, may have disastrous consequences: terrorist use of the Internet will grow; it will become more sophisticated and more manipulative. This, in turn, may lead to harsher counterterrorism measures, enforced by terror-stricken governments and security agencies.

References

[1] From the "Report of the Official Account of the Bombings in London on 7th July 2005", published on 11th May, London: The Stationery Office.

[2] Hoffman, Bruce, 2004. "Redefining counterterrorism: The terrorist leader as CEO", *Rand Review*, Spring 2004, p. 1.

[3] Sageman, Marc. 2004. *Understanding Terror Networks*. Philadelphia: University of Pennsylvania Press.

[4] Hoffman, Bruce, 2003. *Al-Qaida, Trends in Terrorism and Future Potentialities: An Assessment*, Washington, DC: RAND publications.

[5] Weimann, Gabriel, 2004 (a). *WWW.Terror.Net: How Modern Terrorism Uses the Internet*, Special report, Washington DC: United States Institute of Peace.

[6] Weimann, Gabriel, 2004 (b) *Cyberterrorism: How Real Is the Threat?* Special report, Washington DC: United States Institute of Peace.

[7] Weimann, Gabriel, 2006. *Terror on the Internet: The New Arena, The New Challenges*. Washington, DC: USIP Press.

[8] Gadahn, Adam, 2004. "Becoming Muslim", at: http://www.usc.edu/dept/MSA/newmuslims/yahiye.html

[9] Rodriguez, Paul M. 2004. Iranian Group Seeks Recruits for 'Martyrdom Operations', *Insight Online*, posted June 8, 2004, at: www.insightmag.com/global_user_elements/printpage.cfm?storyid=684918

[10] Berger, J.M. 2004, "Online videos shows children training for jihad; young son of Osama bin Laden reads statement", *Intelwire, at:* http://www.intelwire.com/2004_10_13_exclusives.html

[11] Zanini, Michelle and Sean J. A. Edwards, "The Networking of Terrorism in the Information Age," in *Networks and Netwars: The Future of Terror, Crime, and Militancy*, ed. John Arquilla and David Ronfeldt (Santa Monica, CA: RAND Corporation, 2001), 43-430.

[12] Cited by Lawrence Wright, 2004. "The Terror Web", *The New Yorker*, August 2, 2004.

[13] Michelle Malkin, "Trailing Attempted Espionage," *National Review Online* (February 13, 2004), at http://www.nationalreview.com/comment/malkin200402130909.asp.

[14] Niles Lathem, "Al-Qaeda Terror.com," *New York Post* (September 16, 2003), a12.

[15] See SITE at http://www.siteinstitute.org/.

[16] Cited in Laura Rozen, "Forums Point the Way to Jihad," *Wired New* (2003) at http://www.wired.com/news/conflict/0,2100,59897,00.html.

[17] This exchange is cited by Lathem, Niles, 2003, "Al-Qaida Trolls Net", *New York Post*, 15 September, 2003.
[18] Etgar Lefkovitz, "Three Arabs held in Jerusalem cafe poisoning plot", *The Jerusalem Post*, September 10, 2002, at: http://www.jpost.com/
[19] Translated and published online, at: http://www.arabianews.org/english/article.cfm?qid=19&sid=6&printme=1, last accessed: August 2004.
[20] The picture and the text translated to English are included in the report "Educating children for hatred and terrorism: encouragement for suicide bombing attacks and hatred for Israel and the Jews spread via the Internet on Hamas' online children's magazine (Al-Fateh)" posted at: www.intelligence.org.il/eng/sib/11_04/edu.htm
[21] At http://www.worldnetdaily.com/news/article.asp?ARTICLE_ID=31323.

Further References

Lewis, Bernard. 2002. *The Assassins: A radical sect in Islam.* New York: Basic Books.
Stern, Jessica, 2003. The protean enemy, *Foreign Affairs,* August, p. 3.
Tsfati, Yariv and Gabriel Weimann. 1999. Terror on the Internet. *Politika*, 4: 45-64 (in Hebrew).
Tsfati, Yariv and Gabriel Weimann. 2002. WWW.terrorism.com: Terror on the internet. *Studies in Conflict and Terrorism* 25(5): 317-332.
Verton, Dan (2003). *Black Ice: The Invisible Threat of Cyber-Terrorism.* New York: McGraw-Hill Osborne Media
Von Hammer-Purgstall, Joseph. 1968. *The History of Assassins*, New York: Burt Franklin.

Hypermedia Seduction for Terrorist Recruiting
B. Ganor et al. (Eds.)
IOS Press, 2007

Islamist Websites as an Integral Part of Jihad: A General Overview

Yael YEHOSHUA
MEMRI Director of Research

Abstract. The media is viewed by the Islamist terrorist organizations as one of the fronts of jihad against their enemies. The media platform most favored by activists and supporters of Islamist organizations is the Internet, due to its anonymity, availability, low cost and instantaneous communication. The organizations and their supporters disseminate their message via websites and forums in different languages, targeting diverse audiences worldwide. These organizations also utilize the Internet for military and operative purposes in the service of the jihad fighters in the field, but their primary use of this medium is for indoctrination and propaganda. They distribute various messages aimed mainly at glorifying the organizations' activities, spreading their extremist ideology and enlisting support and legitimacy, as well as threatening and influencing their enemies. This Internet activity is described as "propaganda jihad" or "media jihad", waged by those who cannot participate in the actual fighting. Furthermore, Islamist sites constitute an arena of confrontation in their own right. These sites are targeted by hackers trying to disrupt their activity, and in turn encourage Islamist hackers to target "enemy" sites. Islamist organizations also confront the activities of elements suspected of being intelligence agents infiltrating their sites. In addition, the sites constitute an environment for Arab attempts to initiate dialogue with the extremists in order to persuade them to renounce their extremist views. Finally, despite their view of the West as an enemy and of Western culture as corrupt, Islamist organizations do not hesitate to utilize the services of Western Internet companies, and many of their sites are hosted by Western servers. Consequently, it is largely up to the West to come up with solutions, by appointing a body that will issue warnings to the public about sites that disseminate incitement and encourage terrorism, or through legislation that will ban this activity.

Introduction

Alongside military jihad on various battlefields, Islamist organizations are engaged in another type of warfare: jihad on the media front. The media is viewed by Islamist organizations both as an important battlefield and as an effective tool, and they consequently make extensive use of it.

The media platform favored by the Islamist organizations is the Internet, which they prefer for several reasons: first, for the anonymity it allows – anyone can enter and post to a site without divulging personal information; second, due to the medium's availability and low cost – all that is required is a PC and an Internet connection; and third, due to the ability to distribute material to a great number of people over a wide geographic area in a matter of seconds.

The organizations use the Internet mainly for propaganda and indoctrination, but also for operational military needs.

This paper will discuss the distinguishing characteristics of the websites of Islamist organizations and their supporters; the various online activities through which terrorist organizations assist the mujahideen on the ground, both militarily and, especially, with propaganda; and the Internet polemics that these organizations conduct vis-à-vis their enemies.

1. The Internet in the Service of Islamist Organizations

1.1. Main Characteristics of Islamist Websites

Islamist websites operate out of various countries, both Muslim and non-Muslim, and their target audience includes countries and communities all over the world. Accordingly, the websites address their audiences in various languages, from Arabic, Farsi, Urdu, and Turkish to Western languages such as English and French.

Some websites have versions in different languages, both Eastern and Western. Examples are the unofficial Hamas site (www.palestine-info.co.uk) and the site of the British organization Supporters of Shari'a (http://www.supportersofshariah.com), which was headed by Abu Hamza Al-Masri, who is currently serving a prison sentence in London. It is noteworthy, however, that the different versions do not carry the same content, since they are aimed at different audiences.

Islamist sites tend to be transitory and impermanent – new sites appear and others close down or are shut down on a daily basis. An example is the Islamist forum Al-Qal'a (http://www.qal3ati.com), once popular and active, which was shut down after it posted the statement taking responsibility for the July 2005 London bombings. Sites also frequently change their URLs, as in the case of the unofficial Hamas site www.palestine-info.info (previously www.palestine-info.co.uk and www.palestine-info.com).

1.2. Islamist Organizations' Use of Internet Services

Islamist organizations and their supporters use all the usual Internet services, including websites, forums, and discussion groups. According to some media reports, they also use the Internet for communications, instead of telephones and land mail.

1.2.1. Websites

- Official websites of Islamist organizations:

The Islamists' basic use of the Internet is the establishment and operation of websites through which they disseminate their messages. Some Islamist organizations maintain official websites, established by the organization and run by individuals with ties to the organization. It is not always possible to know for certain whether a given website is indeed an official one, but from the website's contents it can often be understood that those running it have direct ties to top organization officials. Since the 2003 invasion of Iraq, many Iraqi terrorist organizations fighting against the Coalition forces have set up official websites. Examples are the site of the Islamic State of Iraq (ISI), an umbrella organization of several Sunni Iraqi terrorist groups led by Al-Qaeda in Iraq (http://islamiciraq.modawanati.com), and the site of the Islamic Army in Iraq (www.iaisite.org) – the largest organization that has not joined the ISI.

- Official websites of religious scholars supporting the Islamist organizations:

Another category is websites of religious scholars who play an active role in setting up the organizations' ideological infrastructure, and in granting religious-legal legitimacy to their activities. Many of these religious scholars are currently serving prison sentences for incitement to terrorism or even for active involvement in terrorism – but this does not prevent them from spreading incitement and supporting terrorism via their websites, which remain active. An example is the website of the Jordanian Sheikh Abu Muhammad Al-Maqdisi (www.almaqdese.net), who was the spiritual mentor of former Al-Qaeda in Iraq commander Abu Mus'ab Al-Zarqawi. The name of Al-Maqdisi's site is Minbar Al-Tawhid Wal-Jihad (The Podium of Monotheism and Jihad), which is reminiscent of the original name of Al-Zarqawi's organization, initially called Jama'at Al-Tawhid Wal-Jihad (The Monotheism and Jihad Group). Al-Maqdisi's website contains numerous writings that provide ideological and religious-legal legitimacy for many Islamist terrorist organizations. [1]

- Unofficial websites supporting the organizations:

Most websites utilized by Islamists are not designated as official websites of a specific organization or individual. Nonetheless, in many instances they promote the aims of certain organizations, regardless of whether members of these organizations are involved in running the site. Al-Qaeda, for instance, currently has no known official website, but there are sites that identify themselves with the organization and its activities, such as Al-Qa'idun (www.qa3edoon.com), whose founders describe themselves as "a group that identifies with the jihad fighters, and whose task is to publish statements from Al-Qaeda in the Arabian Peninsula."

1.2.2. Forums

Internet forums are currently the Internet service most widely used by Islamist organizations for conveying messages related to the organizations' terrorist activities or propaganda. The importance that the organizations attach to these forums as a media venue is evident from the fact that their official statements, such as communiqués or videos, are often posted in forums before they are posted on the websites themselves.

In some forums, such as Al-Hesbah (www.alhesbah.org), most of the rooms are dedicated to the activities of Islamist organizations or support for them. Other forums, such as the Saudi forum Al-Saha (www.alsaha.com), deal with these topics as well as other matters.

Many forums are limited to members only and require prior registration. Some even have an "admissions committee" and assign a different status to different forum members, while other forums are closed and no new members are admitted. While this might appear to run counter to the organizations' desire to spread their messages, registration is possibly aimed at increasing the forum's prestige, keeping tabs on members, as a defensive measure against alleged activity by infiltrators from intelligence services, or – assuming that some sites are actually operated by these services – in order to collect information on the Islamists who enter the site.

The Islamists also use the Internet portal Yahoo for establishing discussion groups to spread their extremist views. One example of this is the Modjahede discussion group (http://groups.yahoo.com/group/modjahede), which included links to guides for the production and use of conventional and nuclear weapons. [2]

2. Islamist Organizations' Internet Activities: An Integral Part of Jihad

Islamists regard their information activities as an integral part of their jihad, and therefore invest great resources – both material and non-material – in Internet activities. In fact, media or information activities on the Internet are described as a kind of jihad that can be carried out by those who cannot take part in the fighting on the battlefield. This kind of jihad is called "propaganda jihad" (*al-jihad al-da'wi*) or "media jihad" (*al-jihad al-i'lami*). The concept is based on the well-known Hadith of the Prophet Muhammad: "One who sees a wrong must correct it with his hand, and if he cannot, then with his tongue, and if he cannot, then in his heart, and this is the weakest level of faith."

This perception is reflected in the media activity of Al-Jabha Al-I'lamiyya Al-Islamiyya Al-'Alamiyya – known in English as the Global Islamic Media Front (GIMF) – an Islamist media group that has posted numerous reports and statements dealing with jihad fighters. GIMF denies that it has a connection to Al-Qaeda, though it has posted several Al-Qaeda statements taking responsibility for terrorist acts.

With the GIMF's establishment, its leader, who calls himself "the second Salah Al-Din," published a statement declaring that "the Media Front's project is a project of media jihad, for jihad is one of the doors to Paradise." Urging Muslims to join GIMF, he said that it was meant for …everyone who believes in Allah, the One and Only, and who believes that Muhammad is his servant and his messenger... [It is meant for] everyone who cannot carry out jihad with his wealth or with his life, for everyone who feels in his soul the jihad of the word and of opinion, for everyone who can reveal the facts to the people of our [Muslim] nation and to those non-Muslims who are not hostile towards us, and for everyone who wants to publish reports on Muslims and to convey their pure words to those who want to take part in spreading awareness by every means."[3]

When, in August 2005, GIMF established the Media Jihad Brigade (Katibat Al-Jihad Al-I'lami), it explained that "the media war is an integral part of the war on the battlefield." The members of the Brigade were described as "a group of Muslims who desire to wage jihad and to bear arms, but until Allah allows them to do so, they take upon themselves to help the jihad fighters by toppling the Zionist hegemony over the media."[4]

Islamists use the Internet as a tool in the military training of jihad fighters by circulating military guidebooks on weaponry, battle tactics, explosives manufacture, and other topics. An example is Al-Qaeda's online military magazine *Mu'askar Al-Battar* (The Al-Battar Training Camp), published by the Military Committee of the *Mujahideen* in the Arabian Peninsula.[5] Some websites also carry courses on manufacturing explosives and even guides for making homemade dirty bombs.[6]

However, Islamist organizations' most prominent use of the Internet is for information and propaganda. They see this use as of the utmost importance, and invest considerable resources in it; thus, it constitutes most of the organizations' online activities. Al-Qaeda, for instance, has media divisions and several production companies, such as Al-Sahab, and is also served by media groups with uncertain affiliation, such as the aforementioned GIMF.

It should be noted that prominent terrorists play an active role in the organizations' online media activities. For example, Fares Al-Zahrani, a top member of Al-Qaeda in the Arabian Peninsula who is currently imprisoned in Saudi Arabia, used to post on various Islamist sites like Minbar Al-Tawhid Wal-Jihad under the pseudonyms Abu Jandal Al-Azdi and Al-Athari.[7] Another senior operative in this organization, 'Abd Al-'Aziz Al-'Anzi, who was likewise arrested after participating in military clashes with Saudi security forces, was also very active on the Internet. Al-'Anzi served as head of the Media Council of Al-Qaeda in the Arabian Peninsula, recruited supporters on Internet forums, and was one of the supervisors of www.alsalafyoon.com, under the pseudonym 'Abd Al-'Aziz Al-Bakri. He also wrote regularly for Al-Qaeda's online magazine *Sawt Al-Jihad* (The Voice of Jihad) under various pseudonyms.[8]

3. Main Goals of Online Propaganda

3.1. Glorifying the Organization's Activities

One of the goals of the organizations' online activities is to praise and exalt their operatives and activities. This is done through statements, speeches, and videos, in which the organization leaders boast of their terrorist activities. A prominent example is speeches by Al-Qaeda leaders like Osama bin Laden and his deputy Ayman Al-Zawahiri, by former Al-Qaeda in Iraq commander Abu Mus'ab Al-Zarqawi, and by his successor Abu Hamza Al-Muhajir. Videos of operations showing preparation and execution of attacks likewise serve to glorify the organization and its fighters.[9]

3.2. Spreading the Organizations' Ideology

In order to spread their extremist ideology, the organizations post essays and fatwas by religious scholars who are active in the organizations or who support their activities. Examples can be found in Al-Qaeda's online magazine *Sawt Al-Jihad* and in the abovementioned website of Sheikh Abu Muhammad Al-Maqdisi, Minbar Al-Tawhid Wal-Jihad, both of which include essays and fatwas by fundamentalist religious scholars supplying ideological justification for terrorist activities.

3.3. Legitimizing the Organizations' Activity

The organizations make an effort to legitimize their activities by bringing various arguments to justify their actions. The issue of legitimizing jihad was discussed explicitly in the magazine of the Information Division of Al-Qaeda in Iraq, *Dhurwat Sanam Al-Islam* ("The Crest of the Summit of Islam" – an expression which is commonly used to refer to jihad). The author of the article, writing under the pseudonym Nur Al-Din Al-Kurdi, wrote: "If we want monotheism to triumph through jihad, we must [conduct] a goal-oriented media [campaign] to explain the aims of the operations carried out by the jihad fighters. [Our] aim is not just to carry out the operation. We must [also] explain why the operation was carried out the way it was. Only then will out mission be complete... Our nation has lost so many campaigns because [it failed on] the media [front]."[10]

3.4. Threatening and Influencing the Enemy

The organizations use the Internet to convey messages to their enemies. A prominent example is a speech by bin Laden which was released by Al-Qaeda's production company Al-Sahab just before the U.S. presidential elections of October 2004. In this speech, bin Laden hinted that a U.S. state that did not vote for Bush would not be targeted. [11] Videos showing small-scale terrorist operations, such as attacks on Hummers and helicopters, or sniper attacks, are likewise aimed at demoralizing the enemy and at influencing public opinion in enemy countries. The organizations assume that showing images of wounded or killed soldiers affects public opinion and may induce the public to pressure its leadership. GIMF's Media Jihad Brigade has declared that its aim is to "affect [the] enemies' morale, thwart their aspirations and dreams, and expose the truth about what goes on in Iraq."[12]

4. The Internet as a Front of Confrontation with the Enemies

In addition to serving as a tool to assist the fighters in combating their enemies on the ground, the Islamist websites are also an environment in which this confrontation takes place.

4.1. Hacker Activities in Both Camps

Islamist websites are subject to hacking attempts by Arab and Western hackers trying to shut down their activity, and the Islamists use the same methods against "enemy" sites, calling it "electronic jihad." GIMF's Media Jihad Brigade claims to have hacked forums and websites used by soldiers in Iraq and their families and to have posted "frightening messages" in them.[13] In fact, Islamist forums such as Al-Firdaws (http://www.alfirdaws.org/vb/) have special areas devoted to electronic jihad, which include instructions for hackers and serve as a platform for exchanging information.

4.2. Alleged Activity by Western and Arab Intelligence Services on Islamist Forums

Some Islamist websites claim that Western and Arab intelligence agents are active on Islamist forums, and accuse various sites of being "espionage sites" run by intelligence services. For example, the Al-Tajdeed forum (www.tajdeed.org.uk), which is affiliated with a London-based Saudi oppositionist organization, has leveled this accusation against Al-Hesbah (www.alhesbah.org), claiming that the latter is operated by Western and Arab intelligence services with the aim of exposing and arresting prominent individuals who post there, such as "Irhabi 007" or "Bin Roma."[14] Such accusations raise important questions regarding the authenticity of some Islamist websites, which is an issue that would require further investigation.

4.3. Attempts to Conduct Dialogue with Individuals on Islamist Sites with the Aim of Reforming Them

A different kind of confrontation is represented by the Saudi Al-Sakinah (Tranquility) campaign, which aims to initiate online dialogue with extremists on the Islamist websites in an effort to lead them to renounce their extremist views. This initiative was launched in 2003 by volunteers and later received the blessing and support of the Saudi Ministry of Religious Endowments and Islamic Affairs.

The initiator and director of the campaign, 'Abd Al-Mun'im Al-Mushawwah, told the Saudi daily *Al-Madina* on July 1, 2005 that the campaign was being conducted by "some 40 ulama and propagators of Islam who have Internet skills. [They] enter extremist websites and forums and converse with the participants in order to bring them to renounce their extremist ideas." He added that the ulama "conduct dialogue [only] with individuals who have expressed solidarity with the operations of violence and terror, but have not participated in them, in order to prevent them from doing so [in the future] – because those who express solidarity [with the terrorists] are likely in the future to turn into collaborators and perpetrators, and this is the most dangerous thing..." He further explained that "after identifying people with extremist views on the various sites, the Al-Sakinah representatives try to draw [them]... into one-on-one dialogue using chat or messenger [programs], since one-on-one dialogue affords more scope for persuasion. It allows you to exchange views far from the hubbub of the public dialogues and without external influences..."

According to reports in the Saudi media, 'Abd Al-'Aziz Al-'Anzi, mentioned above as a terrorist active on the Internet, renounced his extremist views and disappeared from the websites following a conversation with an Al-Sakinah representative.[15]

5. Most Islamist Websites are Hosted in the West

An interesting point is that although they regard the West as an enemy and Western culture as corrupt, Islamist organizations have no qualms about using the services of Western Internet companies, and exploiting the West's freedom of speech. Most of these websites are hosted by servers in the West, and some are even registered there. A two-part MEMRI report titled "Islamist Websites and their Hosts," published in 2004, gave lists of Islamist websites along with their technical specifications. Most of the websites surveyed in this report used servers of Western companies, and there is no reason to assume that the picture has changed significantly since 2004 [16].

The fact that most of the sites utilize Western services means that the West must come up with solutions that do not violate the principle of free speech. There is a need for legislation banning terrorist activities on the Internet. Saudi Arabia, whose tactics in fighting these sites are mentioned above, is currently discussing a new anti-Internet crime bill, which was submitted by the Saudi Communications Committee. Under the proposed legislation, anyone convicted of creating a website affiliated with a terrorist organization will be sentenced to 10 years in prison and fined SAR 5,000,000 (approximately $1,300,000) [17].

Until countries legislate against Islamic terror organizations' online activity, Internet companies themselves must be alert to the content of such sites, and must take steps to prevent the sites' activity. It is further suggested that countries issue warnings about sites that disseminate incitement and encourage terrorism – similar to the travel advisories and warnings issued by the U.S. State Department.

---

[1] Y. Yehoshua, "Dispute in Islamist Circles over the Legitimacy of Attacking Muslims, Shiites, and Non-Combatant Non-Muslims in Jihad Operations in Iraq: Al-Maqdisi vs. His Disciple Al-Zarqawi," MEMRI, *Inquiry and Analysis* No. 239, September 11, 2005.

[2] MEMRI Special Dispatch No. 1039, "Terrorist and Inciting Messages on Yahoo," December 6, 2005; MEMRI Special Dispatch No. 1004: "On Islamic Websites: A Guide for Preparing Nuclear Weapons," October 12, 2005.

[3] Statement by GIMF, posted May 15, 2005 on various Islamist websites. See, for example: http://www.hanein.net/modules.php?name=Forums&file=viewtopic&p=2922.

[4] *Al-Sharq Al-Awsat* (London), August 14, 2005.

[5] MEMRI Special Dispatch No. 637, "The Al-Battar Training Camp: The First Issue of Al-Qa'ida's Online Military Magazine," January 6, 2004.

[6] MEMRI Special Dispatch No. 1004, "On Islamic Websites: A Guide for Preparing Nuclear Weapons," October 12, 2005.

[7] *Al-Sharq Al-Awsat* (London), July 31, 2005.

[8] http://www.alarabiya.net/Articles/2005/10/17/17782.htm.

[9] See, for instance, MEMRI TV Clip No.1131, "An Al-Qaeda Released Video of Attacks in Afghanistan," April 5, 2006; MEMRI TV Clip No.939: "Al-Qaeda Presents Footage of Preparations for Triple Suicide Bombings against Baghdad Hotels," November 26, 2005.

[10] N. Al-Kurdi, "The Fronts of Jihad," *Dhuwat Sanam Al-Islam,* Issue No. 3, pp. 25-27.

[11] MEMRI Special Alert No. 14, "Osama Bin Laden Tape Threatens U.S. States," November 1, 2004; Special Dispatch No. 811, "The Full Version of Osama Bin Laden's Speech," November 5, 2004.

[12] *Al-Sharq Al-Awsat* (London), August 14, 2005.

[13] *Al-Sharq Al-Awsat* (London), August 14, 2005.

[14] D. Hazan, "Al-Tajdeed Versus Al-Hesbah: Islamist Websites & the Conflict Between Rival Arab & Muslim Political Forces," MEMRI, *Inquiry & Analysis* No. 275, May 17, 2006.

[15] Y. Yehoshua, "Reeducation of Extremists in Saudi Arabia," MEMRI *Inquiry & Analysis* No. 260," January 18, 2006.

[16] M. Boccara, "Islamist Websites and Their Hosts Part I: Islamist Terror Organizations," MEMRI, *Special Report* No. 31, July 16, 2004;
M. Boccara & A. Greenberg, " Islamist Websites and their Hosts Part II: Clerics," MEMRI, Special Report No. 35: November 11, 2004

[17] *Al-Watan* (Saudi Arabia), August 29, 2006.

Hypermedia Seduction for Terrorist Recruiting
B. Ganor et al. (Eds.)
IOS Press, 2007

Virtual Communities as Pathways to Extremism

Don RADLAUER
Research Associate
The International Institute for Counter-Terrorism (ICT), Interdisciplinary Center, Herzliya, Israel

Abstract. The Internet is playing an increasing role in terrorism – not only in obvious areas such as command and control, technical instruction, and the publishing of ideological tracts, but also as a social medium in which groups of people form "virtual communities". In some cases, these communities can become progressively radicalized to the point where they eventually commit or support acts of violence. An understanding of "virtual communities" is necessary in order to create means of preventing them from functioning as incubators for terrorism.

Keywords. Virtual communities, online communities, Internet, terrorism, radicalism, extremism

Introduction

One of the notable aspects of much recent terrorism is that the Internet has played a significant role in its preparation and execution: sometimes as a tool for command and control, sometimes as a source of information and technical instructions, sometimes as a medium for publishing ideological manifestos and warped, emotionally manipulative versions of the news of the day, and sometimes as a means for socialization among groups that later carried out or supported terror attacks.

All these uses of the Internet are obviously of concern to those responsible for combating terrorism; but the latter function – the use of the Internet as a social venue – is of special interest. Most, if not all, of the other roles the Internet plays in facilitating terrorism could be accomplished to similar effect without the use of modern communications technologies, albeit at a slower speed; but socialization through computer-mediated communication ("CMC") appears to have some characteristics and potential effects quite different from those of more traditional means of social interaction.

Social groups interacting through various forms of CMC are generally known as "online communities" or "virtual communities". In most cases, of course, such groups are harmless and even beneficial. However, in some cases virtual communities have been known to lead previously moderate participants towards radical political or religious beliefs, and to deepen the commitment and willingness of those already radicalized to carry out violent acts.

In order to explore the virtual-community phenomenon and its implications, it is important first to understand the general phenomenon of community. Equipped with some understanding of what communities are, why they form, and how they function, we can begin to investigate how virtual communities are similar to – and different from – conventional communities.

1. Communities and Why We Need Them

Human beings, in general, need to belong to some form of social group. This is empirically obvious, but has also been well documented in the professional literature; for example, Baumeister and Leary [1] state that "Existing evidence supports the hypothesis that the need to belong is a powerful, fundamental, and extremely pervasive motivation." We are profoundly social animals; and the "lone wolf" among human beings is, in general, just as unhappy as the true lone wolf ejected from his pack.

It has also been observed – notably fifty years ago by Horton and Wohl [2] – that electronic media can provide a substitute for face-to-face group membership. This likely explains the incredible popularity of television shows like "Cheers" (among many others, including daytime dramas), which seem to offer membership in a "parasocial group" [2] as their primary attraction. The viewer of such shows feels him/herself a part of the social group portrayed by the show; s/he can become intensely involved with the characters and their tribulations, even though s/he is well aware of the fictionality of the characters and situations portrayed.

It is evident that for at least some of us, various forms of interpersonal interaction over the Internet can similarly act as an acceptable substitute for face-to-face socialization. Because these Internet-mediated forms of socialization are genuinely interactive (as opposed to the passive parasocial relationship between the viewer and the characters in Cheers, who in fact do not know your name and are not glad you came), they can constitute a genuine, albeit "low-bandwidth", form of social interaction.

1.1. Social-Group Size: Dunbar's Number and the Prisoner's Dilemma

Human social groupings, like those of other animals, appear to be limited in their size. In an attempt to understand the limits of social-group size, British researcher Robin Dunbar correlated neocortex size with social group size in various primate genera, and by extrapolation came up with "Dunbar's Number" as a practical limit to the size of human social groups. Above Dunbar's Number (approximately 150), so much time is spent simply maintaining group cohesiveness that member satisfaction and group stability break down. In practice, and particularly for dispersed societies, the optimal number of members in a functioning community will likely be much smaller [3].

In his investigation of the application of Dunbar's Number to computer-mediated social groupings, Christopher Allen has empirically graphed "group satisfaction" against group size for certain online gaming communities (see Fig. 1).

He found a peak in satisfaction at around seven members for "simple" teams (or cells), and another peak at around fifty members for groups with some hierarchy and specialization. In between is a "chasm", where the group becomes too large to be optimal as a simple social arrangement, and yet isn't large enough for effective hierarchy and specialization to develop [4].

Work by Ernst Fehr and Urs Fischbacher [5] shows that in computerized simulations of the "Prisoner's Dilemma" game [6] with no option for punishment of betrayal, mutual cooperation among imaginary game participants broke down at a size of 8 to 16 group members. This corresponds nicely to the falloff after the first peak of Allen's empirical graph of group satisfaction [5]. When the option of punishing betrayers (at some cost to the punisher) was added to the simulation, intra-group cooperation was more robust – although it still broke down as groups became so large that enforcement became a significant burden for those who chose to do the punishing.

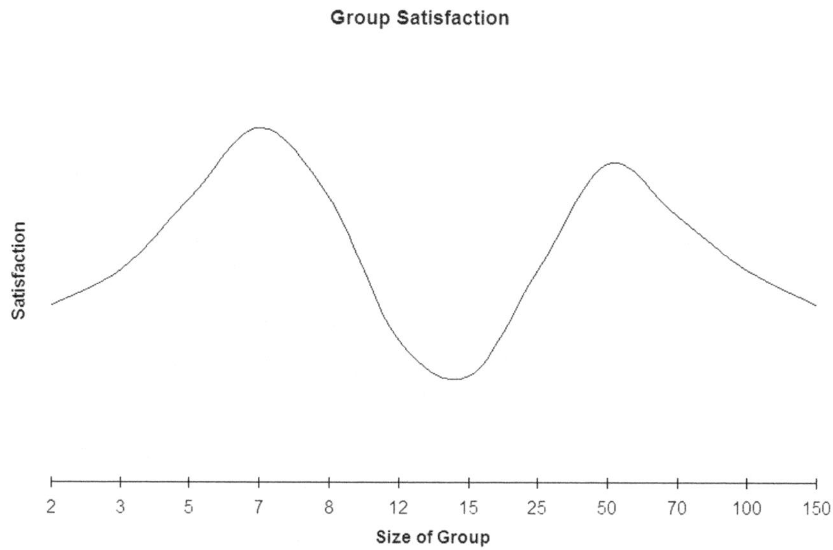

Figure 1. Allen's graph of group satisfaction vs. group size.

Simulated groups that punished both betrayers and non-betrayers who failed to punish betrayers maintained cooperation at much greater sizes. This shows that the addition of a functional hierarchy to the simulation allows larger groups to function effectively. Since Fehr and Fischbacher did not incorporate Dunbar-type limits into their simulation, their fully hierarchical model (the top line in the graph below) does not show a significant decline in mutual cooperation within large groups (see Fig. 2).

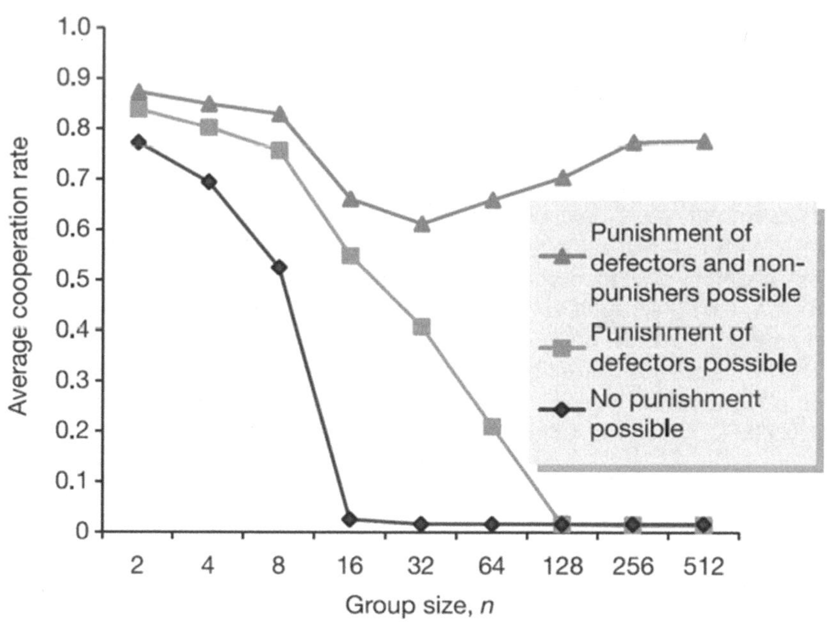

Figure 2. Fehr and Fischbacher's model of functional hierarchy in a Prisoner's Dilemma simulation.

1.2. What Makes a Community Work?

Having established that communities answer a deep need, and that communities of certain sizes are likely to be more stable and satisfying to their members than communities that are too large, too small, or both (in the case of Christopher Allen's "chasm"), we naturally would like to understand more about how communities function. What differentiates a functioning community from a random group of people – the passengers on a bus, for example?

One crucial requirement of a community is almost paradoxically simple: to function as a community, a group of people must feel that they are a community. This sense of community has been further defined as consisting of the following elements [7]:

- Feelings of membership: Feelings of belonging to, and identifying with, the community;
- Feelings of influence: Feelings of having influence on, and being influenced by, the community;
- Integration and fulfillment of needs: Feelings of being supported by others in the community while also supporting them; and
- Shared emotional connection: Feelings of relationships, shared history, and a "spirit" of community.

From the above, it is readily apparent why the relationship between viewers and television-show characters is considered "parasocial" rather than genuinely social. Audience members may feel they are members of the "community" portrayed by their favorite shows, but they have no influence on that "community"; they can "support" their favorite characters only theoretically, and can receive support only with the aid of a good imagination.

In addition to these components of the sense of community, we can add that a functioning community needs something by which to differentiate itself from its surrounding environment. In order to feel fully like "one of us", we need to feel that there is a genuine "us" – and that requires that there be something separating "us" from "them". Traditionally, such differentiation could derive from familial/clan relationships or geographical location: either a fixed location, such as residence in the same village, or, in the case of nomadic groups, a defined range of migration.

But when human agglomerations get large enough (and clan relationships break down sufficiently) so that no available "natural" grouping meets the Dunbar requirements for maximum effective group size, people need to find (or create) other criteria upon which to base a sense of community. Such criteria can take various forms; functional communities (virtual and otherwise) have formed around music and other hobbies, a shared disease, or participation in an online game. In the young, community formation often involves more or less "antisocial" attitudes and behaviors, from wearing "unacceptable" clothes to vandalism to out-and-out crime.

1.3. Interactivity Promotes Sense of Community

As has been shown, part of the sense of community is the feeling that one has an active influence on the community as a whole, and on one's fellow members individually. This applies just as much to virtual communities as to conventional ones. In order to support the formation of a virtual community, an Internet site needs to have appropriate facilities for interaction among members. Such an interactive site has been termed a "virtual settlement" – implying that while an online community may not yet have been formed, an appropriate "place" has been created where such a community can form. It has been shown that highly interactive "virtual settlements" with public exchange of messages are stronger community-builders than less interactive sites [8].

Accordingly, "broadcast" websites may be adequate as an informational and ideological resource for communities to use, but without interactive features they do not function as "virtual settlements" that can support the creation of virtual communities. Chat-rooms, forums, and email lists can function well as "virtual settlements". Blogs, when they include commenting capability, seem to be an intermediate case in which some sense of community can exist, but a strong virtual community seldom if ever comes into existence.

2. How Are Virtual Communities Different from Conventional Social Groups?

The Internet is different from previous social venues in various ways. One particularly important difference is that the Internet can function as both a social venue and as an information source, with one-click linkage between the two. Thus, a virtual community that influences its members to obtain their news and other "facts" preferentially from certain websites can serve as a self-contained, isolated milieu for both socialization and information-gathering. Further, virtual communities are freed from the constraints of geography and of conformity; one's virtual friends can live essentially anywhere in the world, and anyone in need of companionship can find like-minded people without having to conform to the mainstream of one's surrounding non-virtual community. These capabilities, among others, are important in understanding how Internet-based virtual communities can sometimes be a dangerous phenomenon.

2.1. Eliminating Old Isolations and Creating New Ones

In the past, immigrants were effectively isolated from their former countries, and thus faced a high degree of pressure to adopt the language, values, habits, and assumptions of their new home. Information connected to "the old country" was static and essentially "dead", while information connected to the mainstream of the immigrants' new host country was fresh and dynamic. The dominant host-country culture had tremendous cachet; the children of immigrants were eager to cast off the habits and accent of their parents and grandparents, and become as much like their hosts as they could.

The Internet, along with satellite TV, has eliminated much of this isolation, and thus considerably reduced the informational and social pressures leading to assimilation. Immigrants can easily remain in touch with the living, vibrant culture and ideology of their country of origin. The acceptance and promotion of multiculturalism has reduced pressures for assimilation as well; host societies no longer exert the same degree of pressure on immigrant communities to lose their particularist identity. Even the children and grandchildren of immigrants, raised with no strong connection to "the old country", can create such a connection afresh; and so the Internet gives disaffected immigrant-community youth the opportunity of "finding themselves" as "old-country" nationalists.

Because the Internet provides instant, low-cost communications across national boundaries and great distances, Internet-based virtual communities allow small, geographically diffuse groups to maintain and promote their ideological/informational framework undeterred and undetected by outsiders. Like so many features of advanced technology, this can have both positive and negative effects: the same capabilities that allow a madrigal singer in Manila to exchange ideas on performance practice with other madrigal singers in Great Britain also allow extremists of any type to give one other support and encouragement.

This ability to form geographically dispersed virtual communities can lead to the fragmentation of "real" communities: I can feel and believe very different things than my neighbors believe and feel, and yet not feel isolated or discouraged because I have ideological neighbors as close as my computer. The urge to conform is thus considerably weakened. Further, if the virtual community of which I am a member has a major internal dispute, its members are not forced to compromise and continue to live together; it is trivially easy to split off and form a new virtual community. Virtual communities can keep subdividing as long as there are enough new members joining up to keep the resulting communities viable. This "fractal" aspect of virtual communities allows ideological/informational fragmentation to continue almost without limit, profoundly weakening the influence of the mainstream over their members. And since the potential "recruitment space" for a virtual community can be very large, the group does not have to alter or "tone down" its values and characteristics in order to attract members from the limited pool of potential recruits in any particular locale.

The Internet offers privacy, anonymity, a huge recruiting pool, and an unlimited supply of "street corners" on which to meet. All of this reduces the influence of mainstream ideas, cultures, and constraints.

2.2. Internet-Mediated Alienation: Lurkers and Sleepers

As members of a virtual community, terrorists, potential terrorists, and terrorist sympathizers no longer need a local constituency to support them, approve of them, or to inspire them. The global community of true believers (in whatever they believe) is now their community, and their new group identity justifies anything they may do against their host community. By weakening bonds with the local community, membership in a virtual community may permit a loss of normal empathy and inhibitions against killing or injuring one's neighbors, or even against harming one's own family.

Virtual communities often allow the phenomenon of unobtrusive "lurking" – meaning that people can gravitate towards an online community, read what others write, and yet not have to contribute to the discussion or even reveal their presence. (This capability, of course, can also be useful for counter-terrorists!) Analogous behavior is much more difficult in conventional social situations, as "real lurkers" are obvious and annoying. The phenomenon of "lurking" allows a sort of multi-level recruitment strategy for online communities that would be completely impractical for "real" social groupings.

Valdis Krebs [9], a social network analyst, created the following diagram (see Fig. 3) of a typical high-tech interest-based virtual community (which in turn may be made up of several smaller virtual communities for socialization purposes):

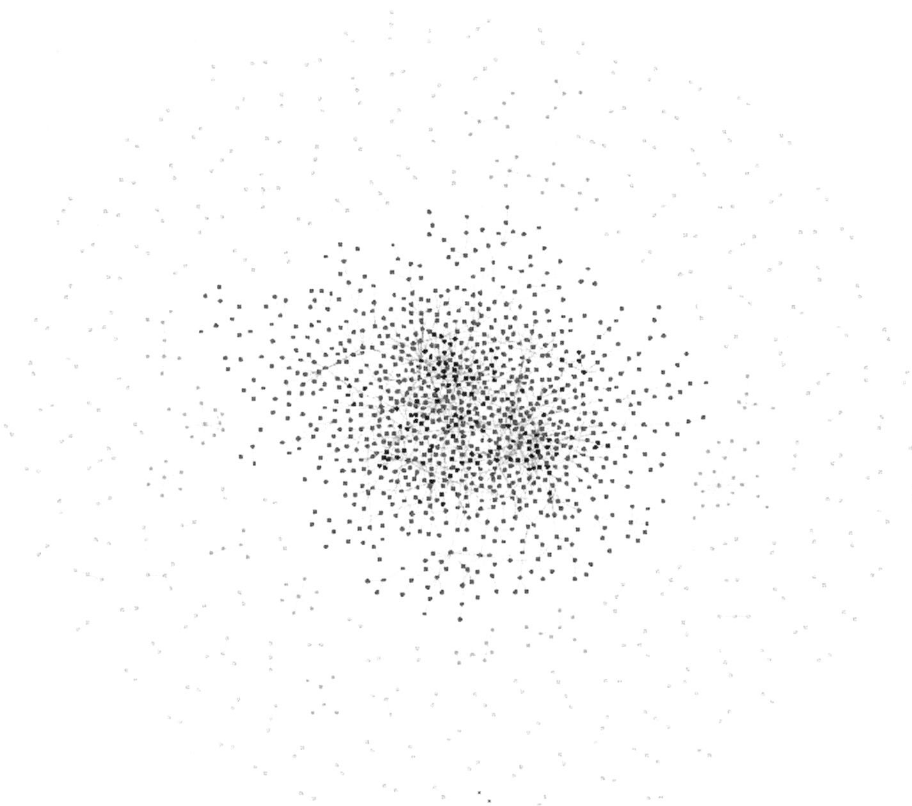

Figure 3. Krebs' depiction of a high-tech interest-based virtual community.

The diagram depicts four types of "nodes", each one representing an individual connected in some way to the community:

- **Black nodes: leadership.** These people are the core leaders of the organization. "They have denser connections to other leaders" and other main network nodes. They keep everything together as the group's connectors.
- **Dark gray nodes: active members.** Active members are tightly connected to the leadership nodes (black). They, in combination with the leaders, are what people refer to as "the group".
- **Medium gray nodes: people actively seeking membership.** These people aren't formally connected to the core group. They are actively working on ways (relationships, credibility, etc.) to connect to "the group". As part of their effort to join the larger group, they have begun to create networks of their own, typically with around 5–15 members.
- **Light gray nodes: lurkers and potential members.** People in this category are not active members of the group, and are only sparsely connected to other lurkers. They may or may not undertake actions that are in line with group goals [9].

The pool of people actively seeking membership and lurking provides substantial potential for recruiting new members and spreading ideological messages.

People participating in Internet-based virtual communities perforce develop some ability to lead a double life: their relations with their "virtual comrades" are largely compartmentalized from their "normal" contacts with those around them, particularly if their online friends would not meet with the approval of parents and other local authority figures. This compartmentalization could conceivably help them to live the double life required of a "sleeper" – an apparently harmless, quiet person who at any time is capable of supporting or carrying out a terror attack. At best, "living online" in political virtual communities enhances members' alienation and isolation from their host communities.

Internet-based virtual communities have a degree of freedom from local observation that is lacking in ordinary "street-corner" gatherings – who knows what Johnny is reading and writing over the Internet? Virtual communities facilitate the phenomenon of the "good boy" who turns out to be a deadly terrorist: until he carries out the attack, his parents, friends, and neighbors think of him as a nice quiet guy who sits in front of his computer a lot.

Because politically-oriented virtual communities are based on ideas and ideologies rather than on geography or family relationships, they are likely to gravitate towards escalating extremism. In general, successful social entities are those that act to strengthen and perpetuate themselves. In political virtual communities, the best "success strategy" consists of deepening members' commitment and extremism, since extremism encourages intra-group solidarity and renders group members increasingly impervious to contrary sources of information. Since the Internet is both an information source and a social medium, there is a synergy between commitment to the group and informational isolation. Loyal group members thus become true believers in extremist ideology, with an understanding of current events based on wildly ideological and inaccurate sources. The combination, obviously, can be deadly.

3. Conclusion

Human beings need to feel that they are members of a community; and yet, for many reasons, modern life often fails to provide opportunities to belong to the kind of community we seem to be "programmed" to need. Given this lack of community in "real life", many people have turned to the Internet, where they have found or created "virtual communities" based around mutually shared interests, beliefs, and ideologies.

Most virtual communities are quite harmless. However, when virtual communities form around extreme political and/or religious beliefs, they can lead to real and severe dangers. Extremist virtual communities are likely to become progressively more extreme; and because they effectively filter the information reaching their hard-core members, they can promote social, ideological, and informational fragmentation and isolation from mainstream society. By providing a sense of "otherness" from ordinary society, extremist virtual communities can also deepen their members' alienation from their surroundings, reducing their normal inhibitions against violence while increasing their ability to perform as "sleepers".

It is imperative that counter-terrorists create some strategy for countering these trends. One approach might be to create and promote alternative "real" and "virtual" social opportunities that do not center around dangerous ideologies. Certainly it would be better to prevent people from slipping into political radicalization in the first place, rather than try to stop them once they have decided to carry out a terror attack.

References

[1] Baumeister, R. F., & Leary, M. R. (1995): "The need to belong: Desire for interpersonal attachments as a fundamental human motivation". *Psychological Bulletin*, 117, 497–529. Accessed at http://tinyurl.com/3dl9xa.

[2] Horton, Donald and R. Richard Wohl (1956): "Mass Communication and Para-social Interaction: Observations on Intimacy at a Distance", *Psychiatry* 19: 215–29. http://www.participations.org/volume%203/issue%201/3_01_hortonwohl.htm.

[3] Dunbar, R. I. M. (1993): "Coevolution of neocortical size, group size and language in humans". *Behavioral and Brain Sciences* 16 (4): 681–735. http://www.bbsonline.org/documents/a/00/00/05/65/bbs00000565-00/bbs.dunbar.html.

[4] Allen, Christopher (2004): "The Dunbar Number as a Limit to Group Sizes", *Life With Alacrity blog*, http://www.lifewithalacrity.com/2004/03/the_dunbar_numb.html.

[5] Ernst Fehr and Urs Fischbacher: "The Nature of Human Altruism", *Nature*, vol. 425, October 2003. http://www.iew.unizh.ch/home/fehr/papers/NatureOfHumanAltruism.pdf.

[6] Kuhn, Steven: "Prisoner's Dilemma", *The Stanford Encyclopedia of Philosophy (Fall 2003 Edition)*, Edward N. Zalta (ed.), http://plato.stanford.edu/archives/fall2003/entries/prisoner-dilemma/.

[7] McMillan, D. W., & Chavis, D. M. (1986): "Sense of community: A definition and theory". *Journal of Community Psychology*, 14(6–23), cited at http://blog.lib.umn.edu/blogosphere/blogs_as_virtual.html. See also Blanchard, Anita L. and Markus, M. Lynne: "The Experienced 'Sense' of a Virtual Community: Characteristics and Processes", *The Data Base for Advances in Information Systems*, Winter 2004, Volume 35, Number 1. http://delivery.acm.org/10.1145/970000/968470/p65-blanchard.pdf?key1=968470&key2=9087614711&coll=GUIDE&dl=ACM&CFID=15151515&CFTOKEN=6184618 or http://tinyurl.com/3bq4ok.

[8] Blanchard, Anita: "Blogs as Virtual Communities: Identifying a Sense of Community in the Julie/Julia Project", *Into the Blogsphere*, University of Minnesota, http://blog.lib.umn.edu/blogosphere/blogs_as_virtual.html.

[9] Robb, John: "Emergent Communities Dedicated to War (London, Iraq, and Al Qaeda)", *Global Guerrillas* blog, 21 July 2005. http://globalguerrillas.typepad.com/globalguerrillas/2005/07/emergent_commun.html. See also Krebs, Valdis E.: "Uncloaking Terrorist Networks", *First Monday*, Volume 7, Number 4 (April 2002), http://firstmonday.org/issues/issue7_4/krebs/index.html.

Hypermedia Seduction for Terrorist Recruiting
B. Ganor et al. (Eds.)
IOS Press, 2007

Common Motifs on Jihadi and Far Right Websites

Michael WHINE
Community Security Trust, UK

Abstract. The far right in the USA and Germany, together with Islamists and jihadists globally, were among the first groups to use the Internet to promote their beliefs, and to direct their activities nationally and transnationally. They recognized that they could publish material which in other formats might be subject to state scrutiny and sanction, and that the medium itself could lend a sense of authority that would otherwise be lacking; furthermore, they could do all of this at a low cost. Both the far right and the Islamists and jihadists share elements in terms of their apocalyptic and messianic, totalitarian belief systems. There has been a mutually reinforcing influence between Islamism, Fascism and Nazism, and these groups are all prepared to use extreme violence against military and civilian targets in pursuit of their aims. They have done so already, with substantial loss of life. There is now evidence that these groups are influencing the other. Some far right extremists have expressed support for jihadi terror, some Islamist sites have republished far right material, and there are many common themes. There is no suggestion, at this stage, of collaboration between them, but law enforcement agencies should be aware of the common links.

Seven uses for websites and other Internet tools can be identified, as used by extremist groups: promotion of ideology; propaganda and incitement; internal communication; recruitment; fundraising; training (including the publication of bomb manuals); command and control (including planning, networking and coordinating action). International organizations which are examining the growth of extremist Internet sites should recognize that the medium has become the primary vehicle for promoting such groups' ideas, and for organizing terrorism and even for training adherents. Counterterrorist initiatives should agree on more effective monitoring and interdiction regimes.

Keywords. Far right, Islamism, neo-Nazi, ideology, propaganda, bomb manuals, training, Internet games.

Introduction

Far right-wing groups in the United States and Germany, plus Islamists, were among the first to recognize the power and benefits that could be derived from the Internet. These groups had common needs and purposes: both wished to transmit material across vast distances. In the case of the American militia movement it was across the breadth of America, while the Islamists needed to reach the increasingly widespread Muslim diaspora.

In Germany, and at approximately the same time, neo-Nazi groups sought to remain small, with independent groups operating in concert rather than forming larger semi-political bodies which might have led to proscription by the state. Their websites and discussion forums enabled this process [1].

The Internet was therefore eagerly seized on as a means to provide communication which could appear authoritative, evade state scrutiny, publish material which in hard form might be illegal and therefore subject to sanction, and provide force enhancement—and all at low cost [2].

Historically, there have been commonalities in the ideologies of Islamism and the far right: the founders of modern Islamist ideologies, Hassan al Banna, Mawlana Maududi and Syed Qutb, were each in turn somewhat influenced by the totalitarian movements of the 1930s and 1940s. It is therefore not surprising that the latter-day adherents of these ideologies have adopted common motifs in their publications, and that these would be reflected in their use of the Internet [3].

The ideological affinities between militant Islam and the contemporary far right have been pointed out by several observers. Chip Berlet has suggested that both seek the overthrow of existing Western governments, in favor of totalitarian monocultural states founded on religious belief or racial exclusivity. Both view the US government as being completely in the power of Jews; many in the far right refer to this government as "ZOG: Zionist Occupied Government". Thus, both groups hate the Jews, who they believe are plotting world domination [4].

Martin Lee [5] has explored the historical links between the Nazis and the Arab world, and the continuing links between neo-Nazis and radical Muslim regimes in Libya and Iran. George Michael [6] has also examined the historical ties and shown how the Internet and globalization have increased and enhanced these groups' interaction. Sometimes this interaction has lead to their establishing global and mutually supportive virtual communities in terms of beliefs, thereby assisting their political extremism, and in turn, terrorism.

The jihadis and elements within the far right add a further common dimension: both promote messianic and apocalyptic ideologies, and both sanction extreme violence in order to bring about redemptive consequences.

In the post 9/11 era, Salafi Jihadi terrorism now relies less on the madrassah and the mosque to promote its doctrines in the West, partly as a result of state surveillance. Increasingly, therefore, its actors are recruited through universities, gymnasia, cyber cafes and bookshops, and the social networks that grow up around, and are encouraged by, such venues. Recruitment takes place via the evangelical sub-terrorist Salafi political groups, such as Hizb ut Tahrir, Al Muhajiroun and its successor groups; via Islamist political groups such as the Muslim Brotherhood and Jamaat e Islami; via the revivalist Tablighi Jamaat; and via the external branches and contact points of Middle East and South Asian jihadi groups, which initiate the radicalization process that offers the springboard or conveyor belt into terrorism [7].

Al-Qaida itself is evolving from the centralized hierarchical organization of the past to the more loosely organized, decentralized and diffuse network of the Global Jihad Movement (GJM), in which the Internet plays a crucial and developmental role. By providing otherwise isolated actors with a medium for communication, the Internet today may therefore be said to serve as a cyber meeting-place to discuss and plan jihad. Not all sites are actually associated with Al-Qaida itself, and Gabriel Weimann points out that only those with certain "fingerprints" in terms of graphics and texts are genuine. But many others offer support, and feed the radicalization process [8].

One site allegedly connected to Al-Qaida was the Azzam Publications site, which issued the following advertisement for its services: "Due to the advances of modern technology, it is easy to spread news, information, articles and other information over the Internet. We strongly urge Muslim internet professionals to spread and disseminate news and information about the Jihad through e-mail lists, discussion groups, and their own websites." [9]

Weimann quotes Thomas Hegghammer of the Norwegian Defense Research Establishment, who noted that "In a sense [the Internet has] replaced Afghanistan as a meeting place for Jihadis." He added, "As the number of terrorist websites has increased the uses to which terrorists put the Internet has diversified. Yet the Internet is not just a tool or a weapon. It is also a target and a weapon." [10]

The German security service recently noted in its annual report that

> The Internet is the key medium of communication for right-wing extremists, who use it to present themselves, make verbal attacks, and carry on internal debates and to mobilize attendance at their rallies and demonstrations. The number of websites run by German right-wing extremists – 950 – remained at the same level as in earlier years. In addition to websites, right-wing extremists increasingly use interactive Internet services for purposes of information or discussion. Along with mailing lists and newsletters, discussion forums are increasingly important for the scene. [11]

In observing the growing links between the American and European far right, the FBI have also noted some links between the American groups and European, Middle East and Asian terrorists. In giving evidence to the Senate Intelligence Committee in February 2002, Dale Watson, FBI Assistant Director for Counterterrorism stated that: "We do see some interaction and communications between groups… With the explosion of the Internet we certainly see white supremacist groups in contact with people in Europe, particularly in Germany." But he added that there was no evidence of any collaboration, so far [12].

The phenomenon of website and discussion groups linked to planned terror has recently been vividly illustrated by the arrests of what appears to be an international group of would-be terrorists in various countries. Most appear not to have met each other, but had communicated via the Internet. Apprehensions in Canada, the USA, the UK and elsewhere were seemingly triggered by the detention of Mirsad Bektasevic (aka Maximus) in Bosnia. At his residence, the authorities found bomb-making materials, a suicide vest, and large quantities of propaganda. Bektasevic, a Swedish-born Muslim of Bosnian origin, was a computer expert who cooperated with Younis Tsouli (aka Irhabi 007), who was subsequently arrested by the British police.

At one end of the transnational conspiracy was a planned attack against the British Embassy in Sarajevo, and at the other was a plan—later abandoned—to attack the Canadian parliament and behead the Canadian Prime Minister. Somewhere in-between was a plan by two Americans, Ehsanul Islam Sadequee and Syed Haris Ahmed, to stage an attack in Washington on the Capitol building, the World Bank, and a fuel depot [13].

Despite their similar use of websites, the nature of jihadi and far right online activities clearly present different levels of threat. Jihadi terrorism constitutes the main contemporary terrorist threat worldwide, whereas far right terrorism barely exists outside the USA at present. The Bologna railway station bombing in 1980, the

Oklahoma City bombing in 1995 and the London nail bombings in 1999 have been the worst acts of far right terrorism. However, it should be noted that many major attacks have been interdicted, without much publicity.

Indeed, US researchers have criticized some agencies of their government for failing to publicly recognize the threat from the far right. According to the Southern Poverty Law Center, far right activists in the USA have produced some sixty terrorist plots since the Oklahoma City bombing. But several of these, it must be stated, were notable for their ineptitude [14].

Among far right terrorist plots, the following two have been the most salient. First, white supremacist William Krar was sentenced to eleven years in prison in 2003, after federal agents in Texas foiled his plot to stage a mass attack—for which he had accumulated enough sodium cyanide to kill 6,000 people, one and a half million rounds of ammunition, and sixty pipe bombs. The FBI found that Krar had forged identity passes to the United Nations and the Pentagon, suggesting these were to have been his targets [15].

Second, David Burgert of Flathead Montana was sentenced to ten years in prison for planning to foment insurrection by assassinating twenty-six local law enforcement officials, and thereafter the National Guard detachment that would be sent to confront him and his co-conspirators. He then expected the US government to call in NATO troops, and it is alleged that he hoped that at this point "true American patriots" would rise up in a ferocious war to reclaim America for white Christians [16].

Europe has not suffered from far right terrorism to the same extent as the USA, but the German authorities did foil a plot in 2003. Eight members of the Kameradschaft Sud were sentenced to up to seven years each for plotting bomb attacks on the Munich Jewish cultural center, which could have killed hundreds of people, including the head of state, who were attending the opening ceremony [17].

To date, far right terrorism has been entirely national, while Islamist terror has sometimes been international, with actors in one country (of which they may or may not be resident) targeting a second country, and sometimes using services within a third country to provide logistical and material support. However, the exchange of ideas and incitement against others is international, and crosses all boundaries.

Clearly there is little room for complacency. The German authorities have recently reported to Europol, the European police agency, that "currently right wing extremist circles reject terrorist violence as a means to achieve their goals." But they added that this may be motivated by tactical considerations, and that "individual terrorist operations by members of the right-wing spectrum, however, cannot be ruled out in the future" [18].

1. Common Motifs

Websites, chat rooms and discussion groups have different uses. Websites portray the users' beliefs and provide a means to promote them, while chat rooms and discussion groups provide interactive forums for the exchange of ideas, and, increasingly, also for command and control functions. Both types of cyber venues may have pass-worded access, and both may contain coded messages in order to protect the secrecy of their communications.

Jihadi and far right users almost universally now resort to offshore Internet service providers, so as to evade sanction; both groups are also prepared to change their service providers frequently and at short notice in order to remain online. US Internet service providers host a wide range of jihadi and far right sites, despite continuing complaints to the service providers and to government agencies. In the US, such sites are less subject to sanction, due to the freedom-of-speech considerations afforded by the First Amendment to the Constitution [19].

Some East European and South East Asian countries allow these sites to be hosted within their countries, and in some cases the local governments are not even aware of the presence of such sites. Often, commercial considerations outweigh political and counterterrorist ones, even when the organizations whose sites are being hosted are proscribed by the United Nations, European Union or USA [20].

For example, Palestinian Islamic jihads operate from five different web hosts; two in the USA, one in Iran, one in Romania and one in Malaysia [21]. Hamas is said to have over twenty websites, controlled from Damascus but hosted in Russia, Indonesia, Malaysia, Pakistan, Eastern Europe and the USA. Hezbollah maintains thirty-six sites, of which eleven are said to be currently active [22].

We can now identify the following seven uses for which jihadi and far right organizations use their websites: promotion of ideology; promotion of propaganda, and incitement (against perceived enemies); internal communication; recruitment; fundraising; training (including the publication of bomb-making and security manuals); command and control (including planning, networking and coordinating action). The following section deals with each of these seven functions in turn. Many examples could be cited to illustrate these uses by both jihadi and far right organizations, but for the sake of brevity I have chosen to refer only to a limited number to illustrate the point being made.

2. Ideology

There are postings on jihadi and far right sites that illustrate similar outlooks, and which demonstrate the value and importance of using the Internet to disseminate and promote ideologies. Some of these postings have become well known far beyond their original target audiences, and have even assumed an importance and notoriety in their own right because they encapsulate the purpose, strategies and tactics of the posters.

Louis Beam's 1983 essay *Leaderless Resistance* defined the concept and modus operandi subsequently adopted by both far right and jihadi terrorists. It was originally published in a small-circulation far right journal, *The Seditionist,* but reached wider audiences when it was republished online in 1992 [23].

In this essay, the author declared that "those who oppose state repression must be prepared to alter, adapt, and modify their behavior, strategy, and tactics as circumstances warrant. Failure to consider new methods and implement them as necessary will make the government's efforts at suppression uncomplicated." [24]

Beam further added that the pyramid scheme of organization was now useless and dangerous for a resistance movement. Electronic surveillance by the state can penetrate the structure and reveal its chain of command, and an infiltrator can compromise and destroy all levels. The cell system, based on the "committees of correspondence" established during the American Revolution, or the post-war Comintern model, Beam suggested, were a more effective model in which small, self-motivated, discrete cells can operate at will [25]. He wrote: "A thousand different small phantom cells opposing them [the Federal Government]… is an intelligence nightmare… A single penetration of a pyramid type of organization can lead to the destruction of the whole. Whereas Leaderless Resistance presents no single opportunity for the Federals to destroy a significant portion of the Resistance." [26]

Osama bin Laden's statement of 23 August 1996, *The Ladenese Epistle: Declaration of War,* was posted on the Internet by the London-based Committee for the Defence of Legitimate Rights, the Saudi Salafi group, directed at that time by Mohammed Al Mas'ari and Saad Al Fagih. It was originally faxed from Afghanistan to Khalid Al Fawwaaz, the director of the London office of the Advice and Reformation Committee [27].

The text is long and rambling, with frequent Koranic and historical references. Bin Laden threatened America, but also referred to "the Zionist-Crusader alliance". He condemned American support for Israel, and the killing of Palestinians. Importantly, he suggested which methods should be employed against the West's military forces: "…a suitable means of fighting… using fast moving light forces that work in complete secrecy". He also praised Muslim youth for their previous efforts, and exhorted them to further individual and collective acts of sacrifice [28].

At another level, it is not difficult to find Islamist and jihadi sites quoting far right activists, and occasionally, the reverse. Hezbollah published the writings of US National Alliance founder, William Pierce; while the Aryan Nations, another major US far right group, has maintained a section on its website for Islamist material [29].

An Aryan Nations leader, August Kreis, publicly proclaimed his group's admiration for Bin Laden in a CNN interview in 2005. In it he declared, "You say they're terrorists, I say they're freedom fighters. And I want to instill the same jihadic (sic) feeling in our peoples' heart, in the Aryan race, that they have for their father, who they call Allah… I don't believe that they were the ones that attacked us… And even if they did, even if you say they did, I don't care." [30]

David Myatt, the British neo-Nazi ideologue who took control of the National Socialist Movement (the successor group to Combat 18) and who subsequently converted to Islam and changed his name to Abdul Aziz ibn Myatt, posted several theoretical tracts on the confluence of jihadi and neo-Nazi ideology [31].

Both groups of extremists use the Internet to promote, discuss and evaluate their strategic and tactical doctrines. Michael Scheur has noted how Al Qaida's websites reviewed these in the light of the insurgencies in Iraq and Afghanistan, suggesting modifications in the use of small groups to wage attacks, and improvised explosive devices. Based on its battlefield experience, Al-Qaida subsequently modified its tactical doctrine and cautioned patience. Scheur quoted military chief Sayf al-Adil, who explained in March 2003 that "there is no harm in delaying action if there is a real interest in doing so. We want to say to those [Muslims] who want a quick victory that this type of war waged by the mujahideen employs a strategy of the long breathe and the attrition and terrorization of the enemy and not holding on to territory." [32]

Statements offering virtually the same advice can be seen on the websites of American far right militia groups, relating to the aftermath of the 1992 Ruby Ridge standoff. In that incident, the wife and child of far right activist Randy Weaver were killed, but this was before the Oklahoma City bombing in 1995. After 1995, the armed militia movement and some within the armed neo-Nazi groups turned their attention away from internal "resistance" to the government and toward other issues, such as immigration for tactical rather than strategic reasons [33].

3. Propaganda and Incitement

Propaganda and incitement against perceived enemies is common on both jihadi and far right websites, and this is where the greatest potential for cooperation lies. The most notable examples occur when far right activists express support for Islamists and jihadi terror. The Anti-Defamation League, however, has reported that right wing propaganda is also being acquired and recycled by Islamists. They have noted the reprinting of material from Lyndon LaRouche by the London-based Al Muhajiroun (its successor groups were proscribed by the British government in July 2006) and the republication of David Duke's articles on some Islamist and jihadi sites [34].

In Britain, in 2000, the Muslim Brotherhood-linked Muslim Association of Britain republished online *The Jewish Threat on American Society,* written by an American Nazi sympathizer in 1934, before he was detained by the American authorities for the duration of the War. The paper was passed off as the work of Benjamin Franklin [35]. In 2003, the London-based International Islamic Federation of Student Organizations republished, on their website, the cover and part of the contents of *The Protocols of the Elders of Zion*, which originally had been published by a US Christian Identity group [36].

The real danger posed by extremist websites, however, is not necessarily that they serve as repositories of inflammatory literature, but that they provide forums and chat rooms for visitors to organize attacks against their enemies.

Recent examples of far right propaganda intended to incite violence might include the posting by Craig Harrison, a neo-Nazi activist from Georgetown, Ontario. He posted calls for the extermination of ethnic minorities and religious groups to *Yoderanium Productions*, a neo-Nazi website. These were so bad that the Forum moderator replaced them with a warning that such comments were illegal. On other occasions Harrison posted repeated calls for genocide to the Freedomsite.org website of Marc Lemire [37].

A posting to the National Socialist Movement Yahoo group website was entitled "Advice for Lone Wolves". In it the author called on supporters to start acting, which means "getting out from behind your computer keyboards and message forums, and actually 'doing' something…" [38]. The author stated:

> The most recent example of this type of activity is the sharp shooter, who wasted that Jew abortionist in New York, but single, and cellular resistance does not necessarily have to include fatality or even illegality. I am not going to bother to put in the usual disclaimers this time, because I think things have reached a point where if the feds want to get me for something, I write they're going to get me, and disclaimers won't matter… I do not give one single remote damn about the 'rights' of any nigger, Jew, abortionist, bureaucrat or whoever. All I am concerned about is the welfare of White people, and that is as it should be.
>
> 1. ACT ALONE, and I do mean ALONE…
>
> 2. DO NOT ROB BANKS or Pizza Huts, or Cookie marts to obtain funds for your activities. The kind of thing you

> will most likely want to do is not really very expensive, and you can raise enough money through working for the items of equipment you will need… act silently, do not send letters or make phone calls to the media boasting or claiming credit for your deeds, calling yourself the "Aryan Revolutionary Strike Force" or nonsense like that.

Internet games are a particular form of incitement common to both jihadis and the far right. They allow the player to attack and even "kill" their enemies; many such games target Jews, the perceived enemy, although Gypsies have also been targets in some games. These games are aimed at children and teenagers, and are intended to incite hatred from an early age.

Rudolf Stefen, the head of the German Federal Office for Inspection of Youth Endangering Materials, outlined the designers' purpose when he stated that: "It's vital for the survival of neo-Nazis to get youth on their side again, and this is how they're trying to do it… The games seek to animate and urge the young to adopt old Nazi ideas as their own and to duplicate Nazi crimes all the way to genocide." [39]

Far right games were created and first distributed in Germany and Austria in the 1980s, with names such as *Hitler Diktator, Stalag* and *Wolfenstein Castle.* They were not directly accessible to members of the public, but were lodged in "mail boxes" which could be called up, and their existence was spread by word of mouth [40]. The German authorities were eventually prevailed upon to prevent the dissemination of these games, and to inform other governments of their dangers, but nonetheless the game sites were publicized rapidly. Their spread would anyway have been difficult to stop as the computer discs were produced covertly and often distributed by hand [41].

One game which was available in the UK was called *Pakibash*; it featured turbaned Asians on flying carpets, and invited players to "wipe the bastards out of existence" [42]. By 1991, the US-based Simon Wiesenthal Center was reporting the existence of over 140 similar games on the market, but their distribution was usually outside normal commercial channels. Titles included *Aryan Test* and *KZ Manager* [43].

Larger organizations entered the market thereafter, and adopted a more open approach. David Pringle of the National Alliance was interviewed by News Channel 4 in 2003, about *White Law*, the game they were marketing. In this game, players tried to kill black people, and then killed Jews in a bonus round. Pringle stated that "We're basically prying the door open, getting in little Jonny and little Janie's minds, just like television." He added that the National Alliance was seeking to capitalize on the success of its predecessor game, *Ethnic Cleansing*, which was so popular that a second pressing was ordered, just a week and a half after the game first went on sale [44].

Hezbollah's interactive video game *Special Force* became available in 2003 in Beirut, and was available at www.specialforce.net. The opening page declared the following:

> To the loyal expectants who endlessly inquired about the launching date of the outstanding game, the first of its kind in the Arab world. The good news is that: The Game is now available in the markets, and includes many surprises. To obtain the game, which includes all that an anxious person dreams of in order to participate in facing the Zionist enemy. Call the

> accredited distributors at your country. Special Force game will render you a partner of the resistance... . Beginning, we must say that we cannot seriously talk about "Arab" Electronic Games, since this sort of games is unavailable in the market... . This gap is one of the gaps that face Arab programming production. This gap must not be underestimated and it heads the top of the Arab programmers' agenda... . This trend was coursed by the Central Internet Bureau of Hizbollah via designing a plan that depends on issuing a number of electronic games that can fill the existing gap within our Arab arena. [45]

The game itself is based on the house-to-house encounter that occurred between Israeli forces and Hezbollah terrorists in Tallousa Village, Southern Lebanon, in May 2000. It allows the player the opportunity to vicariously participate in the fighting, by viewing photos obviously taken by Hezbollah members in combat.

Hamas has also made extensive use of its Internet capability to produce online children's' magazines. These are not interactive games, but they do use stories and poems to glorify terrorism. One, *Al-Fateh*, is an online journal (www.al-fateh.net), which, according to a Center for Special Studies bulletin,

> ...has attractive graphics and contains comics-like drawings and photographs to make it 'friendly' and attractive to its target audience of young children. There are poems and stories ('The Thrush', 'The Troubles of Fahman the Donkey', texts written by children themselves, etc). There are also articles about religious subjects and stories about battles and tales of heroism from Arabic and Islamic history. Side by side with these 'innocent' items are articles preaching the perpetration of terrorist attacks, extolling the suicide bombers and presenting them as role models, and encouraging hatred for Israel and the Jewish people.

The comic has also published a picture of suicide bomber Zaynab Abu Salem, and a photograph of her decapitated head, with the caption "Zaynab Abu Salem, who carried out the suicide bombing attack. Her head was severed from her pure body and her headscarf remained to decorate [her face]. Your place is in heaven in the upper sky, Zaynab... sister (raised to the status of heroic) men." [46].

4. Funding and Recruitment

The SITE Institute reported on a high-tech recruitment drive launched in 2003 to recruit fighters to travel to Iraq and attack US and coalition forces. SITE director Rita Katz stated that "Al-Qaida's use of the Internet is amazing. We know from past cases – from captured Al-Qaida fighters who say they joined up through the Internet – that this is one of the principal ways they recruit fighters and suicide bombers." [47]

Syed Talha Ahsan, a 26 year old British national, was arrested in connection with the US case against Babar Ahmed, and was indicted on 19 July 2006 on charges that he helped to run the Azzam Publications, qoqaz.net and other websites to raise funds and recruit for Al-Qaida. Together with Ahmed, he is accused of managing websites in

Connecticut, Nevada, Malaysia, Ireland, the UK and elsewhere. The indictment alleges that Ahsan was "eulogising dead fighters, for the purpose of recruiting individuals and soliciting donations to support the mujahideen", and that he "possessed various materials, including literature supporting violent jihad throughout the world."

Ahsan's indictment further alleges that "The websites and other internet media AHMAD, AHSAN and others helped create, operate and maintain through their expert advice and assistance were used to support and justify violent jihad. Their expert advice and assistance were directly and integrally linked to ongoing efforts to provide personnel, currency, military items, monetary instruments, and other material support and resources for acts of terrorism."

Concerning Ahmed himself, it was stated in the extradition request submitted by the US authorities that he had "sought and invited and solicited contributions to terrorist causes in Afghanistan and Chechnya" between 1997 and 2004. The best way to help, Ahmed had stated, was "to propagate the cause of jihad among Muslims who are sitting down ignorant… and go to the lands of jihad to fight." [48]

Sami Omar Al-Hussayen, a Saudi computer science doctoral candidate studying in the USA, was charged in March 2004 with running Islamway.com, a Montreal-based Internet site used to recruit and raise funds "for jihad in Israel, Chechnya and elsewhere". According to the indictment, he operated an email group "for people who wanted to participate in violent jihad", and led discussions about potential US targets. The group, which had 2,400 members, "served as a platform for Al-Hussayen's fundraising appeals".

In February 2000, Al-Hussayen "sent a message to all members of the Internet email group urging them to donate money to support those who were participating in violent jihad in order to provide them with weapons and physical strength to carry on the war against those who kill them". He appeared to have been a lone player rather than a member of an organized grouping, and was acquitted of the terrorism charges, but was nevertheless deported from the USA [49].

The domestic terrorism threat is currently greater because of recruitment via the Internet, and because it comes from the far right as well as from left-wing radical environmental groups [50].

Brian Levin and others have noted that the Internet has become the principal form of recruitment, attracting the loners, and the disturbed who boast of finding viable targets. We are likely to see more terrorist attacks by lone wolves, or small cells, according to Levin: "They're in their bedrooms accessing bomb-making information on the Net, and accessing hateful rhetoric which empowers them." [51]

Don Black, the former Ku Klux Klan leader who runs a string of websites from Florida, of which Stormfront is the oldest, most notorious and widely accessed, has stated that white separatists are seeing more Internet activity turn into "real-world activity… . The criticism we've always heard is that people don't do anything but sit behind their computer, and post on message boards… . We are actually turning people out to meetings and getting people involved in activism." [52]

A particularly worrying aspect of recruitment websites, particularly jihadi sites focused on Iraq and the Arab world, is that they are aimed at children as young as seven years. Recent examples include the Death Cult Kiddies' Movie, which shows small children re-enacting the beheading of US hostage Nick Berg. This was available in the UK on Abu Hamza's Supporters of Shariah website, and originated at the Hong Kong-registered www.shareah.org site [53].

5. Command and Control

British groups, along with other European far right groups, increasingly rely on the Internet to organize their activities. For example, the National Front planned a march against the Muslim community in Luton on 1 July 2006, and all arrangements were made via the NF guest site.

> Hope to see as many of you as possible at the Luton march on Sat 1st July 2006, the police have made contact, and have said they may have to change the route a little, because we go to near to an Asian area called Bury Park. Please form up at Luton station at 11.45am, and be ready to move off at 12 noon… . Please bring your own flags, banners, etc… but please remember to watch the wording… transport is coming from around the country email britishnationalfront@hotmail.com for details… [54]

The police subsequently banned the march, on the grounds that it would be likely to provoke violence.

The skinhead "Blood and Honour" group, which originated in the UK and now has offshoots in several countries, spawned the *Redwatch* site, which publishes details of anti-Fascist activists and journalists, including their photographs and home addresses. Sympathizers are invited to attack the targets, and indeed several attacks have taken place, with victims suffering serious injuries. As a consequence, the Polish Foreign Ministry recently requested the authorities in the USA (where the site is hosted) to close down the site, and Britain's Director of Public Prosecutions has asked the Metropolitan Police Service to institute a criminal investigation [55].

Swedish neo-Nazis, who allegedly bombed Iraqi expatriate voters at a polling station just north of Stockholm in December 2005, planned their actions via Internet chat rooms, according to prosecutors [56].

6. Bomb Manuals

According to Gabriel Weimann, in 2000 *The Guardian* newspaper ran a search on the Google search engine for the key words "terrorists" and "handbook". Nearly 4,000 matches were found, including references to guidebooks, manuals, and instruction books. When Weimann himself repeated the search in May 2005, he found 7,900 such sites [57].

Far right and jihadi activists initially may have relied on the same manuals for guerrilla warfare. Some had been copied from military manuals and made available online, and others were compiled by political extremists. Early Al-Qaida terror encyclopedias were based on US Army field manuals, but later examples appear to have drawn on Russian sources [58].

The far right began posting terrorism and bomb-making manuals in the 1990s, although those they originally used appear to have originated in the American anarchist movement. *The Big Book of Mischief – The Terrorist's Handbook* was produced in 1991 by a group calling itself Chaos Industries, based in Chicago [59]. In 1998, David Myatt posted his terrorism manual and ideology handbook, entitled *A Practical Guide to Aryan Revolution – Aryan Resistance Movement* on the web. In the Introduction he wrote:

> A real warrior is someone who acts – who actually does deeds of war – and an elite warrior is someone prepared to undertake whatever actions are necessary in order to bring about, or aid, victory in war. We as Aryans have now to fight a real war against our real enemies in order to win back the freedom which these enemies of ours have taken from us. These enemies tyrannically rule our own Aryan lands, and we are forced to live under a tyrannical government of occupation… . These days, it is necessary for us to fight a *holy war* – we must have fanatical faith in the justice of our cause of freedom. [60]

Chapters in Part One of Myatt's book were devoted to the following topics: methods of covert direct action; escape and evasion; assassination; terror bombing; sabotage; racial war. Part Two outlined "The strategy of revolution", and included chapters on "How to create a revolutionary situation and direct action groups". The Appendices covered the following areas: rules of engagement; the conduct of soldiers of the Aryan Liberation Army; disciplinary offences; and court martial offences.

Other chapters included: aims and tactics; racism; the will of nature; our revolutionary task; why carrying weapons and vigilante action are right; national socialism: the rule of excellence; what is our life for? A National-Socialist answer; in league with warrior Gods; why the police are our enemies [60].

Myatt was arrested and his site was taken down immediately after his apprehension, but not before it had been mirrored elsewhere, including by jihadi sites. He was, however, not charged, as his posting had been hosted abroad and was therefore beyond British jurisdiction. The hosting was done by FTC:NET, a Canadian far right site, which also hosted the French Charlemagne Hammerskin group. Their leader, Herve Guttoso, was subsequently extradited by the UK to France [61].

The London nail bomber, David Copeland, who joined the National Socialist Movement after leaving the British National Party, revealed at his 1999 trial at the Central Criminal Court in London that he had learned to make bombs from Chaos Industries' *The Terrorists Handbook.* The book contained almost one hundred pages of bomb designs [62].

In 2001, it was noted that among Al Qaida members fleeing Afghanistan, "Every second Al-Qaida member (was) carrying a lap top computer along with a Kalashnikov." Five years later, it has been observed that the Global Jihad Movement has migrated from physical space to cyber space, and that its adherents have "sought to replicate the training, communication, planning, and preaching facilities they lost in Afghanistan with countless new locations on the Internet." [63]

Al-Qaida started publishing its virtual training manual, *Al Battar*, in 2004. In 2000, a different Al-Qaida manual had been found at the home of Anas Al Liby in Manchester. This was *Military Studies in the Jihad Against the Tyrants.* The format was similar to that of the latter publication, with an Introduction in which the author/s laid out their theological justification for terrorism:

> After the fall of our orthodox caliphates on March 3, 1924 and after expelling the colonialists, our Islamic nation was afflicted with apostate rulers who took over in the Moslem nation… . Those apostate rulers threw thousands of the Haraka Al-Islamiya [Islamic Movement] youth in gloomy jails and detention centers… . They [the rulers] tried using every means and [kind of] seduction to produce a generation of young men that did not know [anything] except what they [the rulers] want. These young men realized that an Islamic government would never be established except by the bomb and rifle. Islam does not coincide or make a truce with unbelief, but rather confronts it. [64]

Again this introduction was followed by technical chapters on principles of military organization; necessary qualifications and characteristics for the organization's member; means of communication and transportation,; and so on. As with Myatt's book, many of the instructions were accompanied by elaborate "theological" justifications, which in this case included verses from the Koran and the sayings of the Prophet.

Abdul Hameed Bakier, a Jordanian researcher, recently reviewed three terrorism manuals available online: *The Great Jihad Encyclopaedia*, Abu Yahia al-Libi's letter on US interrogation techniques, and a posting on how to confront and cope with intelligence agency interrogators.

He states that the first of these was originally released in 2002, and is a manual on security and intelligence techniques, collected from different sources, and from experienced jihadi militants. The seventy-page document is broken up into different sections, and Bakier suggests that these are taken from the archives of Maktab Al-Khidamat (Afgan Services Bureau), Al-Qaida's predecessor organization. It reviews intelligence skills, such as espionage, counter-terrorism, sabotage, the importance of human intelligence, and agent handling. Bakier notes that the contents are of a high quality, and clearly some are based on Soviet techniques [65].

The second document, Al-Libi's letter, outlines in four major parts the US arrest and prisoner transfer procedures, as well as interrogation techniques. This content is based on the author's personal experience. Al-Libi (aka Mohammed Hassan) is a senior Al-Qaida member of Libyan origin, who also posted a video tape on the Internet recently, in which he incited Muslims to attack France, Norway, and Denmark in the wake of the cartoons issue.

The third document covers different aspects of interrogation by security services, and is a compilation of postings from three different jihadi sites prepared by an individual named Al-Kandahari.

Bakier's assessment is that these postings were prepared from non-Arabic and government sources. The contents of the *Encyclopaedia* are organized in a logical, sequential training order, albeit interspersed with Koranic verses and stories of intelligence practice in Muslim history. Bakier further suggests that many of the intelligence lessons were drawn from the Eastern school style of training, and that the information comes from ex-Iraqi intelligence officers who joined Al-Qaida in Iraq [66].

Following the August 2006 foiled London plane bombing plot, Bakier reported that the website of the al-Aqsa Martyrs Brigade (linked to Hamas), at http://www.kataebaqsa.org, carried links to "a colossal amount of data on creating [liquid] explosives, part of a more extensive training document called *The Preparation Encyclopedia – All The Mujahid Needs*" [67].

Website downloads of bomb manuals and "militant Islamic" journals were discovered in the Melbourne home of Aimen Joud and Shane Kent, who allegedly had connections to the Ben Brika group arrested in Sydney and Melbourne in 2005. According to the prosecutors, they also had graphic video footage of the execution and decapitation of a man said to have come from Chechnya in 1998 [68].

7. Assessment

Despite some common historical links between their founding ideologues, there were few remaining links between militant Islam and the far right immediately after the 1980s. The post-war Arab nationalism of Syria and Egypt was waning, and most Nazi fugitives who had sought refuge in the Arab world were in retirement or had died. The Palestinian groups that emerged after the Six Day War in 1967 had no history of cooperation with the far right (although they subsequently did train some German neo-Nazis in Lebanon), and the Muslim Brotherhood had been effectively, though temporarily, quashed.

The telecommunication revolution that followed the end of the Cold War, the Soviet invasion of Afghanistan and the call for a jihad to repel the communist invaders changed everything. The far right abandoned the struggle against the communist threat as its chief enemy, and replaced it with the struggle against the New World Order that had been announced by President Bush after Coalition forces repelled the Iraqi invasion of Kuwait.

The American and European far right also increasingly define their struggle as one against secret cabals which control the world, which they characterize as the New World Order or ZOG. Their conceptualization of this new enemy closely parallels that adopted by militant Islam and the jihadis, in particular, who emerged from Afghanistan, many of whom have sought refuge in Europe.

Comparing the far right and militant Islam, George Michael observed that "the extreme right and Islam share a similar eschatology, in which the old order is viewed as incorrigibly corrupt, something that must be totally effaced in order to build a new order. For these reasons, new opportunities for cooperation began to emerge by the late 1990s." [69]

As terrorists continue to move from hierarchical formats toward information-age networked designs, it can be expected that some effort will go into building arrays of transnational groups linked by the Internet, rather than into building stand-alone groups. The would-be terrorists arrested in Canada, the USA, UK and elsewhere (referred to at various earlier points of this chapter), and even those apprehended in the UK at the beginning of August 2006 in connection with the planned plane bombings, may be evidence of this trend

Investigations into the 7 July 2005 London bombings and the recently intercepted plane bombings have already shown strong connections between London, Pakistan, and Germany. The Internet appears to have played a major role in these connections, as a communications medium, as a source of information regarding the bombs' manufacture, possibly as the means by which the players were radicalized, and possibly as a medium for command and control [70].

The Internet is particularly aimed at and accessed by young people, and several national police authorities have recently reported on the youthfulness of terrorist suspects currently coming before their countries' courts. Some suspects are only just over the age where they can be held criminally liable.

An Al-Qaida website, *Global Islamic Media Front* (GIMF), recently announced that the Internet now serves as the best alternative to broadcast television for jihadi activists, since the Western media frequently obfuscates the facts. The site called on Muslims to join Western forums to voice their concerns about Iraq, Afghanistan, Palestine and Chechnya. Several months later, a second GIMF posting called on "People of Jihad to create a media war that goes parallel to the military war" [71].

Europol has noted, in the context of the Netherlands—although the observation could equally apply worldwide—that terrorism is increasingly "bottom up rather than top down"; and that it is young people who constitute the driving force among today's terrorists, even if many of the ideologues and "father figures" are of an older generation. Evidence given recently to the US House Select Committee on Intelligence noted that the target audience for terrorist propaganda, and specifically for recruitment, is aged seven to twenty-five, and that the report-writers had even seen products aimed at children younger than seven [72].

Former US security analysts Daniel Benjamin and Steven Simon have drawn attention to the recruitment by jihadists within universities, where they can more easily access the technological skills they need to mount operations, and where they can use the Internet in place of training camps [73].

The danger, therefore, is that extremist websites serve to radicalize, incite hatred, and provide the manuals to build bombs; in addition, they are a medium through which to organize attacks and terrorism. The danger is exacerbated by the fact that it is young people, particularly, who are accessing the sites and becoming the players.

Jihadis and the far right share many common perceptions about globalization, existing Muslim regimes, the West, Zionism, and the Jews. The founding ideologues whose works have guided these extremist groups' activities certainly borrowed from, and were influenced by, each other's writings. Many people in both arenas are driven by messianic and apocalyptic world outlooks, although it has been the jihadis who have proved to be far more threatening. It is therefore unsurprising that the websites of far right and militant Islamic groups share some common motifs, and that there are some crossovers and shared postings.

8. Recommendations

There should be more effective and concerted effort by governments and international security agencies to pressurize Internet Service Providers in the West to close down the most violent jihadi and far right websites. Some such sites are hosted in the USA, where free speech and commercial considerations outweigh security concerns. Others are hosted in Eastern Europe, or in South East Asia, where legal controls do not operate, or where there is little legal oversight.

Both the Organization for Security and Cooperation in Europe (OSCE) and the United Nations have held conferences in recent years, to focus international attention on the relevant issues; these organizations are in a position to recommend action, including sanctions.

The Internet is emerging as the primary vehicle by which extremist and sub-terrorist ideologies, as well as terrorist ideologies themselves, are disseminated. National security and law enforcement agencies should take greater account of the wider context and of the ideologies—which are so freely disseminated online—that provide the conveyor belt to terrorism. This development also presents opportunities to track extremist and terrorists and map their connections.

Particular attention should be paid to the increasing sophistication of users, and their use of proxies, anonymizers and other means designed to hide their tracks. International security forums should particularly address the issue of terror recruitment and the dissemination of Internet games. Their ban should be made the subject of a United Nations convention.

References

[1] Right-wing extremism on the Internet, *Federal Office for the Protection of the Constitution*, Cologne, January 2000 http://www.verfassungsschutz.de/publikationen/gesamt/page13.htm

Michael Whine, "The Use of the Internet by Far Right Extremists", chapter in *Cybercrime: Law, Security And Privacy in the Information Age*, eds. Brian Loader and Douglas Thomas; Routledge, London, January 2000. Also available at http://www.ict.org.il/articles/articledet.cfm?articleid=413

[2] Gabriel Weimann, *Terror on the Internet*, United States Institute of Peace Press, Washington DC, 2006

Michael Whine, Cyberspace – A New Medium for Communication, Command, and Control by Extremists, *Studies in Conflict & Terrorism*, Volume 22 Number 3, July – September 1999, RAND/Taylor & Francis, USA; also available at http://www.ict.org.il/articles/cyberspace.htm

[3] Michael Whine, Islamism and Totalitarianism: Similarities and Differences, *Totalitarian Movements and Political Religions*, Vol 2 Number 3, Autumn 2001, Frank Cass, London

Michael Whine, Eine unheilige Allianz. Internationale Verbindungen zwischen Rechts-extremismus und Islamismus, chapter in *Globalisierter Rechts-extremismus?*, VS Verlag fur Sozialwissenschaften, Berlin, January 2006

[4] Racial Nationalism, the Third Position, and Ethnoviolence, Political Research Associates, Somerville MA, November 2005, http://www.publiceye.org/fascist/third_position.html

Peter Finn, "Unlikely Allies Bound by a Common Hatred", *Washington Post*, 29 April 2002.

[5] Martin A. Lee, *The Swastika and the Crescent*, Intelligence Report, Southern Poverty Law Center, Montgomery Ala, Spring 2002, http://www.splcenter.org/intel/intelreport/article.jsp?aid=132&printable=1

[6] George Michael, *The Enemy of My Enemy – The Alarming Convergence of Militant Islam and the Extreme Right*, University Press of Kansas, 2006

[7] Marc Sageman, *Understanding Terror Networks*, pp.137 – 173, University of Pennsylvania Press, Philadelphia, 2004

[8] Gabriel Weimann, *Terror on the Internet*, United States Institute of Peace Press, Washington DC, 2006, p 66,

[9] Ibid, p 66

[10] Ibid, p 145

[11] *2004 Annual Report of the Office for the Protection of the Constitution*, Federal Ministry of the Interior of the Federal Republic of Germany.

[12] John Solomon, "US Extremists, Terror Groups Eyed," *Associated Press,* Washington DC, 28 February 2002.

[13] Worldwide web of terrorism suspects exposed, *The Australian*, 8 June 2006, http://www.theaustralian.new.com.au/printpage/0,5942,19400650,00.html

"Canadian Arrests Portray the Value of the Internet in Global Networks", *SITE Institute*, USA, 6 June 2006, www.siteinstitute.org

Canada holds terrorist suspects, *RUSI/Jane's Homeland Security & Resiliance Monitor*, London, Vol.5 No.7, September 2006.

Tony Allen-Mills, "The nest of city bombers Canada found in time", *The Times*, London, 11 June 2006

Rita Katz and Michael Kern, "Terrorist 007 Exposed", *Washington Post*, 26 March 2006.

United States of America against Ehsanul Islam Sadequee, *Affidavit in Support of Arrest Warrant*, United States District Court Eastern District of New York, 28 March 2006.

[14] Andrew Blejwas, Anthony Griggs and Mark Potok, "Terror From the Right", Intelligence Report, Southern Poverty Law Center, 2006. http://www.splcenter.org/intel/intelreport/article.jsp?aid=628

[15] Lois Romano, "Domestic Extremist Groups Weaker but Still Worrisome", *Washington Post*, 19 April 2005, http://www.washingtonpost.com/ac2/A64058-2005Apr118?language=printer

Julian Borger, "US Extremists to be sentenced over bomb plot", *The Guardian,* London, 8 January 2004

[16] "David Burgert sentenced in U.S. District Court", News Release, United States Attorney, District of Montana, 12 November 2004, http://www.usdoj.gov/usao/my/PressRelease/20041112141900.htm

[17] "European Jihadists: case studies", *Jane's Terrorism & Security Monitor,* London, July 2005

[18] *Terrorist Activity in the European Union-Situation and Trends Report*, Europol, October 2004-October 2005, The Hague,2 May 2006; p 18

[19] See for example, "The users of the alnusra.net forum", Society for Internet Research, 16 May 2006, . http://www.sofir.org/reports/2006-15-05-alnusra.html

[20] "The Palestinian Islamic Jihad Internet infrastructure and its Internet Webhosts", Intelligence and Terrorism Information Center at the Center for Special Studies (CSS), Israel, December 2005

[21] ibid

[22] "Terrorism and Internet: an examination of Hamas's websites and the hosting providers used by them", CSS, Israel, June 2006

The websites of Hizbollah, Internet Haganah, 18 July 2006, http://internet-haganah.com/hmedia/17jul06-hizbollah_sites.html

[23] Louis Beam, "Leaderless Resistance", *The Seditionist*, Issue 12, February 1992, http://www.louisbeam.com/leaderless.htm

[24] ibid, p.2

[25] ibid, p.3

[26] ibid, p.5

[27] MSANews: "The Ladenese Epistle: Declaration of War", http://www.mynet.net/-msanews/MSANEWS/199610/19961012.3.htm, downloaded 26 June 1997

Also available at: http://www.ummah.net/sos/laden_fatwa.htm, downloaded 15 December 1998

[28] ibid

[29] http://www.aryan-nations.org/islam/index.htm

[30] Henry Schuster, "An unholy alliance – Aryan Nation leader reaches out to Al-Qaida", CNN.com, 29 March 2005, http://cnn.usnews.printthi.clickability.com/pt/cpt?action=cpt&title=CNN.com++An+unholy+alliance

[31] See; David Myatt, "Islam and National Socialism", http://www.geocities.com/davidmyatt88?islamandns.htm;

David Myatt, "National-Socialism and Muslims", http://www.geocities.co/davidmyatt88/ns_muslims.htm;

David Myatt, "My Conversion to Islam", http://www.davidmyatt.portland.co.uk/texts/my_conversion_to_islam.htm

[32] Michael Scheur, "Al-Qaeda's Tactical Doctrine for the 'Long War' ", *Terrorism Focus*, Volume III, Issue 10, The Jamestown Foundation, Washington DC, 14 March 2006.

[33] See for example, "Improvised Weapons of the American Underground", (http://www.holt.org/archives/weopons/weopons.html) , and many other online publications and websites from that time.

[34] "Jihad Online: Islamic Terrorists and the Internet", Anti-Defamation League, New York, 2002

[35] Select Committee on Religious Offences in England and Wales Minutes of Evidence – Supplementary memorandum from the Home Office (4), 12 June 2002, http://publications.parliament.uk/pa/1d200203/1drelof/95/2061201.htm,12

[36] "Protocols of the Learned Elders of Zion", http://churchoftrueisrael.com/protocols/dreammap.html, downloaded 16 January 2003

[37] Amy Brown-Bowers, "Hate message too extreme for neo-Nazis", CanWest News Service, *National Post*, 15 June 2006

Richard Warman and the Canadian Human Rights Commission and Craig Harrison, *Canadian Human Rights Tribunal*, 2006 CHRT 30, 2006/08/15,www.chrt-tcdp.gc.ca/earch/files/t1072_5305ed15aug06.pdf

[38] "Advice for Lone Wolves", National Socialist Movement Yahoo Group, 8 June 2006.

[39] Video Nazis, *The Press & Journal Aberdeen*, 14 Feb 1989

[40] "Nazi video nasties haunt Germany", *Sunday Times*, London, 26 July 1987

[41] "Echoes of Nazi Germany, Racist, anti-Semitic computer games gain popularity", *The Guardian*, London, 24 Nov 1988

[42] "Racist Game", *The Observer*, London, 26 February 1989

[43] Response – *The Wiesenthal Center World Report* ,Vol 12, No 2, USA, Summer 1991

[44] "Are Some video Games Promoting Hate?", WNBC, New York, 5 May 2003

[45] http://download.specialforce.net/english/begin/start/htm, downloaded 3 September 2003

[46] "Educating children for hatred and terrorism: encouragement for suicide bombing attacks and hatred for Israel and the Jews spread via the Internet on Hamas' online children's magazine (Al-Fateh)", Intelligence and Terrorism Information Center at the Center for Special Studies, Israel, October 2004, http://www.intelligence.org.il/eng/sib/11_04/edu.htm

[47] Gabriel Weimann , *Terror on the Internet*, United States Institute for Peace Press, Washington DC, 2006, p 120

[48] United States of America v. Syed Talha Ahsan, United States District Court, District of Connecticut, July 2006

"British man accused of aiding terrorists", *Associated Press*, 20 July 2006, http://www.boston.com/news/local/connecticut/articles/2006/07/21/british_man_a

United States of America v. Babar Ahmad, "Affidavit in support of request for extradition of Babar Ahmed", No.3:04M240(WIG), September 2004

[49] United States of America vs. Sami Omar Al-Hussayen, United States District Court for the District of Idaho, March 2004.

http://en.wikipedia.org/wik/Sami__Omar_Al-Hussayen

[50] Lois Romano, "Domestic Extremist Groups Weaker but Still Worrisome", *Washington Post,* 19 April 2005, http://www.washingtonpost.com/ac2/A64058-2005Aprl18?language=printer

[51] "Secret FBI Report Highlights Domestic Terror", ABC News, USA, 18 April 2005

[52] Chuck McCutcheon, Newhouse News Service, USA, 13 July 2004

[53] Steven Donald Smith, "Terrorists Use Internet for Propaganda, Defense Officials Say", *American Forces Press Service*, 5 May 2006, http://www.defenselink.mil/cgi-bin/dlprint.cgi?

[54] Terry Blackham, National Front members' emailing, 26 June 2006

[55] "Polish Foreign Ministry seeks U.S. help to close down nationalist website", *People's Daily*, 17 June 2006, http://english.people.com.cn/200605/29/eng20060529_269372.html

"Another arrest linked to Nazi Internet site", Poland.pl, 6 June 2006, http://www.poland.pl/news/article,Another_arrest_linked_to Nazi_Internet_site,id,22

"Un site neo-fasciste menace des journalistes: RSF alerte les authorities", International Freedom of Expression Exchange, 26 May 2006, http://www.ifex.org/en/content/view/full/74672/

"Far right website has local targets", *Bolton Evening News*, 24 June 2006
2006

Author's correspondence with Director of Public Prosecutions, 2006.

[56] Karl Ritter, "3 Men charged with planning terror attack against Swedish church," *Associated Press*, 3 May 2006

[57] Gabriel Weimann, *Terror on the Internet*, United States Institute for Peace Press, Washington DC, pp.124 – 125.

[58] Jerrold M Post, *Military Studies in the Jihad Against the Tyrants – The Al-Qaida Training Manual*, Frank Cass, London, December 2002.

[59] *The Big Book of Mischief - The Terrorist's Handbook*, Chaos Industries, Chicago, 1991, also posted to rec.pyrotechnics Newsgroup, 1992

[60] "A Practical Guide to Aryan Revolution", http://www.ftcnet.com/-chs/readm18.htm, downloaded 8 January 1998

[61] Ben MacIntyre, "Internet neo-Nazi suspect arrested in Britain", *The Times*, 19 February 1998

Susannah Herbert, "France calls on Britain to extradite 'neo-Nazi' ", *The Daily Telegraph*, London, 19 February 1998

"Un reseau neonazi mis au jour", *le Figaro*, Paris, 18 February 1998.

[62] p.73, Graeme McLagen and Nick Lowles, *Mr Evil –The Secret Life of Racist Bomber and Killer David Copeland*, John Blake, London, 2000

[63] Steve Coll and Susan B Glasser, "Terrorists Turn to the Web as Base of Operations", *Washington Post*, 7 August 2005

[64] Jerrold M Post, *Military Studies in the Jihad Against the Tyrants: the Al-Qaeda Training Manual*, Frank Cass, London, December 2002

[65] Abdul Hameed Bakier, "Jihadis Adapt to Counter-Terror Measures and Create New Intelligence Manuals", *Terrorism Monitor*, IV (14) The Jamestown Foundation, Washington DC, 13 July 2006.

[66] ibid

[67] Abdul Hameed Bakier, "Liquid Explosive Training Manual Easily Attainable on Jihadi Forums", *Terrorism Focus*, III (32) The Jamestown Foundation, Washington DC, 15 August 2006.

[68] Katie Lapthorne, Herald Sun, Australia, 2 August 2006, http://www.news.com.au/heraldsun/story/0,21985,19988180-662,00.html

[69] p.131, George Michael

[70] p.31, Report of the Official Account of the Bombings in London on 7th July, HC 1087, House of Commons, The Stationery Office, London.

[71] "The Global Islamic Media Front Announces the Initiation of Infiltrating Western Internet Forums, and issues a Call to Able Muslims to Join Information Jihad", *SITE Publications*, 12 January 2006

"Global Islamic Media Front Encourages Online Jihadists to Attack America Through the Written and Visual Media", *SITE Publications*, 12 January 2006.

[72] pp19 – 30, Europol

[73] Daniel Benjamin & Steven Simon, *The Next Attack*, Regnery Publishing Inc, USA, 2005

Hypermedia Seduction for Terrorist Recruiting
B. Ganor et al. (Eds.)
IOS Press, 2007

Using the Internet to Uncover Terrorism's Root Causes

Dr. Joshua SINAI
Program Manager, Counterterrorism Studies and Education,
The Analysis Corporation, McLean, VA

Abstract. Monitoring terrorists' use of the Internet provides an opportunity to collect information about the root causes driving terrorist warfare. Terrorist groups and their supporters are extensively active on Internet websites and in chat rooms and forums. Tracking such sites yields information about their grievances, ideologies, ambitions, and other factors that motivate them to conduct terrorist activities. The methodology presented in this chapter aims to enable counterterrorism analysts to hierarchically decompose the underlying factors driving a terrorist insurgency, by focusing on the content of terrorist-related websites.

Keywords. Insurgency, root causes, ideology, modernization, revivalist, relative deprivation, radical subculture.

Introduction

The widespread use of the Internet by terrorist groups and their supporters offers myriad opportunities for the counterterrorism community. Such opportunities include monitoring and tracking terrorist activities in cyberspace, so as to gather intelligence information about their motivations, agendas, and ongoing and future warfare activities. Although it falls below the threshold of terrorist warfare *per se*, such monitoring also provides an opportunity to understand the root causes driving terrorist warfare.

Thus, the counterterrorism community can track terrorists' use of the Internet to uncover insights about their communications, information gathering, training and education, fundraising, operational planning, command and control, publicity and propaganda, and the radicalization and recruitment of potential operatives. In addition, "official" postings on their websites, and discussions in their forums and chat rooms, also provide information about some of the root causes driving their grievances, ambitions and other factors that motivate them to conduct terrorist activities.

This chapter's objective is to provide a methodology to enable counterterrorism analysts to hierarchically decompose the underlying factors driving a terrorist insurgency. Such an analysis would focus on the content of official terrorist websites and their accompanying forums and chat rooms, in order to formulate appropriate responses—whether coercive or conciliatory—that will be capable of resolving such conflicts.

1. Root Causes of Insurgencies

Terrorism is a type of insurgent warfare by sub-national groups against the state [1]. Groups are driven to commit acts of terrorism, in which they target innocent civilians or armed military, by a spectrum of motivations. These include real or perceived grievances, psychological dispositions, ambitions, and other factors. Terrorism does not emerge within a vacuum, but is the product of a confluence or coalescence of factors, whether in the societies where terrorist groups originate or in those of their targeted adversaries. Terrorism is not necessarily the product of a single causal factor, but is often rather a convergence of many interrelated factors and causes. These causes vary from one conflict to another, and change dynamically over time.

One must also consider how new sets of underlying causes that emerge over time sustain a terrorist insurgency. It is important to analyze all of these underlying factors, because, at least in theory, addressing such factors or drivers would solve problems that have identifiable root causes. Uproot the cause and the problem is solved. By contrast, leave the roots intact and the terrorism problem will persist.

Nevertheless, it must be recognized that, due to a variety of factors [2], while some root causes may be resolvable through conciliatory measures, such as a negotiated settlement, others may only be resolvable through continued military means.

As mentioned above, root causes are not static, but are dynamic and constantly changing. In fact, even when a government begins to address what it considers to be the primary root cause underlying a conflict, it is always possible for the terrorist group to claim that another, yet unresolved, root cause needs to be attended to. Thus, to attain a complete picture of the underlying causes driving an insurgency, it is crucial to examine them at levels that are both general and specific. These levels include the individual, group, societal and governmental levels.

At the individual level, moreover, different root causes may influence different types of operatives within a terrorist group, including their supporters. Thus, different drivers may influence members of a group's combat unit and its supporting "civilian" infrastructure. Such key members include those responsible for funding and logistical arrangements, as well as individuals who sympathize with a group's cause, and whose views can be gleaned from their Internet discussions.

Root causes also need to be examined and synthesized from the divergent points of view of the insurgents, the targeted governments, and academic researchers, the latter of whom may offer a more objective and independent assessment of such underlying drivers.

Any conceptual framework designed to explain the causes underlying a terrorist insurgency needs to overcome what Marc Sageman calls the "problem of specificity" [3]. Thus, while it may be possible to identify some of the necessary conditions for such outbreaks, Sageman argues that "such an approach is overly broad and leads to the fundamental problem of specificity. Many people share the identified characteristics of such an analysis, but very few become terrorists. What accounts for this vast difference?" [3].

Thus, the first step is resolve these theoretical and empirical issues regarding the correlation of factors characteristic of aggrieved populations (or populations that perceive themselves to be aggrieved) and factors characteristic of those who ultimately become terrorists. Thereafter, a conceptual framework can be applied to investigate how terrorist-related websites can reveal the underlying conditions of specific conflicts. Such an examination should take into account the factors driving terrorists to attack particular adversaries, and the choice or use of particular weapons in such warfare.

The framework we propose for understanding the root causes of terrorist insurgencies is based on seven interrelated steps. These are presented below in successive order.

2. Towards a Conceptual Framework

First, one needs to identify a terrorist insurgency's "physical" and "virtual" manifestations. These take the form of a spectrum of activities on the ground, ranging from the type of terrorist warfare being waged against a targeted country (or community) to political warfare, such as the radicalization and recruitment of operatives. Type of warfare includes, for example, the use of tactics involving conventional low impact, conventional high impact, or CBRN weapons and devices[1]. Radicalization and recruitment of operatives takes place among supporting communities, including mosques, schools, and other facilities.

In "virtual" space, evidence of political warfare can be found in terrorist-related websites, forums and chat rooms, where a group's propaganda is intended to radicalize supporters and recruit new members. Interestingly, cyberspace provides terrorist groups with unlimited geographical reach, unconstrained by national borders.

---

[1] In this framework, conventional low impact (CLI) refers to the use of conventional means to cause relatively few casualties; conventional high impact (CHI) to the use of conventional means to cause catastrophic damages, such as crashing airliners into the World Trade Towers; and chemical, biological, radiological or nuclear (CBRN) warfare to the use of "unconventional" means to inflict mass casualties (although chemical and radiological devices may also cause a relatively low number of casualties). This is not intended to be a rigid trichotomy, and it is possible for groups to use a combination of these types of weapons and devices in their warfare.

For each of these "physical" and "virtual" manifestations, a root cause would then be identified. For example, one would attempt to explain why a group had chosen to embark on conventional low impact, conventional high impact, or CBRN warfare, or "conventional" versus suicide martyrdom tactics in order to achieve its strategic objectives.

Second, a terrorist insurgency's ideological drivers need to be identified. Ideas and, in the contemporary period, especially radical religious ideologies, are among the major drivers that mobilize individuals and groups into committing acts of terrorism, and provide them with a cultural and religious underpinning and guide for action.

The problems facing individuals whose societies are transitioning from traditional to modern ways of life are an important determinant of terrorist insurgencies. This is a particular concern with terrorist groups that espouse fundamentalist religious ideologies. The drive towards modernization produces socio-economic and political dislocations in society, and individuals are left to cope with the challenges ushered in by modernization. This change fosters the rise of new elites, who represent democratization, secularization, industrialization and technocracy. These elite individuals emerge into positions of influence, replacing their more traditional counterparts. This changeover results in a series of conflicts between the advocates and opponents of modernity.

According to Arie Kruglanski, radical ideologies appeal to individuals experiencing psychological uncertainty because such ideologies are "formulated in clear-cut, definitive terms" and provide "cognitive closure." [4] Thus, among those who cling to traditional religious values and ways of life, some will respond by espousing revivalist ideologies that promote resentment and even xenophobia towards those who believe in the benefits of modernity. Often, this perceived opponent includes foreigners, especially as represented by Western Europe and the United States. Such revivalist ideologies, which affect not only Muslim societies but Christian and Jewish ones as well, promote a vision that advocates restoration of their own societies to a state of religious "purity", which supposedly existed in previous centuries. These kinds of revivalist movements provide fertile ground for the support of, and recruitment into, terrorist organizations.

Thus, in examining terrorist websites and accompanying chat rooms, it is important to identify the ideologies, whether political or religious, that motivate their adherents to embark on terrorism. For example, the Palestinian Hamas is driven by a combination of Palestinian nationalism and Muslim Brotherhood ideologies; Al-Qaida and its affiliated groups are influenced by Salafi Jihadi ideologies; while the Provisional IRA, during its prime, was motivated by a blend of Irish Catholic nationalism and certain socio-economic beliefs. On these groups' websites, their ideologies are represented by what the groups consider to be their traditional religious, historical or secular nationalistic symbols.

Third, it is crucial to examine the structural preconditions underlying a terrorist insurgency. Structural preconditions "focus on social conditions ('structures') that affect group access to services, equal rights, civil protections, freedom, or other quality-of-life measures." [5] Examples of structural preconditions include government policies, actions by security forces, access (or lack thereof) by a local population to social institutions, and so on [5]. The state or government is the key focus in structural theories of terrorism, because of its role in serving as the precipitating factor for a terrorist uprising. According to this theory, societal injustice, popular discontent, the alienation of elites, and a sense of societal crisis are key ingredients for a terrorist eruption in society [6].

In addition to such structural preconditions which depend on the central role of the state and its instruments of power in causing terrorist rebellions, other preconditions depend on the confluence of factors below the state level. Relative deprivation (RD) theory, which was developed by Ted Robert Gurr, focuses on the relationship between frustration and aggression [7]. As applied to the terrorism milieu, according to Gurr, feelings of frustration and anger underlie individual decisions to engage in collective action against the perceived source of their frustration, and constitute one of the necessary conditions for joining a terrorist group. The motive for such individuals to engage in political violence is a perception that they are *relatively* deprived, vis-à-vis other groups, in an unjust social order. When rising expectations are met by governmental resistance in the form of sustained political repression, low ranking socio-economic or political status, or lack of educational opportunities, a group is likely to turn to political violence.

As a corollary to relative deprivation, absolute deprivation theory holds that when a group has been deprived of the basic necessities for survival by a government or social order (which includes being subjected to physical abuse, poverty or starvation), it turns to political violence [8]. Thus, the difference between relative and absolute deprivation is based on the degree of discrepancy experienced by an individual or group. Absolute deprivation is the discrepancy between what people have and what they need for daily sustenance, while relative deprivation refers to what they have versus what they believe they deserve vis-à-vis others in society. Both types of deprivation are capable of driving those believing themselves to be aggrieved into carrying out acts of political violence.

Moving from these general theories to more specific indicators of terrorist outbreaks, there are underlying structural factors that need to be examined at the societal, group, and individual levels. At the societal level, countries that are vulnerable to terrorism are those which experience economic and social inequality, poverty, low levels of social services, a lack of political or civil rights, low literacy rates and a lack of education, and ethnic conflict. These factors are likely to serve as springboards for terrorism, or at least some manifestation of active discontent that may fall below the threshold of violence.

The group level of analysis intervenes between the societal and individual levels. Thus, at the group level, agents of mobilization, such as charismatic leaders or radical movements and their political and religious ideologies, serve as causal factors for terrorist outbreaks.

At the individual level, susceptibility to radicalization and actual recruitment into terrorist organizations are additional causal factors.

The ecological framework for examining terrorist insurgencies therefore comprises at least three levels—individual, group and societal, with additional explanatory value provided by theories at more general levels.

As demonstrated by these approaches, no single theory or structural (or sub-structural) factor provides a sufficient explanation for terrorism. Links between poverty, lack of education and other socio-economic variables, including the clash between tradition and modernity, and terrorism, are complex, as are the links between state repression, lack of political opportunities and terrorism. Other political factors may also contribute to terrorism, such as the ability of terrorist groups to exploit political disorder or a lack of political order in weak or failed states. At other times, however, groups may actually receive political support from governments, such as support by the Iranian and Syrian governments to the Lebanese Hezbollah.

Theories or models for understanding terrorism therefore need to focus on the causal mechanisms or processes in which multiple factors, working together in specific social and other contexts, influence and drive terrorist insurgencies. Causal mechanisms alone are insufficient to bring about a terrorist insurgency; another intervening variable is crucial to initiating such an outbreak.

In this fourth interrelated step, certain types of enablers—in the form of leaders, radical subcultures, and susceptible individuals—are necessary to mobilize and empower a disaffected population to embark on a terrorist insurgency.

While effective leadership traits may be contextually dependent [9], certain management traits are necessary for an effective terrorist leader. Leaders with charismatic personalities are able to transform grievances and frustrations into a political agenda for violent action, to radicalize supporters, and recruit and mobilize operatives to sacrifice their lives for the cause. They also attract a coterie of dedicated and capable associates to help manage and sustain the group, as well as imposing strict discipline over the rest of the group.

Radical subcultures are another crucial enabler in driving a terrorist insurgency, through the provision of new operatives. The relationship between terrorist groups and their sympathizer and support communities can be viewed as a pyramid, with the terrorist groups at the apex and the sympathizers and supporters at the base. The higher levels of the pyramid are characterized by increasing levels of mobilization, commitment, and engagement in terrorist activities. For a terrorist group to grow and sustain itself over time, it requires a continuous and ever-expanding connection to the larger pyramid for support, and a pool of recruits to replace operatives who are arrested or killed in action. Terrorist groups could not survive without community support, which is often based on ethnic or religious ties.

In addition, radical subcultures, whether they exist in immigrant communities or prisons, can be termed "susceptible" or "feeder communities", because of the ease with which they can be infiltrated by terrorist movements for potential recruits and supporters.

Susceptible individuals constitute another type of enabler. For example, using social network analysis, Marc Sageman found that a significant proportion of terrorists who join the Global Salafi Jihadist movement are drawn into such groups by social bonds. Some are linked through marriage and family connections, and others through friendships acquired at a local mosque. He argues that such social interactions are a crucial factor in explaining why, for example, upwardly mobile, middle class, and yet lonely and disaffected men turn to certain mosques, where they become radicalized and recruited into terrorism. For other individuals, such as Muhammad Atta (one of the 9/11 operatives), after leaving their home country in the Middle East to be educated in West, religion provides them with a means to find friends, gain a sense of spiritual fulfillment and restore their sense of self [10].

The fifth interrelated step in causing a terrorist insurgency or influencing the decision to carry out major attacks are precipitants. These take the form of triggering events. Such events might include outrageous acts committed by a group's adversary; political failure (e.g. the breakdown of the June 2000 Camp David negotiations was used as a pretext for instigating violence); certain provocative events (e.g. Ariel Sharon's visit to the Temple Mount in Jerusalem on September 28, 2000); or the publication of controversial cartoons. There are other types of events which can act as precipitants too, and such events inevitably receive extensive coverage in terrorist-related websites.

Terrorist insurgencies also persist over time. In the sixth step, it is crucial to examine the factors that sustain such outbreaks over time [11]. These factors include unresolved root causes that continue to drive the insurgency—new and different factors that have little or nothing to do with the initial root causes. These include cycles of revenge, the need for a group to provide for its members or for the survival of the group itself, profitable criminal activities, and a feeling by a group's operatives that they have no choice but to continue with their armed struggle because it offers the only alternative to imprisonment or death at the hands of the authorities.

In the seventh step, in ideal cases, it is hoped that the previous six steps of mapping a conflict's root causes will produce the knowledge and insight on the part of governments to formulate appropriate responses to terrorism. Decisions must be made regarding which response would be most effective in terminating a terrorist insurgency, whether peacefully, militarily, or through a combination of such measures. By incorporating an understanding of a conflict's underlying causes into a government's "combating terrorism" campaign[2], such response strategies and tactics can be effectively calibrated to address the specific challenges and threats.

---

[2] Combating terrorism (CbT) is an umbrella concept incorporating anti-terrorism, which is defensively oriented, and counter-terrorism, which is offensively oriented.

3. Conclusion

Addressing a conflict's root causes may not necessarily or automatically lead to conflict termination. First, there may not always be a direct correlation between a specific root cause and a terrorist rebellion, because of the myriad alternative forms of action, ranging from non-violent to violent, that may be available to a group to express its underlying grievances and demands. In fact, a terrorist rebellion is likely to occur only when certain significant propitious circumstances—in the form of political, economic, social, military, and other underlying trends—coincide and coalesce. But even these trends may not be sufficient to launch such rebellions, unless they are buttressed by the availability of effective leaders, organizational formations (including a willing cadre), access to particular types of weaponry, and the logistical and other covert capabilities to carry out an operation against an adversary.

Second, root causes should not be viewed as necessarily static. Some of the root causes that might play a significant role in the initial phase of a conflict may later on become peripheral, while other root causes may emerge as paramount at a later phase in a conflict.

Once the spectrum of a conflict's underlying root causes are mapped and identified—initially, in most cases, at the academic level, and then at the governmental level—it will be up to governments and their security and military organizations to formulate the appropriate response measures to resolve these underlying problems and combat terrorism. For the underlying factors to be resolved, however, it is also up to the insurgents to incorporate into their demands grievances and other objectives that are amenable to the "give and take" of compromise and negotiations. Otherwise, even addressing a conflict's root causes may not succeed in terminating the insurgency.

Hopefully, the seven-step framework offered in this chapter provides a methodology for counterterrorism analysts to derive insight from the content and discussions on terrorist websites and their accompanying chat rooms. This insight would focus on the underlying factors driving the terrorist insurgencies that are being promoted by the relevant sponsoring groups. With terrorist groups and their supporters increasingly resorting to cyberspace to establish their "virtual" communities (while continuing to operate in "physical" space), the Internet provides an important medium to glean valuable information. This information pertains to both the underlying causes driving terrorist insurgencies, as well as the measures that are necessary to terminate them.

References

[1] See Gabriel Weimann, *Terrorism and the Internet* (Washington, DC: United States Institute of Peace, 2006).

[2] For initial examinations of root causes, see Walter Reich, editor, *Origins of Terrorism: Psychologies, Ideologies, Theologies, States of Mind* (Washington, DC: Woodrow Wilson Center Press, 1998); and Tore Bjorgo, editor, *Root Causes of Terrorism* (New York: Routledge, 2005).

[3] Marc Sageman, "Threat Convergence: The Future of Terrorism Research" [unpublished paper prepared for the workshop on "Threat Convergence: Possible New Pathways to Proliferation – Terrorism, Weapons of Mass Destruction, and Weak and Failing States", April 7, 2006, Washington, DC; cited by permission].

[4] Arie Kruglianski, "Inside the Terrorist Mind," paper presented to the National Academy of Science annual meeting, Washington, DC, April 29, 2002.

[5] Gus Martin, *Understanding Terrorism: Challenges, Perspectives, and Issues* (Thousand Oaks, CA: SAGE Publications, 2003), p. 67. Martin based his analysis on the works of Steven E. Barkan and Lynne L. Snowden, *Collective Violence* (Boston: Allyn & Bacon, 2001) and Jack A. Goldstone, "Introduction: The Comparative and Historical Study of Revolutions," in Jack A. Goldstone, editor, *Revolutions: Theoretical, Comparative, and Historical Studies* (San Diego, CA: Harcourt Brace Jovanovich, 1986).

[6] *Ibid.*, p. 68.

[7] Ted Robert Gurr, *Why Men Rebel* (Princeton, NJ: Princeton University Press, 1970).

[8] See Charles Y. Glock, "The Role of Deprivation in the Origin and Evolution of Religious Groups," in *Religion and Social Conflict* , R. Lee and M. E. Marty, editors (New York: Oxford University Press, 1964), pp. 24-36

[9] John Horgan, "The Search for the Terrorist Personality," pp. 3-27, in Andrew Silke, editor, *Terrorists, Victims and Society: Psychological Perspectives on Terrorism and its Consequences* (New York: John Wiley & Sons, 2003).

[10] Marc Sageman, *Understanding Terror Networks* (Philadelphia: University of Pennsylvania Press, 2004).

[11] The discussion of factors responsible for sustaining a terrorist insurgency is based on Tore Bjorgo, "Introduction," in *Root Causes of Terrorism*, Tore Bjorgo, editor (New York: Routledge, 2005).

Hypermedia Seduction for Terrorist Recruiting
B. Ganor et al. (Eds.)
IOS Press, 2007

Hypermedia Communication as a Modern Means for the Creation of Terrorist and Counterterrorist Consciousness

Luibomyr TOKAR

National Institute of International Security Problems National Security and Defense Council of Ukraine- Hypermedia
tokar@niisp.gov.ua

Abstract. The classical definitions for terrorism are related to its characteristics of disorder and insecurity, and are inappropriate for "post-9/11" thinking. The current situation is caused by the chaos of a changing world, which has created modern types of terrorism. But the "new" terrorist dynamic must be considered as a continuation of the "old", and also as the "continuation of politics by other means", to use the words of Carl von Clausewitz. Certain new elements and new directions have added to the effectiveness of modern terrorism, including the latest media means; the "new" terrorism and the media share a symbiotic relationship. This form of terrorism is itself a media-psychological weapon, which damages the communicative sphere and creates massive threats to the wider society.

Key words. "New" terrorism, war on terrorism, media-psychological weapon, critical thinking.

Introduction

The history of the world, according to the great German philosopher Hegel, is none other than the progress of the "consciousness of freedom." However, contrary to this optimistic statement, terrorist forces depend upon the consciousness of fear and violence. In current times, rogue states and the forces behind global terrorism are building weapons that are intended to disrupt healthy, non-violent consciousness. Often, these forces do succeed in destroying the consciousness of freedom.

The most powerful ideologies of the twentieth century era, namely communism and capitalism, were both concerned with the vision of a superior world order. Each sought ways of translating this utopian ideal into reality, and developed ideas about how to organize economic and political life.

The new digital telecommunication technologies, or hypermedia (hypertext and multimedia), have strongly influenced the redistribution of political power in the twenty-first century. In our hypermedia age, technologies such as the Internet, encryption, and high-resolution satellite imaging favor non-territorial communities; the same technologies also shift policy-making from individual nations to transnational corporations, non-governmental organizations, and activists. These include not only legitimate organizations, but also illegal, criminal ones.

1. Mass Media and "New Terrorism"

Historically, the term "media" refers to all existent methods and channels of information and entertainment. The media encompasses not only newspapers, radio and television, but also other important forms of communication, such as books, films, music, theatre and the visual arts. The late twentieth century has seen the globalization of the mass media culture, along with many informal methods of communication (such as gossip in taverns, streets and marketplaces). These informal channels coexist with the latest multimedia technology in contemporary societies.

The term "terrorism", as used in this chapter, denotes a particular type of violence; it is a synonym for politically-motivated violence in general, which includes such characteristics as the creation of a climate of extreme fear. Terrorism may be seen as a propagation of policy by violent means, in line with Carl von Clausewitz's notion that "war is the continuation of policy by other means" [1].

There are symbiotic relationships between terrorism and the media, and terrorists' exploitation of the media plays a crucial role in terrorist propaganda. Thus there is a need for voluntary self-regulation by the media. Media personnel should be assessing the impact of mass communications on policy options within a democratic society, as well as monitoring the media's response to terrorism.

The political weapon of terrorism does not serve individual television producers and journalists directly, but it would be foolish to deny that many modern terrorists and certain sections of the mass media do become locked into relationships of considerable mutual benefit. When one says "terrorism" in a democratic society, one is effectively saying "media". Terrorism itself is a psychological weapon which depends upon communicating a threat to the wider society. This, in essence, is why terrorism and the media enjoy a symbiotic relationship. We could conclude that before we can understand what gives rise to new forms of violence and terrorism, we first have to understand the impact of new digital telecommunication technologies on contemporary societies.

2. A Reassessment of Old Approaches

Now we encounter a basic dilemma. Terrorist "all-out war" against modern civilization as whole, and against particular states, is a cruel reality; whereas counterterrorism in its modern forms is inappropriate for the new situation. Violent, non-state actors pose a pressing challenge to human and national security across the globe. The new waves of terrorism, against this background, might be interpreted as the exacerbation of chaotic behavior, both within and beyond existing orderly systems. Robert Cooper, bestselling author, scholar, and British diplomat, has written that "We may not be interested in chaos, but chaos is interested in us" [2]. This idea is the subject of the modern "world risk society" theories, with their many applied aspects and many practical consequences.

There are two contrasting approaches to the question of national and international security. Both approaches share a common goal, namely to combat the violence that exists in the modern world. But the two methods are very different, with one relying on "soft" tactics and the other on "hard" strategies.

The traditional approach to international security problems, during the cold war era, has been one of offensive or defensive interventionism. This approach is based on the effort to control the foreign policies of threatening states (rogue or failing states) through military force. The logical conclusion in this direction would be a doctrine of "pre-emptive wars", which equates to a "hard" security policy.

The "soft" security-related policy aims to minimize challenges proceeding from threatening societies and their states. Such challenges are minimized by strengthening political and economic relations, via cultural penetration and absorption, and by limiting the existence of socio-cultural "ghettos".

The synthesis of the soft and hard approaches appears to be impossible at the current time; however, such a synthesis, should it become possible, would allow for a more effective effort in confronting international terrorist threats.

Modern media (that is, the communicative aspects of the media) are creating a breeding ground for modern terrorism. This is a cyber-environment, where de-realization and de-personification are commonplace. However, modern media have an equally great potential to create the preconditions for effective antiterrorist actions, in a gentle or "synergistic" style.

Troubling questions abound, perhaps the most salient of which is: Could terrorism exist without the media? This issue creates a permanent tension between security, on the one hand, and various liberties on the other.

The quasi-revolutionary terrorist activity of yesterday, together with the stereotype of a terrorist who is motivated by a "progressive" ideology of national or social "liberation", and who is overtly backed by state sponsors, has quite disappeared nowadays. The ultimate goal of previous terrorist organizations was social and political chaos. These terrorist efforts invariably resulted in one of the following: a highly controlled society, with the emergence of totalitarian or authoritarian governance ("states of national security"), such as the former USSR and some Third World countries; a "pre-revolutionary" state, which entails great social disorder and anomie; and finally, emerging from the pre-revolutionary state, a "new order", which lacks the chaotic nature of its predecessor.

In contrast, the major distinguishing feature of the new style of terrorism is its lack of a center of gravity, as we might commonly understand it. Instead, there is a bazaar of violence. The extremely vivid nature of the terrorist network implies that it lives on the edge of chaos. This is a network whose very viability depends upon its ability to respond rapidly to incoming information.

The new terrorism very actively exploits the social chaos connected with globalization, "violent" modernization, and the numerous social freedoms and human rights which are prominent in a Western democratic society. Such terrorism is connected with the activity of neotribal warlords in the global market place ("bazaar").

The most striking examples of this "marketization" of terrorism, with its invisible hands and barbarous markets of death, are the situations in Iraq, Chechnya and Kosovo, and their surrounds. Many violent non-state actors, with close connections between them, control this terrorism in "bazaar" style. Such terrorist acts often occur when a stateless terrorist group seeks resources from a terrorist (rogue) state, and is either successful or unsuccessful in obtaining those resources.

Terrorist weapons of mass disruption (deconsolidation) inflict much physical damage, but the more important impact occurs on the psychological level. People begin to regard each other as a possible terrorist threat, and this increases the need to guard against any future attacks via greater security; this, in turn, creates a threat to democracy.

The best weapon against modern terrorism is mass consolidation, coupled with anti-terrorist education. These factors allow for the emergence of new type of mass consciousness, which can play a powerful and positive role in altering the environment in which terrorism breeds.

The free media in an open society are particularly vulnerable to exploitation and manipulation by ruthless terrorist organizations. According to Paul Wilkinson [3], terrorists generally have four main objectives in using media channels:

1. To create propaganda about their deeds and thus sow extreme fear among their target groups;
2. To mobilize wider support for their cause among the general population, and to sway international opinion by emphasizing such themes as the righteousness of their cause and the inevitability of their victory;
3. To frustrate and disrupt the response of governments and security forces, for example by suggesting that all practical anti-terrorist measures are inherently tyrannical and counterproductive;
4. To mobilize, incite and boost their constituency of actual and potential supporters, and in so doing to increase recruitment, raise more funds and inspire further attacks.

But the threat of terrorism should not be used as an excuse to impose restrictions on the freedom of expression, or on the freedoms of the media and of information. Specifically, the following rights must be preserved: the protection of confidential sources of information; accessing of information held by public bodies; freedom of movement; and privacy of communications. It is a common view in free societies that media outlets; journalists', publishers' and broadcasters' associations; academic institutions and other civil society organizations should take measures to enhance the capacity of the media to report professionally on terrorism.

The media should also promote tolerance, including through training and by providing opportunities to discuss ethical issues relating to the reporting on terrorism [4]. But it is also true that terrorist threats can be used in post-totalitarian countries as a convenient justification for repressing the freedom of expression.

3. The "New Terrorism" and the New Media

In *How Modern Terrorism Uses the Internet,* Professor Gabriel Weimann indicates that terrorism has a close connection with the new communication technologies, and is constantly changing its character and modes of operation. Thus, today's post-modern terrorism has a new face: it is "less centralized, less structured, and less organized, yet far more dangerous than the terrorism of the late twentieth century" [4].

According to Weimann, terrorists have established a sophisticated and dynamic presence on the Internet, one that has completely transformed the way they communicate, obtain information, conduct propaganda and issue threats. Weimann points out that the Internet provides terrorists with numerous advantages. First, just as multimedia conglomerates control newspapers, radio and television stations, terrorist groups are exploiting the Internet to create their own multimedia synergies. These are managed by media departments that develop videos, chat rooms, links to sister sites, downloadable posters and screen savers. Like an ever-expanding conglomerate, they have even established their own on-line "universities" that serve as jihad academies.

Second, the Internet gives terrorist groups the capability to extend their virtual community to any part of the world. In such a way, Al-Qaida's dream of establishing a pan-Islamic caliphate, which so far has failed to be implemented on the ground, is taking shape in cyberspace.

The success of terrorist forces depends upon their ability to recruit new members and passive sympathizers, and for this they need ever more effective hypermedia communications. It is a fundamental truth that without such communications, there would be no phenomenon of global terrorism. Terrorist communications have undergone a revolution, and are becoming increasingly more sophisticated in terms of quality, content, and transmission capacity.

The Internet has significantly expanded the opportunities for terrorists to secure publicity. Until the advent of the Internet, terrorists' ability to win publicity for their causes and activities depended on their attracting the attention of television, radio, or the print media. Now, many terrorists have direct control over their media messages. This offers further opportunities to shape how they are perceived by different target audiences, and to manipulate their own public images and the images of their enemies. Terrorist communication has evolved to a point where the terrorists themselves can now control the entire production process, determining the content, context and medium over which their message is projected, towards the precise audience (or multiple audiences) whom they seek to reach.

The implications of this information revolution are enormous. Terrorist movements now have the ability to shape and disseminate their own messages in their own way, which enables them to bypass completely the more traditional, established media outlets. As Tina Brown, the doyenne of post-modern media, has pointed out, the "conjunction of 21st-century Internet speed and 12th-century fanaticism has turned our world into a tinderbox."[5]

Thus the Internet, once regarded as an engine for education and enlightenment, has now become an essential means for the dissemination of various kinds of violent terrorist consciousness. It has become one of the most important tools in terrorist strategy.

The other side of the coin, namely the Internet's potential for public dissemination of antiterrorist information, has not received sufficient attention. There has been an inadequate effort to develop antiterrorist operations in this realm; to date, little use has been made of the Internet for counterterrorist initiatives. This is despite the Internet's potential as a virtual sanctuary, where the propaganda, conspiracy theories, and messages of hate, intolerance and violence that are purveyed through the same hypermedia could be challenged effectively.

In order for counterterrorist information operations to become more effective, a far better understanding is needed of terrorist groups and of the constituencies which they have already proven effective in reaching and motivating. This knowledge would allow for counterterrorist initiatives that are more specifically focused and better tailored to the variety of audiences which one seeks to inform and influence

"If you know the enemy and know yourself," Sun Tzu famously advised centuries ago, "you need not fear the results of a hundred battles" [6]. Indeed, what remains missing, four and a half years since the current terrorist war began, is a thorough, systematic understanding of the enemy. Such an understanding must encompass motivation as well as mindset, decision-making processes as well as command and control relationships, and ideological constructs as well as organizational dynamics.

4. Terrorist Websites, their Audiences, and their Typical Content

There are several rhetorical devices which one commonly encounters on terrorist websites. The first is the claim that terrorists have no choice other than to turn to violence. Violence is presented as a necessity, foisted upon the weak as their only means by which to respond to an oppressive enemy. While such websites avoid mentioning how the terrorists victimize others, the forceful actions of governments and regimes which combat terrorists are heavily emphasized. The terrorist organization is depicted as being constantly persecuted, its leaders subject to assassination attempts and its supporters massacred, its freedom of expression curtailed, and its adherents arrested. This tactic, which portrays the organization as small, weak, and hunted down by a strong power or a strong state, turns the terrorists into the underdog.

A second and related rhetorical ploy used to justify violence is the demonizing and delegitimization of the enemy. Members of the terrorist movement or organization are presented as freedom fighters, forced against their will to use violence because a ruthless enemy is crushing the rights and dignity of their people or group. Descriptions of the actions of the enemy are characterized by emotive terms such as "slaughter", "murder", and "genocide". The enemy of the movement or organization is the real terrorist, many sites insist; "Our violence is tiny in comparison to his aggression" is a common argument. Once again, this kind of rhetoric tries to shift the responsibility for violence from the terrorist to the adversary, which is accused of displaying its brutality, inhumanity, and immorality.

A third rhetorical device makes extensive use of the language of non-violence in an attempt to counter the terrorists' violent image. Although these are violent organizations, many of their sites claim that they seek peaceful solutions, and that their ultimate aim is a diplomatic settlement achieved through negotiation and international pressure brought to bear against a repressive government.

The revolution in electronic communications has led to a situation where any and all of the following may be used as terrorist weapons: videotapes, CD-ROMs, DVDs, laptop and desktop computers, CD burners, e-mail accounts, and the Internet and worldwide web access in general. Virtually every terrorist group in the world today has its own Internet website, and, in many instances, a single group may maintain multiple sites in different languages, with different messages tailored to specific audiences. Public access to terrorist websites, like any other websites, is easy.

According to Weimann [4], some 4,800 terrorist or terrorist-related websites currently exist; other sources estimate that more than forty active terrorist organizations operate worldwide. As the following list shows, these organizations and groups appear in all corners of the globe:

- ***Middle East:*** Hamas (the Islamic Resistance Movement); Lebanese Hezbollah (Party of God); Al Aqsa Martyrs Brigades; Fatah Tanzim; Popular Front for the Liberation of Palestine (PFLP); Palestinian Islamic Jihad; Kahane Lives Movement; People's Mujahedin of Iran (PMOI—Mujahedin-e Khalq); Kurdish Workers' Party (PKK); the Turkish-based Popular Democratic Liberation Front Party (DHKP/C); and Great East Islamic Raiders Front (IBDA-C).

- ***Europe:*** Basque ETA Movement; Armata Corsa (Corsican Army); Irish Republican Army (IRA).

- ***Latin America:*** Peru's Tupak-Amaru (MRTA) and Shining Path (Sendero Luminoso); Colombian National Liberation Army (ELN-Colombia); Armed Revolutionary Forces of Colombia (FARC).

- ***Asia:*** Al-Qaida; Japanese Supreme Truth (Aum Shinrikyo); Ansar al Islam (Supporters of Islam) in Iraq; Japanese Red Army (JRA); Hizb-ul Mujehideen in Kashmir; Liberation Tigers of Tamil Eelam (LTTE); Islamic Movement of Uzbekistan (IMU); Moro Islamic Liberation Front (MILF) in the Philippines; Pakistan-based Lashkare-Taiba; and the rebel movement in Chechnya.

The ability to communicate in real time via the Internet, using a variety of compelling electronic media (including video, digital photographs, and audio clips, accompanied by visually stimulating web design), has enabled terrorists to reach their potential audiences faster, more pervasively, and more effectively. The Internet has become indispensable to terrorists and their sympathizers and supporters worldwide, as it allows for a myriad of opportunities for radicalization, recruitment, propaganda and training. The role that the Internet plays in this specific case of the radicalization of mass consciousness is beyond doubt.

Which audiences do terrorists target through their websites? According to Prof. Weimann, the content of the websites suggests three different audiences.

- ***Current and potential supporters.*** Terrorist websites make heavy use of slogans, and offer items for sale, including T-shirts, badges, flags, videotapes and audiocassettes, all evidently aimed at sympathizers. Often, an organization will target its local supporters with a website written in the local language, and will provide detailed information about the activities and internal politics of the organization, its allies, and its competitors.

- ***International public opinion.*** The international public, who are not directly involved in the conflict but who may have some interest in the issues involved, are courted with websites using languages other than the local tongue. Most sites offer several different languages. ETA's website, for instance, offers information in Castilian, German, French, and Italian; while the MRTA site offers Japanese and Italian in addition to its English and Spanish versions; and the IMU site uses Arabic, English, and Russian. For the benefit of their international audiences, the sites present basic information about the organization, and extensive historical background material.

- ***Specific audiences.*** Judging from the content of many of the sites, it appears that foreign journalists are also targeted. Press releases are often placed on the websites in an effort to get the organization's point of view into the traditional media. The detailed background information which is provided is also very useful for international reporters and politicians. One of Hezbollah's sites specifically addresses journalists, inviting them to interact with the organization's press office via e-mail.

- ***The enemy public:*** Efforts to reach the enemy public (i.e. citizens of those states against which the terrorists are fighting) are not as clearly apparent from the content of many websites. However, some sites do seem to make an effort to demoralize the enemy, by threatening attacks and by fostering feelings of guilt about the enemy's conduct and motives.

What is the content of terrorist websites? Typically, a site will provide the history of the organization and its activities; a detailed review of its social and political background; accounts of its notable exploits; biographies of its leaders, founders, and heroes; information on its political and ideological aims; fierce criticism of its enemies; and up-to-date news. Nationalist and separatist organizations generally display maps of the areas in dispute. For example, the Hamas website shows a map of Palestine, the FARC website shows a map of Colombia, and the LTTE site presents a map of Sri Lanka.

Most websites refrain from referring to the terrorists' violent actions or their fatal consequences, with a reticence that is presumably based on propagandist and image-building considerations.

The Internet can be used not only to solicit donations from sympathizers, but also to recruit and mobilize supporters to play a more active role in support of terrorist activities or causes.

Weimann [7] states that "In addition to seeking converts by using the full panoply of website technologies (audio, digital video, etc.) to enhance the presentation of their message, terrorist organizations capture information about the users who browse their websites. Users who seem most interested in the organization's cause or well suited to carrying out its work are then contacted." Weimann [7] also cites Zanini and Edwards as saying, "Recruiters may also use more interactive Internet technology to roam online chat rooms and cyber cafes looking for receptive members of the public, particularly young people." Electronic bulletin boards and user nets (issue-specific chat rooms and bulletins) can also serve as vehicles for reaching out to potential recruits.

5. Towards a Counterterrorist Media Strategy

Given the proliferation of terrorist and radical websites, the efforts of antiterrorist forces—in terms of Internet-based public diplomacy and information operations—have been elementary if not inadequate. Counterterrorist organizations need to plan and implement programs to develop new information sources and means of communication. The traditional media sources (e.g. television, radio, and print media) have been superseded by the Internet, and it is the Internet which must now be utilized in the effort to create a mass consciousness of antiterrorism.

In this regard, any communication efforts by counterterrorist organizations which rely on standard or traditional information outlets are becoming increasingly anachronistic. Television, radio and printed media are less relevant, or are even completely irrelevant, as far as terrorist radicalization processes are concerned. They fail to reach a vital audience segment, one which does need to be influenced, namely youth and young adults. Such young people often no longer obtain their news from established television, cable or radio stations, but may do so exclusively from Islamists and Marxists, nationalists and separatists, racists and anarchists—all of whom find the Internet alluring.

Without knowing the enemy, counterterrorist organizations cannot successfully penetrate terrorist cells, and cannot sow discord and dissension in their ranks and thus weaken them from within. And the most basic requirement of an effective counterterrorist strategy, namely to pre-empt and prevent terrorist operations and deter their attacks, cannot be fulfilled.

Terrorist movements throughout history have presented themselves as monoliths, united and in agreement over their fundamental objectives, aims, strategies, tactics and targets. All too often, their opponents succumb to such fictions, and therefore fail to seize and exploit critical opportunities for weakening the terrorist organization. Such opportunities would center on the deepening of existing schisms, while also creating new sources of dissension. Emergent ideological fault-lines could be widened, and wedges could be driven into the heart of terrorist movements, based on their own internal disagreements.

This approach of undermining terrorist groups from within has arguably been missing in the current war on terrorism. Until counterterrorist forces recognize the importance of this vital prerequisite, they will remain perennially on the defensive. They will be inherently reactive rather than proactive, deprived of the capacity to recognize—much less anticipate— important changes in the enemy's modus operandi, recruitment and targeting. For while terrorist communications continue to change and evolve, so will the nature of terrorism itself.

A second, related aspect of a potential solution to terrorism concerns the efforts of the relevant governments to organize and engage in information operations, and to develop resources, especially with respect to the Internet. For example, given the increase in politically motivated crimes, the German government has stated that it will intensify education programs in schools to prevent young people from drifting into extremist movements. Also, many modern schools are using an educational multimedia program to combat extremism and terrorism [8].

These are appropriate responses to a very real danger. According to the annual report of Germany's Office for the Protection of the Constitution[1], there was a rise of 27% in politically motivated crimes by far-right groups between 2004 and 2005. The actual figures of such crimes were 12,051 in 2004 and 15,361 in 2005. Over the same period, the number of neo-Nazis in Germany rose by 300 to 4,100, and the number of right-wing extremists prepared to engage in violence rose by 400 to 10,400. The number of radical Islamic groups active in Germany had increased from 24 to 28.

Last year in Germany, during the federal parliamentary elections, 5% of young men between the ages of 18 to 25 in the Western states voted for the far-right wing National Party. In Eastern Germany, the figure was 10 %, which is probably due to the fact that these states have experienced specific problems since the fall of the Berlin Wall in 1989. The very high unemployment rate, for example, has meant that young people are susceptible to the fellowship offered by extremist groups. Thus the Eastern states of Germany have been attractive recruiting grounds for far-right parties.

How should those societies respond? This is not the place to attempt anything like a definitive answer, but two things seem clear. First, the relevant parties must be better informed about the uses to which terrorists put the Internet, and better able to monitor their activities. As noted at the start of this chapter, journalists, scholars, policymakers, and even security agencies have tended to focus on the exaggerated threat of cyber-terrorism, and have paid insufficient attention to the more routine uses made of the Internet. Those uses are numerous and, from the terrorists' perspective, invaluable. Hence, it is imperative that security agencies improve their ability to study and monitor terrorist activities on the Internet, and to explore measures to limit the use of this medium by modern terrorists.

Second, while democratic societies must thus better defend themselves against terrorism, they must not, in the process, erode the very qualities and values that make those societies worth defending. The Internet is in many ways an almost perfect embodiment of the democratic ideals of free speech and open communication; it is a marketplace of ideas unlike any that has existed before.

Unfortunately, as this chapter has shown, the freedom offered by the Internet is vulnerable to abuse from groups that, paradoxically, are themselves often hostile to uncensored thought and expression. But if, through fear of further terrorist attacks, democratic societies were to circumscribe their own freedom to use the Internet, then they would be handing the terrorists a victory, and dealing democracy a blow. One should not forget that fear inflicted by terrorism has, in the past, been manipulated by politicians to pass legislation that undermines individual rights and liberties.

The use of advanced techniques to monitor, search, track, and analyze communications carries inherent dangers. Although such technologies might prove very helpful in the fight against cyber-terrorism and Internet-savvy terrorists, they would also hand participating governments tools with which to violate civil liberties, both domestically and abroad. This danger is intensified in the case of authoritarian governments and agencies that have little public accountability. It does not take much imagination to recognize that the long-term implications of such interventions could be profound and damaging for democracies and their value systems. This would add a heavy price, in terms of diminished civil liberties, to the high toll exacted by terrorism itself.

---

[1] The OPC is the German federal domestic intelligence service that monitors all forms of extremism, espionage and counter-sabotage.

References

[1] Carl von Clausewitz, *On War: The complete translation* by Colonel J.J. Graham. Published by N. Trübner, London, 1873. Posted to the web by Clausewitz Homepage. Book VIII. Chapter VI.B. "War is an Instrument of Policy". www.clausewitz.com/CWZHOME/VomKriege2/Bk8ch06.html
[2] Robert Cooper, "The Breaking of Nations. Order and Chaos in the Twenty-First Century". *Atlantic Monthly Press*, New York, 2004. p.3. (Robert Cooper is former Director-General for External and Politico-Military Affairs at the General Secretariat of the Council of the European Union).
[3] Paul Wilkinson, "The Media and Terrorism". *Terrorism and Political Violence.* Jg.9, Nr.2, 1997, p. 51-64
[4] Gabriel Weimann, www.terror.net: "How Modern Terrorism Uses the Internet". // United States Institute of Peace. Special Report 116. March 2004. p. 12.
[5] Tina Brown, "Death by Error", *Washington Post*, 19 May 2005.
[6] Konrad N., Sun Tzu. *Traktat o voennom iskusstvie.* Moscow., Voenizdat, 1950.
[7] Weimann, G. (2007). "Using the Internet for Terrorist Recruitment and Mobilization"; see Chapter 5 in this publication.
[8] *International Herald Tribune*, May 23, 2006.

Mining the Jihadist Network

Hypermedia Seduction for Terrorist Recruiting
B. Ganor et al. (Eds.)
IOS Press, 2007

Computer-Assisted "Dark Web" Website and Forum Collection, Analysis and Visualization

Hsinchun CHEN, Ph.D.
McClelland Professor of Management Information Systems
Director, Artificial Intelligence Lab and Hoffman E-Commerce Lab
Management Information Systems Department
Eller College of Management, University of Arizona, USA

Abstract. In this chapter we discuss technical issues regarding intelligence and security informatics (ISI) research towards accomplishing the critical missions of international security and counterterrorism. We propose a research framework addressing the technical challenges facing counterterrorism and crime-fighting applications, with a primary focus on the knowledge discovery from databases (KDD) perspective. We also present several Dark Web-related case studies in collecting, analyzing and visualizing open-source terrorism information. Using a web spidering approach, we have developed a large-scale, longitudinal collection of extremist-generated Internet-based multimedia and multilingual contents. We have also developed selected computational link analysis, content analysis, and authorship analysis techniques to analyze the Dark Web collection.

Keywords. Intelligence and security informatics, terrorism informatics, Dark Web

1. Introduction

The tragic events of September 11 2001 and the subsequent anthrax contamination of letters had a drastic effect on many aspects of American society. Terrorism has become the most significant threat to national security, because of its potential to cause massive damage to infrastructures, economies, and people.

In response to this challenge, federal authorities are actively implementing comprehensive strategies and measures to achieve the three objectives identified in the "National Strategy for Homeland Security" report [1]. Those objectives are: (1) preventing future terrorist attacks; (2) reducing the nation's vulnerability; and (3) minimizing the damage and recovering from attacks that occur. State and local law enforcement agencies, likewise, have become more vigilant about criminal activities which might harm public safety or threaten national security.

Academics in the fields of natural sciences, computational science, information science, social sciences, engineering, medicine, and many others have also been called upon to help enhance the government's ability to fight terrorism and other crimes. Science and technology have been identified in the "National Strategy for Homeland Security" report as key in winning the new counterterrorism war [1]. In particular, it is believed that information technology will play an indispensable role in making our nation safer [2]. This role is concerned with supporting intelligence and knowledge discovery, through collecting, processing, analyzing, and utilizing terrorism- and crime-related data [3, 4]. Based on this criminal and intelligence knowledge, the federal, state, and local authorities can make timely decisions to select effective strategies and tactics. They can also allocate the appropriate amount of resources to detect, prevent, and respond to future attacks.

2. Problems and Challenges

Currently, intelligence and security agencies are gathering large amounts of data from various sources. Processing and analyzing such data, however, has become increasingly difficult. By treating terrorism as a form of organized crime, we can categorize these challenges into three types:

Characteristics of criminals and crimes. Some crimes may be geographically diffused and temporally dispersed. In organized crimes such as transnational narcotics trafficking, criminals often live in different countries, states, and cities. Drug distribution and sales occur in different places and at different times. Similar situations exist with regard to other organized crimes (e.g. terrorism, armed robbery, and gang-related crime). As a result, the investigation must cover multiple offenders who commit criminal activities in different places at different times. This can be fairly difficult, given the limited resources of intelligence and security agencies. Moreover, as computer and Internet technologies advance, criminals are utilizing cyberspace to commit various types of cyber-crimes, under the disguise of ordinary online transactions and communications.

Characteristics of crime and intelligence-related data. A significant challenge concerns information stovepipe and overload, which results from using diverse data sources, multiple data formats, and large data volumes. In many domains, such as marketing, finance, and medicine, data can be collected from particular sources, such as sales records from companies, or patient medical history from hospitals. But the intelligence and security domain does not have any such well-defined data source. Both authoritative information (e.g. crime incident reports, telephone records, financial statements, immigration and custom records) and open source information (e.g. news stories, journal articles, books, Web pages) need to be gathered for investigative purposes. Data collected from these different sources are often in diverse formats, ranging from structured database records to unstructured text, image, audio, and video files. Important information, such as criminal associations, may be available but contained in unstructured, multilingual text, which remains difficult to access and retrieve. Moreover, as data volumes continue to grow, extracting valuable, credible intelligence and knowledge becomes increasingly problematic.

Characteristics of crime and intelligence analysis techniques. Current research on the technologies for counterterrorism and crime-fighting applications lacks a consistent framework with which to address the major challenges. Some information technologies (including data integration, data analysis, text mining, image and video processing, and evidence combination) have been identified as being particularly helpful [2]. However, the question of how to employ these technologies in the intelligence and security domain remains unanswered. It is not clear how they might best be utilized to effectively address the critical mission areas within national security.

With regard to the critical missions of national security and various data and technical challenges, we believe there is a pressing need to develop the science of "Intelligence and Security Informatics" (ISI) [3, 4]. The main objective of this science is the "development of advanced information technologies, systems, algorithms, and databases for national security related applications, through an integrated technological, organizational, and policy-based approach" [3].

3. An ISI Research Framework: Techniques and Caveats

We believe that knowledge discovery and data mining (KDD) techniques can play a central role in improving the counterterrorism and crime-fighting capabilities of intelligence, security, and law enforcement agencies. These techniques reduce cognitive and information overload.

Knowledge discovery refers to non-trivial extraction of implicit, previously unknown, and potentially useful knowledge from data. Knowledge discovery techniques promise an easy, convenient, and practical way of exploring very large collections of data for organizations and users. They have been applied in marketing, finance, manufacturing, biology, and many other domains (e.g. predicting consumer behaviors, detecting credit card frauds, or clustering genes that have similar biological functions) [5]. Traditional knowledge discovery techniques include association rules mining, classification and prediction, cluster analysis, and outlier analysis [6]. As natural language processing (NLP) research advances, text mining approaches that automatically extract, summarize, categorize, and translate text documents have also been widely used [7, 8].

Many of these KDD technologies could be applied in ISI studies [3, 4]. Using the special characteristics of crimes, criminals, and crime-related data, we can categorize existing ISI technologies into six classes:

- *information sharing and collaboration*
- *crime association mining*
- *crime classification and clustering*
- *intelligence text mining*
- *spatial and temporal crime mining*
- *criminal network mining*.

These six classes are grounded on traditional knowledge discovery technologies, but with a few new approaches added, including spatial and temporal crime pattern mining, and criminal network analysis. These recent additions are particularly relevant to counterterrorism and crime investigation. Although information sharing and collaboration are not data mining *per se*, they help to prepare, normalize, warehouse, and integrate data for knowledge discovery, and thus are included in the framework.

We present in Figure 1 our proposed research framework, with the horizontal axis representing crime type and the vertical axis the six classes of techniques [3]. The shaded regions on the chart show promising research areas; that is, a certain class of techniques is relevant to solving a certain type of crime. Note that more serious crimes may require a more complete set of knowledge discovery techniques. For example, the investigation of organized crimes such as terrorism may depend on criminal network analysis technology, which requires the use of other knowledge discovery techniques, such as association mining and clustering.

An important observation about this framework is that the high-frequency occurrences and strong association patterns of severe and organized crimes, such as terrorism and narcotics, present a unique opportunity and potentially high rewards for adopting such a knowledge discovery framework.

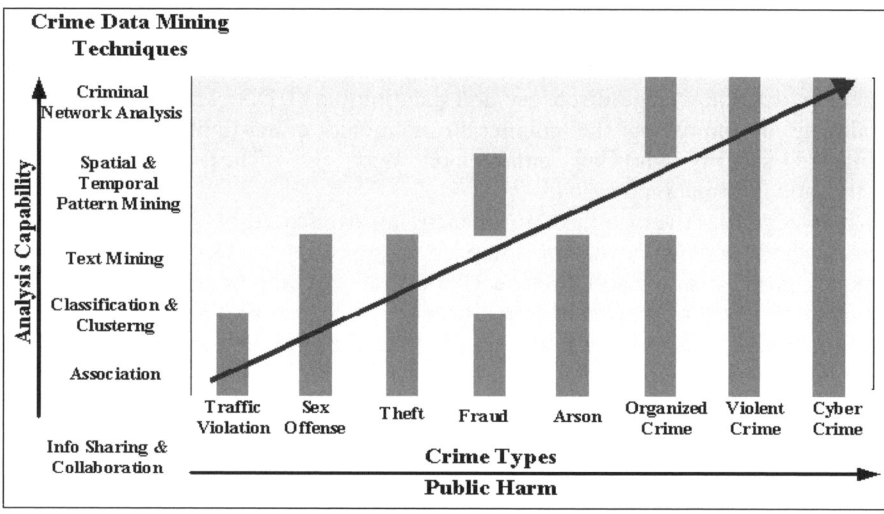

Figure 1. A knowledge discovery research framework for ISI

Several unique classes of data mining techniques are of great relevance to ISI research. *Text mining* is critical for extracting key entities (people, places, narcotics, weapons, time, etc.) and their relationships, as presented in voluminous police incident reports, intelligence reports, open source news clips, and so on. Some of these techniques need to be multilingual in nature, and may include possibilities for machine translation and cross-lingual information retrieval (CLIR).

Spatial and temporal mining and visualization is often needed for geographic information systems (GIS) and temporal analysis of criminal and terrorist events. Most crime analysts are well trained in GIS-based crime mapping tools. However, automated spatial and temporal pattern mining techniques (e.g. hotspot analysis) have not been adopted widely in intelligence and security applications. Organized criminals (e.g. gangs and narcotics dealers) and terrorists often form inter-connected covert networks for their illegal activities. Often referred to as "dark networks," these organizations exhibit unique structures, communication channels, and resilience to attack and disruption. New computational techniques, including social network analysis, network learning, and network topological analysis (e.g. random network, small-world network, and scale-free network) are needed for the systematic study of those complex, covert networks. We broadly consider these techniques under *criminal network analysis* in Figure 1.

Before we review in detail relevant ISI-related data mining techniques, applications, and literature (see section 4 of this chapter), we wish briefly to discuss the legal and ethical caveats regarding crime and intelligence research.

The potential negative effects of intelligence gathering and analysis on the privacy and civil liberties of the public have been well publicized [9]. There exist many laws, regulations, and agreements governing data collection, confidentiality, and reporting, which could directly impact the development and application of ISI technologies. We strongly recommend that intelligence and security agencies and ISI researchers be aware of these laws and regulations with regard to their research.

Moreover, we suggest that a hypothesis-guided, evidence-based approach be used in crime and intelligence analysis research. That is, there should be probable and reasonable parameters for causes and evidence when selecting particular individuals or data sets for analysis. Proper investigative and legal procedures need to be strictly followed. It is neither ethical nor legal to "fish" for potential criminals from diverse and mixed crime, intelligence, and citizen-related data sources.

The well-publicized Defense Advanced Research Program Agency (DARPA) Total Information Awareness (TIA) program, and the Multi-State Anti-Terrorism Information Exchange (MATRIX) system, for example, have recently been shut down by the US Congress due to their potential misuse of citizen data. Such misuse represented an impairment of civil liberties [10].

In an important recent review article by Strickland, Baldwin, and Justsen [11], the authors provide an excellent historical account of government surveillance in the United States. The article reviews new surveillance initiatives in the age of terrorism (including the passage of the USA Patriot Act) and discusses—in great depth—the impact of technology on surveillance and citizen rights. Finally, it proposes a balance between the necessary secrecy and oversight. We believe this to be one of the most comprehensive articles to have been written on the topic of civil liberty issues, within the context of national security research.

We now summarize some of the key points made in the above-mentioned article, as applied within the context of our proposed ISI research. Readers are strongly encouraged to refer to the original source [11] for more details.

Framed in the context of domestic security surveillance, Strickland et al.'s paper considers surveillance an important intelligence tool, which has the potential to contribute significantly to national security—but also to infringe on civil liberties. As faculty members of the University of Maryland Information Science department, the authors believe that information science and technology has drastically expanded the mechanisms by which data can be collected, and knowledge extracted and disseminated through automated means

An immediate result of the tragic events of September 11 2001 was the extraordinarily rapid passage of the USA Patriot Act in late 2001. The legislation was passed by the Senate on October 11 2001, and by the House on October 24 2001. It was signed by the President on October 26 2001. But the legacy of the consensus at that time, and the concomitant lack of detailed debate and consideration, has been a bitter ongoing national argument as to the proper balance between national security and civil liberties. The Act contains ten titles within 131 pages. It amends numerous laws, including, for example, the following:

- expansion of electronic surveillance of communications in law enforcement cases
- authorized sharing of law enforcement data with intelligence
- expansion of the acquisition of electronic communications as well as commercial records for intelligence use, and
- creation of new terrorism-related crimes.

However, as new data mining and/or knowledge discovery techniques have matured and become potentially useful for national security applications, there have been great concerns about violating civil liberties. Both the DARPA's TIA Program and the Transportation Security Administration's (TSA) Computer Assisted Passenger Prescreening Systems (CAPPS II) were cited as failed systems that faced significant media scrutiny and public opposition. Both systems were based on extensive data mining of commercial and government databases collected for one purpose but shared and used for another purpose. Both systems were sidetracked by a widely perceived threat to personal privacy.

Based on many of the debates generated, Strickland et al. suggest that data mining using public or private sector databases for national security purposes must proceed in two stages. First, the search for general information must ensure anonymity; second, the acquisition of specific identity, if required, must be court authorized under appropriate standards (e.g. in terms of "special needs" or "probable causes").

In their concluding remarks, the authors caution that secrecy within any organization could pose a real risk or constitute abuse, and must be constrained through effective checks and balances. They also state that professionals within the fields of information science and technology are ideally situated to provide the tools and techniques by which the necessary intelligence is collected, analyzed, and disseminated. Finally, they comment that civil liberties are protected through established laws and policies.

4. University of Arizona Artificial Intelligence Lab Research

In response to the challenges of national security, the Artificial Intelligence Lab at the University of Arizona has developed many research projects over the past decade. These projects have been conducted together with the affiliated National Science Foundation (NSF) COPLINK Center for law enforcement and intelligence research.

Research has addressed the six critical mission areas identified in the "National Strategy for Homeland Security" report [1], namely:

- intelligence and warning
- border and transportation security
- domestic counter-terrorism
- protection of critical infrastructure and key assets

- defense against catastrophic terrorism, and
- emergency preparedness and responses.

The main goal of the Arizona lab is to develop information and knowledge management technologies appropriate for capturing, accessing, analyzing, visualizing, and sharing law enforcement and intelligence related information [12].

In this chapter we present four case studies of relevance to intelligence and warning based on our Dark Web research. In Case Studies 1 and 2, we report on the Dark Web Portal project, which collects open source terrorist website information via select spidering and portal techniques. A (limited access) web portal has been developed to support retrieval and analysis of these extremist-generated contents. Case Study 3 reports how US domestic extremist groups used the Web to disseminate their ideology, recruit members, and support communications. Case Study 4 reports on a novel Arabic language model for authorship identification of Dark Web online forums.

4.1 Case Study 1: The Dark Web Portal

The Internet has become a global platform for anyone to disseminate and communicate information. Terrorists take advantage of the freedom of cyberspace, and construct their own websites to propagate terrorist beliefs, share information, and recruit new members. Websites of terrorist organizations may also connect to one another through hyperlinks, forming a "Dark Web".

At the Arizona lab we are building an intelligent web portal, called Dark Web Portal, to help terrorism researchers collect, access, analyze, and understand terrorist groups [13, 14]. This project consists of three major components: Dark Web testbed building, Dark Web link analysis, and Dark Web Portal building.

4.1.1 Dark Web Testbed Building

Utilizing reliable governmental sources such as the Anti-Defamation League (ADL), FBI, and United States Committee for a Free Lebanon (USCFL), we identified 224 USA domestic terrorist groups and 440 international terrorist groups. For USA domestic groups, group-generated URLs can be found in FBI reports and Google Directory. For international groups, we used the group names as queries to search major search engines such as Google, and manually identified the group-created URLs from the result lists.

To ensure that our testbed covered the major regions of the world, we sought the assistance of language and domain experts in English, Arabic, and Spanish to help us collect URLs in several major regions. All URLs collected were manually checked by experts to make sure that they had been created by terrorist groups. After the URL of a group was identified, we used the SpidersRUs toolkit, a multilingual digital library building tool developed by our lab, to collect all the web pages under that URL and store them in our testbed. Table 1 shows a summary of web pages collected from three rounds of spidering (performed bi-monthly).

Region		U.S.A. Domestic			Latin-America			Middle-East		
Batch #		1st	2nd	3rd	1st	2nd	3rd	1st	2nd	3rd
# of seed URLs	Total	81	233	108	37	83	68	69	128	135
	From literature & reports	63	113	58	0	0	0	23	31	37
	From search engines	0	0	0	37	48	41	46	66	66
	From link extraction	18	120	50	0	32	27	0	31	32
# of terrorist groups searched		74	219	71	7	10	10	34	36	36
# of Web pages	Total	125,610	396,105	746,297	106,459	332,134	394,315	322,524	222,687	1,004,785
	Multi-media files	0	70,832	223,319	0	44,671	83,907	0	35,164	83,907

Table 1. Summary of URLs identified and web pages collected in the Dark Web collection

4.1.2 Dark Web Link Analysis and Visualization

Terrorist groups are not atomized individuals but actors linked to each other through complex networks of direct or mediated exchanges. Identifying how relationships between groups are formed and dissolved within the terrorist group network would enable us to decipher the social milieu and communication channels among terrorist groups across different jurisdictions.

Previous studies have shown that the link structure of the Web represents a considerable amount of latent human annotation [15]. Thus, by analyzing and visualizing hyperlink structures between terrorist-generated websites and their content, we could discover the structure and organization of terrorist group networks. We could also capture those networks' dynamics, and understand their emerging activities.

4.1.3 Dark Web Portal Building

Using the Dark Web Portal, users are able quickly to locate specific Dark Web information in the testbed, through keyword searches. To address the information overload problem, the Dark Web Portal is designed with post-retrieval components. A modified version of a text summarizer, called TXTRACTOR, which uses sentence-selection heuristics to rank and select important text segments [16], is added into the Dark Web Portal. The summarizer can flexibly summarize web pages using three or five sentences, such that users can quickly get the main idea of a web page without having to read though it.

A categorizer organizes the search results into various folders, which are labeled according to the key phrases extracted by the Arizona Noun Phraser (AZNP) [17] from the page summaries or titles. This facilitates an understanding of different groups of web pages. A visualizer clusters web pages into colored regions using the Kohonen self-organizing map (SOM) algorithm [18], thus reducing the information overload problem when a large number of search results are obtained. Post-retrieval analysis could also help to reduce the information overload problem.

However, unless the language barrier problem is overcome, researchers are limited to data in their native languages, and cannot fully utilize the multilingual information in our testbed. To address this problem, we added a cross-lingual information retrieval (CLIR) component into the portal. Based on our previous research, we have developed a dictionary-based CLIR system for use in the Dark Web Portal. It currently accepts English queries, and retrieves documents in English, Spanish, Chinese, and Arabic. Another component that will be added to the Dark Web Portal is a machine translation (MT) component, which will translate the multilingual information retrieved by the CLIR component back into the users' native languages.

Figure 2 shows a sample search session. Suppose the user is interested in the terrorist group "Ku Klux Klan" and uses it as a search term. Two types of search forms are available: simple search and advanced search. Our user chose to use the simple search first (the advanced mode gives users more options to refine their search). For example, he can specify that he wants web pages containing the exact search phrase. In addition, he can restrict the results to a few terrorist categories, or he can choose to search a particular file type, such as PDF or Word files.

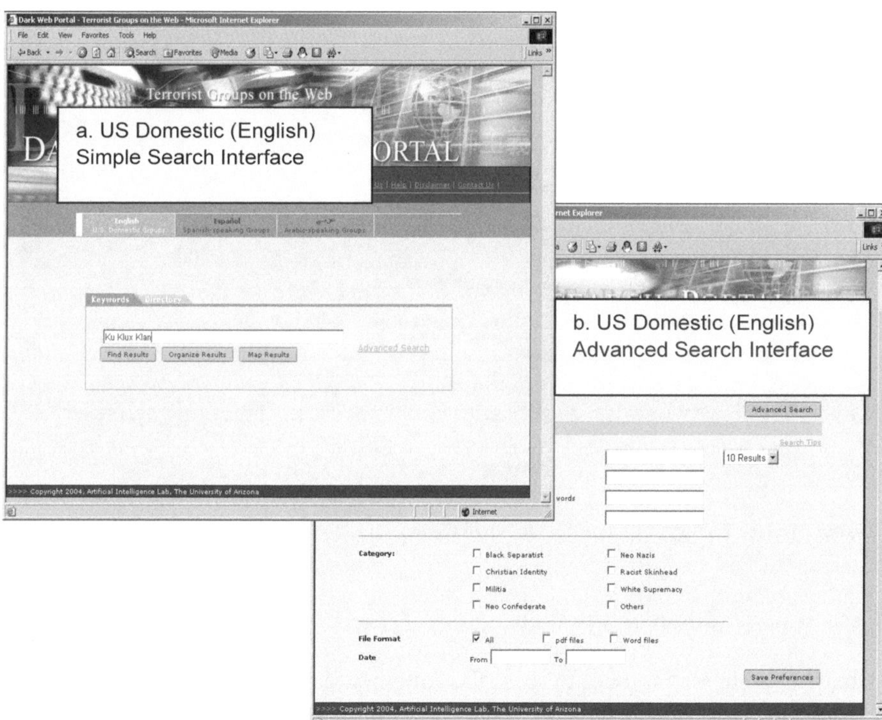

Figure 2. Dark Web Portal interfaces: simple search and advanced search

By hitting the "Find Results" button, the top twenty results are displayed; see Figure 3. On the top of the result page a list of "suggested keywords" appears; in this case with terms such as "Aryan Nations" and "David Duke". These help the user to expand or refine his query.

Along with the web page result display, our portal also presents the terrorist group name and the corresponding group category. As terrorist group web pages may often disappear, "Cached Pages" for each web page collected at different time periods are provided (e.g. 2004/03). Additionally, the user can view web pages, PDF files, or Word files by clicking the corresponding links.

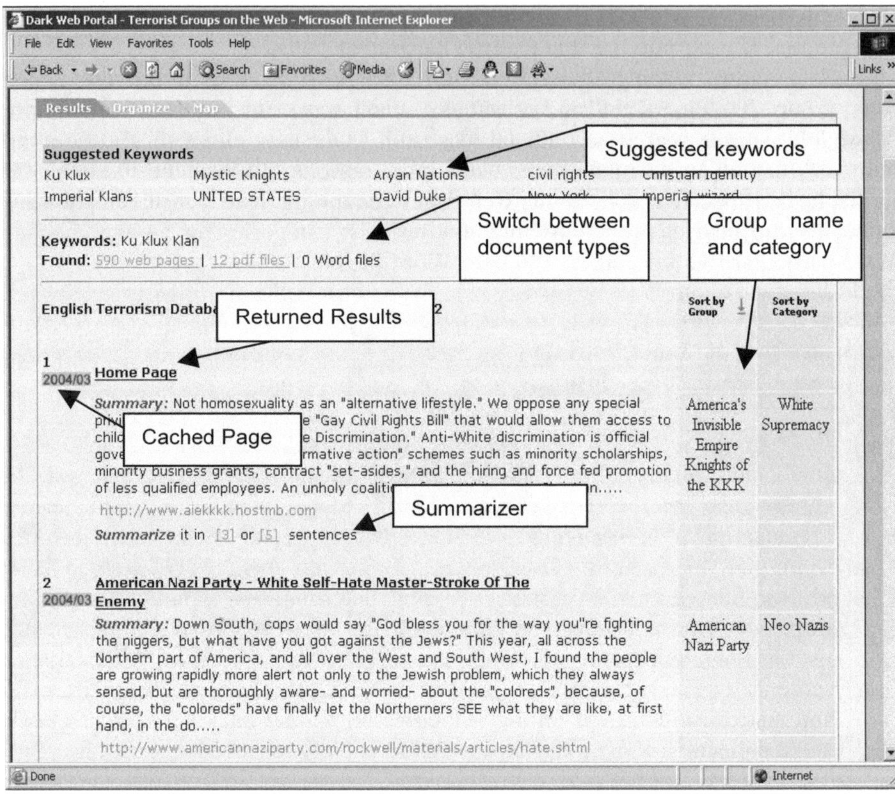

Figure 3. Dark Web Portal interfaces: returned results

4.2 Case Study 2: Jihad on the Web

As terrorist groups continue to use the Internet as their main communications, recruiting, and propaganda tool, a systematic and system-aided approach to studying their presence on the Web is critically needed. Weekly, news coverage includes excerpts from videos that have been produced and webcasted by terrorists. It has become clear that terrorists have exploited the Internet beyond routine communication and propaganda operations, so as to better influence the outside world [19].

Some terrorism researchers posit that terrorists have used the Internet as a broadcast platform for the "terrorist news network", which is an effective tactic because they can reach a broad audience with relatively little chance of detection [20, 21, 22]. Although this alternate side of the Internet (referred to as the Dark Web) has recently received extensive government and media attention, systematic understanding of how terrorists use the Internet for their campaign of terror is very limited. In this case study, we explore an integrated computer-based approach to harvesting and analyzing websites produced or maintained by Islamic jihad extremist groups or their sympathizers. The purpose of this study is to deepen our understanding of how jihad terrorists use the Internet, especially the World Wide Web, in their terror campaigns.

More specifically, we built a high-quality jihad terrorism web collection using a web harvesting approach; we conducted hyperlink analysis on this collection to reveal various facets of jihad terrorism web usage. We hope, in this manner, to supplement the existing high-quality but manually-driven terrorism research with a systematic, automated web spidering and mining methodology.

4.2.1 Building the Jihad Web Collection

To guarantee that our collection is comprehensive and representative, we take a three-step systematic approach in constructing it:

1. *Identifying seed URLs and backlink expansion:* The first task is to find a small set of high-quality jihad websites. To identify terrorist groups, we relied completely on the US Department of State's list of foreign terrorist organizations. In particular, we selected only Middle-Eastern organizations from that list for this study. After identifying the terrorist groups in the Middle-East region, we manually searched major search engines to find websites of these groups. Our goal here was not to construct a comprehensive list of URLs, but merely to compile a small list of high-quality URLs that can serve as the seeds for backlink expansion. The backlinks of these URLs were automatically identified through Google and Yahoo backline search services, and a collection of 88 websites was automatically retrieved.

2. *Manual collection filtering:* Because bogus or unrelated terrorist sites can make their way into our collection, we developed a manual filtering process based on evidence and clues in the websites. In addition to sites which explicitly identify themselves as the official site of a terrorist organization or one of its members, a website that contains praise of—or adopts ideologies espoused by—a terrorist group is included in our collection.

3. *Extending search:* To ensure the comprehensiveness of our collection, we augment it by means of expanded searches. Based on the 26 websites identified in the previous step, we constructed a small lexicon of commonly-used jihad terms with the help of Arabic language speakers. Examples of highly relevant keywords included in the lexicon are: "حرب صليبية" ("Crusader's War"), "المجاهدين" ("Moujahedin"), "الكفار" ("Infidels"), and so on. This lexicon is then utilized to perform expanded searches. The same rules used in the filtering process are used here to discern fake and unrelated

websites. Our final jihad web collection contains 109,477 jihad web documents; these include HTML pages, plain text files, PDF documents, and Microsoft Word documents.

4.2.2 Hyperlink Analysis on the Jihad Web Collection

We believe the exploration of hidden jihad web communities can give insight into the nature of real-world relationships and communication channels between terrorist groups [22]. Uncovering hidden web communities involves calculating a similarity measure between all pairs of websites in our collection.

We define similarity as a function of the number of hyperlinks in Website A that point to Website B, and vice versa. In addition, a hyperlink is weighted in proportion to how deep it appears in the website hierarchy. The similarity matrix is then used as input to a Multi-Dimensional Scaling (MDS) algorithm [23], which generates a two dimensional graph of the websites. The proximity of nodes in the graph reflects their similarity level.

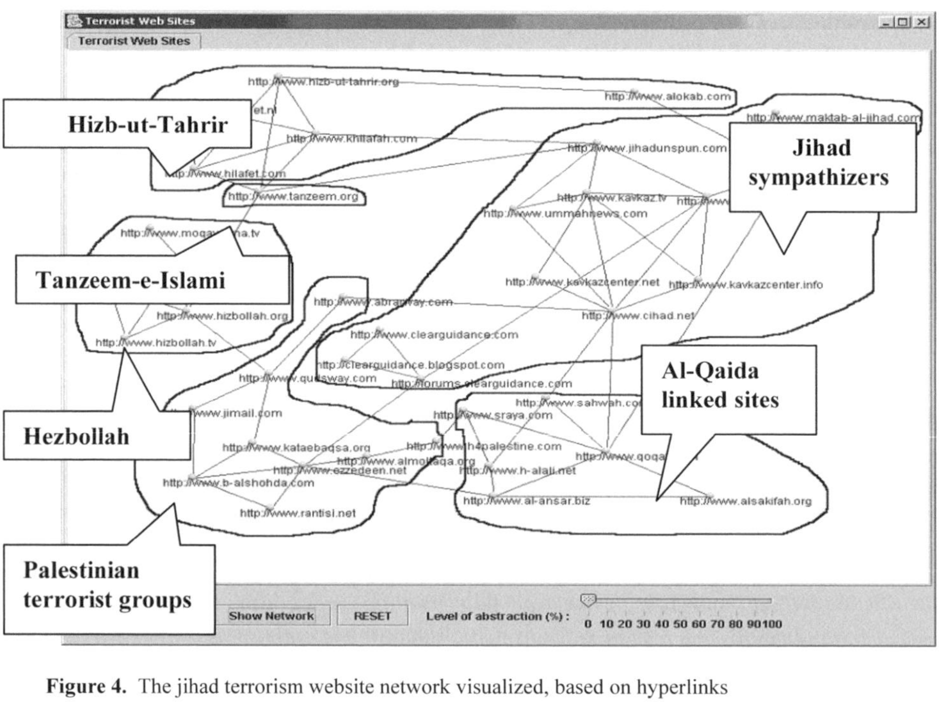

Figure 4. The jihad terrorism website network visualized, based on hyperlinks

As shown in Figure 4, domain experts recognized six clusters representing hyperlinked communities in the network. On the left side of the network resides the Hezbollah cluster; Hezbollah is a Lebanese militant organization. Established in 1982 during the Israeli invasion of Lebanon, the group routinely attacked Israeli military personnel until their pullout from south Lebanon in 2000.

A cluster of websites of Palestinian organizations occupies the bottom-left corner of the network, including Hamas, Al-Aqsa Martyr's Brigades, and the Palestinian

Islamic Jihad. An interesting observation here is the close proximity of the Hezbollah community and the Palestinian militant groups' community. Hezbollah has traditionally sympathized with the Palestinian cause.

On the top-left corner sits the Hizb-ut-Tahrir cluster. Hizb-ut-Tahrir is a political party with branches in many countries across the Middle-East and in Europe. Although the group is believed to be associated with Al-Qaida, this apparent relation between the two groups has not been verified.

Looking at the bottom-right corner, one can see a cluster of Al-Qaida affiliated sites. This cluster has links to two radical Palestinian websites. Al-Qaida sympathizes with Palestinian groups, and some Palestinian Islamist groups, such as Hamas and Islamic Jihad, share the same Salafi ideology with Al-Qaida.

In the top-right corner, the Jihad Sympathizers web community includes websites maintained by sympathizers of the Global Salafi movement. For example, "kavkazcenter.net" and "clearguidance.com" are two websites maintained by sympathizers of the Chechen rebels. As expected, the sympathizers' community does not have any links to Hezbollah's community, since the two groups follow radically different ideologies.

Visualizing hyperlinked communities in this manner can lead us to a better understanding of the jihad web presence. Furthermore, it helps to foretell likely relationships between terrorist groups.

4.3 Case Study 3: US Domestic Extremist Groups on the Web

Although not as well-known as some of the international terrorist organizations, the extremist and hate groups within the United States also pose a significant threat to our national security. Recently, these groups have been intensively utilizing the Internet to advance their causes. Thus, understanding how these domestic extremist and hate groups develop their Web presence is very important in addressing any domestic terrorism threats.

This case study proposes the development of systematic methodologies to capture domestic extremist and hate groups' website data, and to support subsequent analyses. We aim to answer the following research questions: What are the most appropriate techniques for collecting high-quality web pages of domestic extremist and hate groups? What systematic procedures can be used to analyze and visualize the content of these individual websites?

We propose a sequence of semi-automated methods to study these domestic sites. First, we employ a semi-automatic procedure to harvest and construct a high-quality domestic terrorist website collection. We then perform hyperlink analysis, based on a clustering algorithm, to reveal the relationships between these groups. Lastly, we conduct an attribute-based content analysis to determine how these groups use the Web for their purposes. Because the procedure adopted in this study is similar to that reported in Case Study 3, "Jihad on the Web", we summarize only selected, interesting results here.

4.3.1 Collection Building

We manually extracted a set of URLs from the relevant literature. In particular, the websites of the "Southern Poverty Law Center" (SPLC, www.splcenter.org) and the Anti-Defamation League (ADL, www.adl.org) are authoritative sources of information on domestic extremists and hate groups. A total of 266 seed URLs were identified in the SPLC and ADL websites, together with the Google directory.

A backlink expansion of this initial set was performed, and the count increased to 386 URLs. The resulting set of URLs was validated through an expert filtering process, and a total of 97 URLs were deemed relevant. We then spidered and downloaded all the web documents within the identified websites. The result was a final collection that contains about 400,000 documents.

4.3.2 Hyperlink Analysis

Using the MDS algorithm [23], we visualize the hidden hyperlinked communities among 44 websites randomly retrieved from our collection. Three communities are identified in the network shown in Figure 5.

The left side of the network shows the "Southern Separatists" cluster, which mainly consists of the websites of new confederate organizations in the Southern states. These groups espouse a separatist ideology which promotes the idea of establishing an independent state in the south. In addition, they share elements of white-supremacy ideas with other non-neo-confederate racist organizations, such as the KKK.

A cluster of the websites of white supremacists occupies the top-right corner of the network, including Stormfront and White Aryan Resistance (www.resist.com).

Neo-Nazi groups occupy the bottom portion of Figure 5.

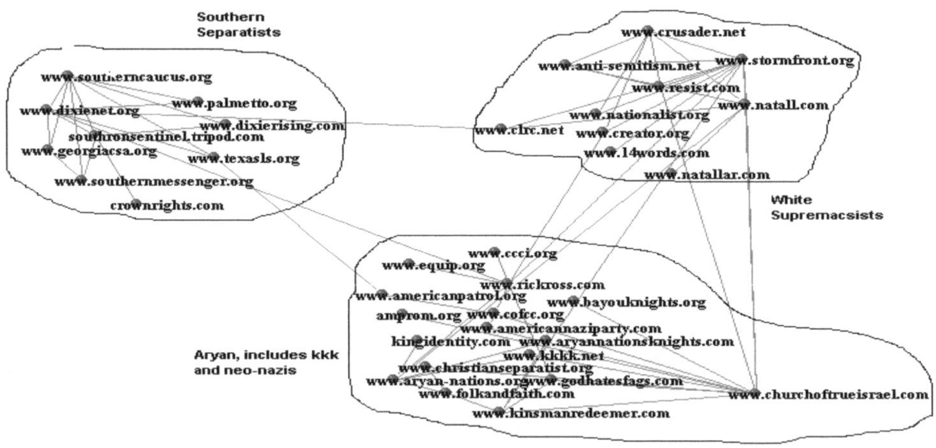

Figure 5. Web community visualization of selected domestic extremist and hate groups

4.3.3 Content Analysis

We asked our domain experts to review each website in our collection, and to record the presence of low-level attributes based on an eight-attribute coding scheme. The eight attributes were as follows: Sharing Ideology, Propaganda (Insiders), Propaganda (Outsiders), Recruitment and Training, Command and Control, Virtual Community, Fundraising, and Communications.

For instance, the webpage of "Nation of Islam" contains recordings of the organization's leaders, intended for their followers. The presence of these recordings contributes to the website's content richness, and is coded under the "Propaganda (Insiders)" attribute. Our web coding scheme is similar in nature to the one developed by Demchak et al. [24] for coding government website characteristics.

The manual coding of the attributes in a single website takes about 45 minutes. After coding the websites in our collection, we compared the content of six categories of domestic extremist and hate groups, as shown in Figure 6. "Sharing Ideology" is the attribute with the highest frequency of occurrence across all categories. This attribute encapsulates all communication media devoted to portraying the goals of the extremist group, defining its general policies, and presenting the foundational ideology.

In addition, the attributes of "Propaganda (Insiders)" and "Recruitment and Training" are widely used by all groups on their websites.

Another interesting observation is the low presence of "Propaganda (Outsiders)", with the exception of Eco-terrorism/Animal Rights groups, which are considered to have a much wider audience than the racist groups, who have a more targeted audience. Much research is still needed for the systematic understanding of how domestic extremist and hate groups use the Web to promote their causes.

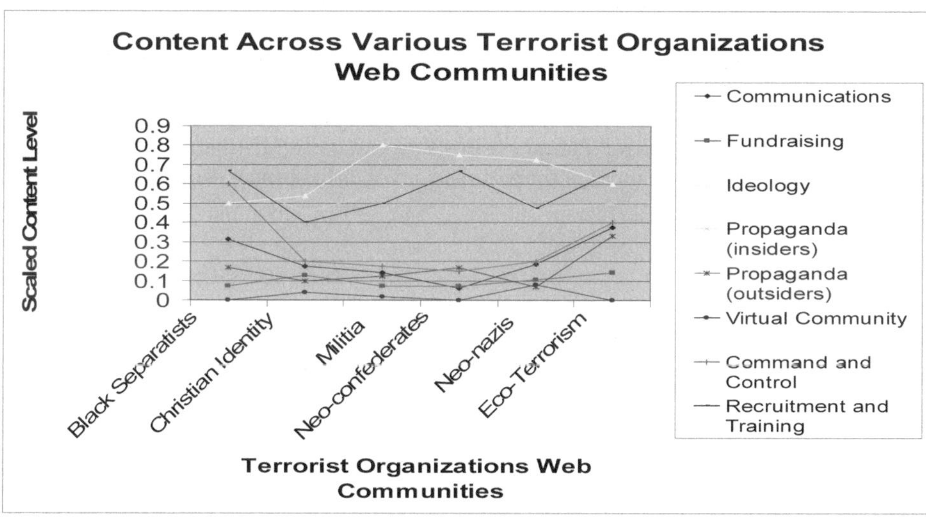

Figure 6. Content coding for various domestic extremist organizations web communities

4.4 Case Study 4: Developing an Arabic Authorship Model for Dark Web Forums

The evolution of the Internet as a major international communication medium has spawned the advent of a multilingual dimension. Applying authorship identification techniques across multilingual web content has become important, due to increased globalization and the ensuing security issues that are created.

Arabic is one of the six official languages of the United Nations, and the mother tongue of over 300 million people. The language is attracting interest due to its socio-political importance and its distinctiveness from Indo-European languages. In terms of our work, the morphological challenges pertaining to Arabic pose several critical problems for authorship identification techniques. These problems could be partially responsible for the lack of previous authorship analysis studies relating to Arabic.

In this case study, we apply an existing framework for authorship identification to Arabic web forum messages. Techniques and features are incorporated to address the specific characteristics of Arabic, resulting in the creation of an Arabic language model. We also present a comparison between English and Arabic language models.

Most previous authorship studies have focused only on English, with a few studies done on Greek and Chinese. Stamamatos et al. applied authorship identification to a corpus of Greek newspaper articles [25]. Peng et al. conducted experiments on English documents, Chinese novels, and Greek newspapers using an n-gram model [26]. Zheng et al. performed authorship identification on English and Chinese web forum messages [27]. In all of these previous studies, the results for English have been better than results for the other languages.

Applying authorship identification features across different languages is not without its difficulties. Since most writing style characteristics were designed for English, they may not always be applicable or relevant for other languages. Structural and other linguistic differences can create feature extraction nightmares.

Arabic is a Semitic language, meaning that it belongs to the group of Afro-Asian languages, which includes Hebrew. It is written from right to left, with letters in the same word being joined together in a way similar to that of English cursive writing. Semitic languages have several characteristics that can cause difficulties for authorship analysis. These challenges include properties such as inflection, diacritics, word length, and elongation.

4.4.1 Inflection

Inflection is the derivation of stem words from a root. Although the root has a meaning, it is not a word in itself, but rather a class that contains stem instances (words). Stems are created by adding affixes (prefixes, infixes, and suffixes) to the root, using specific patterns. Words with common roots are semantically related. Arabic roots are three to five letter consonant combinations, with the majority being three letters.

Al-Fedaghi and Al-Anzi believe that as many as 85% of Arabic words are derived from a tri-lateral root, suggesting that Arabic is highly inflectional [28]. Inflection can cause feature extraction problems for lexical features, because high levels of inflection increase the number of possible words, since a word can take on numerous forms.

4.4.2 Diacritics

Diacritics are markings above or below letters, used to indicate special phonetic values. An example of diacritics in English (or French) would be the little markings found on top of the letter "e" in the word résumé. These markings alter the pronunciation and meaning of the word. Arabic uses diacritics in every word to represent short vowels, consonant lengths, and relationships between words.

4.4.3 Word Length:

Arabic words tend to be shorter than English words. This shortness of Arabic words reduces the effectiveness of many lexical features. The short-word count feature, used to track words of three letters or smaller, may have little discriminatory potential when applied to Arabic. Additionally, the word-length distribution feature may also be less effective, since Arabic word length distributions have a smaller range.

4.4.4. Elongation:

Arabic words are sometimes stretched out or elongated. This is done for purely stylistic reasons, using a special Arabic character that resembles a dash ("-"). Elongation is possible because Arabic characters are joined during writing. Table 2 shows an example of elongation. The word MZKR ("remind") is elongated with the addition of four dashes between the "M" and the "Z." Although elongation provides an important authorship identification feature, it can also create problems.

Elongated	English	Arabic	Word Length
No	MZKR	مذكر	4
Yes	M----ZKR	مـــــذكر	8

Table 2. An example of Arabic elongation

Our testbed consisted of English and Arabic datasets. The English dataset was adapted from Zheng et al.'s study, and consists of messages from USENET newsgroups [27]. The dataset identifies twenty authors engaged in potentially illegal activities relating to computer software and music sale and trading. The data consists of twenty messages per author, a total of 400 messages.

The Arabic dataset was extracted from Yahoo groups, and is also composed of twenty authors and twenty messages per author. These authors discuss a broader range of topics, including political ideologies and social issues in the Arab world.

Based on previous studies, there are numerous classification techniques that can provide adequate performance. In this research, we adopted two popular machine learning classifiers, namely ID3 decision trees (called C4.5) and Support Vector Machine (SVM). The Arabic feature set was modeled after the English feature set. It includes 410 features, with the key differences highlighted in Table 3.

Feature Type	Feature	English	Arabic
Lexical, F1	Short Word Count	Track all words 3 letters or less	Track all words 2 letters or less
	Word Length Distribution	1-20 letter words	1-15 letter words
	Elongation	N/A	Track number of elongated words
Syntactic, F2	Function Words	150 words	250 words
	Word Roots	N/A	30 roots
Structural, F3	No Differences	-	-
Content Specific, F4	Number of words	11	25

Table 3. Differences between English and Arabic feature sets

The results for the comparison of the different feature types and techniques are summarized in Table 4 and Figure 7. For both datasets, accuracy was increased proportionally with the addition of more feature types. The maximum accuracy was achieved with the use of SVM and all feature types for both English and Arabic. Using all features with the SVM classifier, we were able to achieve an accuracy level of 85.43% for the Arabic data set; this was a level lower than the 96.09% achieved for the English data set.

Accuracy (%)	English Dataset		Arabic Dataset	
Features	**C4.5**	**SVM**	**C4.5**	**SVM**
F1	86.98%	92.84%	68.07%	74.20%
F1+F2	88.16%	94%	73.77%	77.53%
F1+F2+F3	88.29%	94.11%	76.23%	84.87%
F1+F2+F3+F4	89.31%	**96.09%**	81.03%	**85.43%**

Table 4. Accuracy for different feature sets across techniques

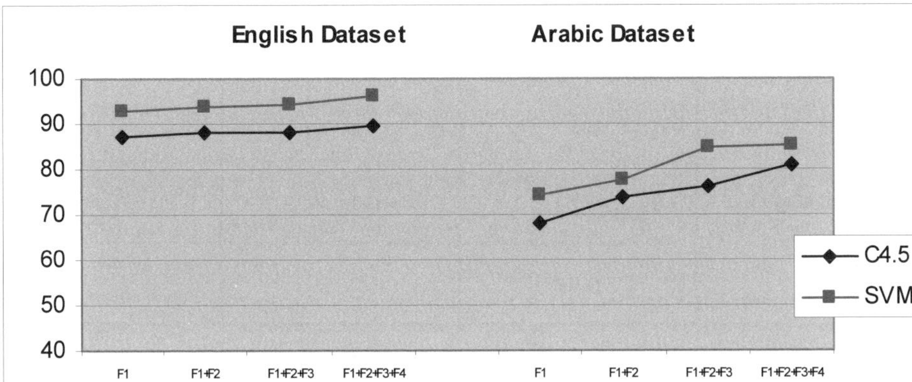

Figure 7. Authorship identification accuracies for different feature types

A comparison of C4.5 and SVM revealed that SVM significantly outperformed the decision tree classifier, in all cases. This result is consistent with the results of previous studies, which have also showed SVM to be superior. The difference between the two classifiers was consistent across English and Arabic, with English accuracies being about 10% higher.

In the future, we would like to analyze authorship differences at the group-level within a specific language. Identification of unique writing style characteristics for speakers of the same languages, across different geographic locations (e.g. Iraq vs. Palestine), cultures (e.g. Sunni vs. Shiite), and interest groupings (e.g. terrorist) could prove to be an interesting endeavor.

5. Conclusions and Future Directions

In this chapter we have discussed technical issues regarding intelligence and security informatics (ISI) research, as it relates to the critical missions of international security. We have proposed a research framework to address the technical challenges facing counterterrorism and crime-fighting applications, with a primary focus on the knowledge discovery from databases (KDD) perspective.

We have also presented several Dark Web-related case studies in open-source terrorism information collection, analysis, and visualization. As this new ISI discipline continues to evolve and advance, several important directions need to be pursued. These include technology development, testbed creation, and social, organizational, and policy studies. We hope that active ISI research will help to improve knowledge discovery and dissemination, and enhance information sharing and collaboration among academics; local, state, and federal agencies; and industry. This would result in positive impacts on all aspects of our society.

References

[1] Office of Homeland Security. (2002). National Strategy for Homeland Security. Washington D.C.: Office of Homeland Security.

[2] National Research Council. (2002). *Making the Nation Safer: The Role of Science and Technology in Countering Terrorism*. Washington, DC: National Academy Press.
[3] Chen, H., Miranda, R., Zeng, D. D., Demchak, C., Schroeder, J., and Madhusudan, T. (Eds.). (2003). *Intelligence and Security Informatics: Proceedings of the First NSF/NIJ Symposium on Intelligence and Security Informatics*. Berlin: Springer.
[4] Chen, H., Moore, R., Zeng, D., and Leavitt, J. (Eds.). (2004). *Intelligence and Security Informatics: Proceedings of the Second Symposium on Intelligence and Security Informatics*. Berlin: Springer.
[5] Fayyad, U. M., and Uthurusamy, R. (2002). "Evolving Data Mining into Solutions for Insights." *Communications of the ACM, 45*(8), 28-31.
[6] Han, J., and Kamber, M. (2001). *Data Mining: Concepts and Techniques*. San Francisco, CA: Morgan Kaufmann.
[7] Chen, H. (2001). *Knowledge Management Systems: A Text Mining Perspective*. Tucson, Arizona: Knowledge Computing Corporation.
[8] Trybula, W. J. (1999). "Text Mining." In M. E. Williams (Ed.), *Annual Review of Information Science and Technology (ARIST)* (vol. 34, pp. 385-419). Medford, NJ: Information Today, Inc.
[9] Cook, J. S., and Cook, L. L. (2003). "Social, ethical and legal issues of data mining." In J. Wang (Ed.), *Data mining: Opportunities and challenges* (pp. 395-420). Hershey, PA: Idea Group Publishing.
[10] American Civil Liberties Union. (2004). *MATRIX: Myths and Reality*. Retrieved July 27, 2004, from the World Wide Web: http://www.aclu.org/Privacy/Privacy.cfm?ID=14894&c=130
[11] Strickland, L. S., Baldwin, D. A., and Justsen, M. (2005) "Domestic Security Surveillance and Civil Liberties", In B. Cronin (Ed.), *Annual Review of Information Science and Technology (ARIST),* Volume 39. Information Today, Inc., Medford, New Jersey.
[12] Chen, H., Zeng, D., Atabakhsh, H., Wyzga, W., and Schroeder, J. (2003). "COPLINK: Managing Law Enforcement Data and Knowledge". *Communications of the ACM, 46*(1), 28-34.
[13] Chen, H., Qin, J., Reid, E., Chung, W., Zhou, Y., Xi, W., et al. (2004). "The Dark Web Portal: Collecting and Analyzing the Presence of Domestic and International Terrorist Groups on the Web." In *Proceedings of the 7th Annual IEEE Conference on Intelligent Transportation Systems (ITSC 2004).*
[14] Reid, E. O. F., Qin, J., Chung, W., Xu, J., Zhou, Y., Schumaker, R., et al. (2004). "Terrorism Knowledge Discovery Project: A Knowledge Discovery Approach to Address the Threats of Terrorism". In H. Chen, R. Moore, D. Zeng and J. Leavitt (Eds.), *Proceedings of the Second Symposium on Intelligence and Security Informatics (ISI'04)* (pp. 125-145). Berlin: Springer.
[15] Gibson, D., Kleinberg, J., and Raghavan, P. (1998). "Inferring Web Communities from Link Topology". In R. Akscyn, D. McCracken, and E. Yoder (Eds.), *Proceedings of the 9th ACM Conference on Hypertext and Hypermedia* (pp. 225-234). New York: Association for Computing Machinery.
[16] McDonald, D., and Chen, H. (2002). "Using Sentence-selection Heuristics to Rank Text Segments in TXTRACTOR". *Proceedings of the 2nd ACM/IEEE-CS Joint Conference on Digital Libraries* (pp. 28-35).
[17] Tolle, K. M., and Chen, H. (2000). "Comparing noun phrasing techniques for use with medical digital library tools". Journal of the American Society for Information Science, 51(4), 352-370.
[18] Kohonen, T. (1995). *Self-Organizing Maps.* Berlin: Springer-Verlag.
[19] Arquilla, J. & Ronfeldt, D. F. "Advent of Netwar". (1996). *Rand Report,* http://www.rand.org/.
[20] Elison, W. (2000) Netwar: "Studying Rebels on the Internet". *The Social Studies* 91, 127-131.
[21] Tsfati, Y. & Weimann, G. www.terrorism.com: *Terror on the Internet. Studies in Conflict & Terrorism,* 25, 317-332 (2002).
[22] Weimann, G. (2004). www.terrorism.net: "How Modern Terrorism Uses the Internet". *Special Report* 116, U.S. Institute of Peace.
[23] Torgerson, W. S. (1952). "Multidimensional scaling: Theory and method". *Psychometrika, 17*, 401-419.
[24] Demchak, C., Friis, C., and La Porte, T. M. (2000). "Webbing Governance: National Differences in Constructing the Face of Public Organizations". In G. David Garson (Ed.), *Handbook of Public Information Systems*. New York: Marcel Dekker.
[25] Stamatatos, E., Fakotakis, N., and Kokkinakis, G. (2001). "Computer-based Authorship Attribution without Lexical Measures". *Computers and the Humanities, 35*(2), pp. 193-214.
[26] Peng, F., Schuurmans, D., Keselj, V., and Wang, S. (2003). "Automated Authorship Attribution with Character Level Language Models." Paper presented at the 10th Conference of the European Chapter of the Association for Computational Linguistics.
[27] Zheng, R., Qin, Y., Huang, Z., and Chen, H. (2003). "Authorship Analysis in Cybercrime Investigation". In H. Chen, R. Miranda, D. Zeng, C. Demchak, et al. (Eds.), *Proceedings of the First NSF/NIJ Symposium on Intelligence and Security Informatics (ISI'03)* (pp.59-73). Berlin: Springer
[28] Al-Fedaghi Sabah S. and Al-Anzi, F. (1989) "A New Algorithm to Generate Arabic Root-pattern Forms". *Proceedings of the 11th National Computer Conference*, King Fahd University of Petroleum & Minerals, Dhahran, Saudi Arabia, (pp. 4-7).

Further references

Adderley, R., & Musgrove, P. B. (2001). "Data mining case study: Modeling the behavior of offenders who commit serious sexual assaults". In F. Provost & R. Srikant (Eds.), *Proceedings of the 7th ACM SIGKDD International Conference on Knowledge Discovery and Data Mining* (pp. 215-220). New York: Association for Computing Machinery.

Chen, H. & Xu, J. (2005) "Intelligence and security informatics for national security: a knowledge discovery perspective", In B. Cronin (Ed.), *Annual Review of Information Science and Technology (ARIST),* Volume 40. Information Today, Inc., Medford, New Jersey.

Fayyad, U. M., Djorgovshi, S. G., & Weir, N. (1996). "Automating the analysis and cataloging of sky surveys". In U. Fayyad, G. Piatetsky-Shapiro, P. Smyth & R. Uthurusamy (Eds.), *Advances in knowledge discovery and data mining* (pp. 471-493). Menlo Park, CA: AAAI Press.

O'Hara, C. E., & O'Hara, G. L. (1980). *Fundamentals of Criminal Investigation* (5th ed.). Springfield, IL: Charles C. Thomas.

Hypermedia Seduction for Terrorist Recruiting
B. Ganor et al. (Eds.)
IOS Press, 2007

The Internet as a Tool for Intelligence and Counter-Terrorism

Yael SHAHAR
Director, Database Project
The International Institute for Counter-Terrorism (ICT), Interdisciplinary Center, Herzliya, Israel

Abstract. The internet is crucial to the daily operations of the radical groups making up the global jihad. The internet supplies the jihad movement with its recruiting and propaganda interface, as well as the means for ideological growth and the exchange of ideas. Without free and open communication, a movement of this size breaks down. The jihadi online presence is literally the physical brain of the global jihad movement. The very openness and accessibility of this medium provides the intelligence community with a wealth of material for foundation intelligence and analysis. This resource has been neglected in recent years due to lack of qualified researchers and linguists. The key to countering these problems may lie in harnessing the power of the private and academic sectors as unofficial research arms of the counter-terrorism community.

Keywords. Open source intelligence, internet intelligence monitoring, global jihad movement

Introduction

The uses made of the internet by terrorist organizations and extremists groups is nothing new. The terrorists use the net for the same reasons—and in the same ways—that we all use the net: for marketing, for communications, command and control, for intelligence gathering and datamining. Analysts have been bemoaning this fact for almost a decade, noting all the while that intelligence agencies consistently lag behind the terrorists in the use of the web. These complaints are legitimate, and highlight a central difference in both attitude and aptitude between terrorists and their more institutional opponents.

At present, intelligence agencies are focusing most of their efforts on the threat posed by the larger jihad movement, also referred to as the salafist jihad movement or the global jihad. For the purposes of this paper, I will use the later term, with the stipulation that the global jihad refers not only to the loose network of radical organizations and cells currently under scrutiny by the intelligence community, but also to the larger social movement espousing the jihad ideology. Without this social movement, the global jihad would not be the threat that it is.

Moreover, without the internet, the radical groups making up the global jihad's cadre of militants would remain a widely dispersed and isolated group of cells that happened to claim the same historical roots. It is the internet which has "globalized" the jihad movement. The network of global jihad is a product of the communications revolution. But of course, the same can be said of Microsoft. The internet has changed the way large organizations operate, be they multinational corporations or political movements. The internet is a facilitator of globalization and the global jihad is by definition a global movement.

The internet supplies the jihad movement with more than its recruiting and propaganda interface. It also provides the organization with the means for ideological growth and the exchange of ideas. Without free and open communication, a movement of this size breaks down. The jihadi online presence is literally the physical brain of the global jihad movement.

The very openness and accessibility of this medium provides the intelligence community with a wealth of material for foundation intelligence and analysis. This resource has been neglected in recent years due to lack of qualified researchers and linguists. The key to countering these problems may lie in harnessing the power of the private and academic sectors as unofficial research arms of the counter-terrorism community.

1. The Internet as Training Camp: To shut down or to tune in?

A common refrain over the past few years has been the constant complaint that the terrorists are "winning the battle in cyberspace." They're technologically creative, highly mobile, unfettered by either moral or bureaucratic constraints. They've got websites out there to fight their battle for the hearts and minds of supporters and potential supporters, to terrify their enemies, and in general "win friends and influence people." And of course, you can't shut them down. If you close down a website, assuming that you can persuade the service provider to do this, then you can be sure it will just pop up again somewhere else.

It was some time before the analysts began suggesting that, instead of engaging in the futile effort to close down websites, we should make full use of the potential to learn about our enemy from what they say about themselves, and even more importantly, among themselves. After all, a good intelligence agency should make a point of reading everything that the enemy writes, squeezing every drop of information on the opponent's psychological state, tactical capabilities, and strategic planning. This is what intelligence agencies do. Only it wasn't being done—or wasn't being done very well—with regard to the vast publishing empire of the internet. And with good reason. The sheer number of websites out there dealing with stuff that we should know about is daunting. Its content is ever changing. And of course there are linguistic issues, as well as cultural issues. To really make sense of it all would require the services of a veritable army of qualified personnel, and would be far beyond the budgetary—not to mention the bureaucratic—capabilities of most intelligence agencies.

But such an army of personnel does exist, even if it's a modest and rather irregular army. Independent research bodies and individuals have been monitoring terrorist websites and listening in on jihadist forums for years. Some of these people are contributors to the current volume—people like Gabi Weimann, who catalogued terrorist websites long before it became fashionable to bemoan the terrorists' subversion of the internet; or the folks at MEMRI, who've been translating relevant documents for years. Aharon Weisburd has been sleuthing into the identities of the jihadi webmasters and forum posters. The Arizona Artificial Intelligence Lab has pioneered new tools for finding out who's who on the "jihadnet," and analyzing their methods of persuasion.

Note that all of this has come from the private sector. Smart governments realize the usefulness of these skills early on, and have taken some steps to make effective use of them. Arizona's AI Lab has been the beneficiary of more than $20M in research funding from federal agencies.

And of course, government think tanks and institutes are also dealing with jihadi materials more and more. Of particular note, the Combating Terrorism Center (CTC) at West Point has done some outstanding analysis work in applying what is learned on the web to concrete tactical operations. Some the examples cited here of intelligence gathering from Islamist online materials come from *Stealing al-Qaida's Playbook*, by Jarret M. Brachman and William F. Mccants [1] of the CTC. Just to give a taste of what such analysts have come up with over the past two years, I would like to present some examples of how open source datamining can clue us into how the "bad guys" think, and what to do about it.

2. Intelligence from the Web

Before going further, I would like to review briefly what good intelligence is and is not.

Intelligence is not "news". Governments often press intelligence agencies into service as open source news monitoring agencies, insisting on getting up-to-date reports on things that are covered by the mainstream wire services. This kind of pressure is often detrimental to the agencies' primary tasks, which is providing an insight into fundamental processes. Yes, an intelligence agency should be up to date with what's going on in a particular sphere of interest. But intelligence gathering is not meant to be news reportage.

At its best, intelligence leads to understanding. Properly understood, intelligence-gathering should lead to a fundamental understanding of what is going on in the sphere of interest, who the main players are, and who their friends and enemies are. With regard to a specific opponent, this intelligence should include what the opponent is doing and what the opponent is thinking. This is called "foundation intelligence", and without it, an intelligence agency is no more than an under-funded and under-staffed news agency.

Understanding leads to predictions. Proper foundation intelligence can lead to a better understanding not only of how things are likely to unfold in the sphere of interest, but also what effect contingencies outside of that sphere may have on events. Extrapolation from foundation intelligence is the work of professional analysts, who provide the crucial link from *foundation intelligence* to *tactical intelligence.*

Essentially, good foundation intelligence helps to formulate answers to the following questions regarding the opponent

- What are they saying?
- What can we learn from it?
- What should we do about it?

While the Internet is unlikely to be a lucrative source of reliable tactical intelligence, it is a goldmine for foundation intelligence. This kind of intelligence deals with the following basic issues regarding the opponent:

- Ideology. What are their core beliefs? What divisions are there in the movement?
- Strategy. What are their ultimate goals and overall game plan?
- Tactics. By what methods do they propose to reach these goals?
- Structure. How do they make decisions, and who makes them?

3. What we Learn from Jihadist Websites and Forums

The real heart of the global jihad is expressed online. This is the venue where ideas are hashed out, dissent is either neutralized or accommodated, and strategies and tactics are discussed. For the most part, we don't expect to find online discussions of actual attack planning. But we can learn how ideology and opinion are shaping in the larger Muslim community who form the jihad's target constituency and who supply the movement's pool of recruits.

In essence, Al-Qaida's flight into cyberspace was necessitated by the destruction of jihadi training camps following the September 11 attacks. The movement's leaders have had to turn to cyberspace as a way of maintaining contact with a geographically dispersed constituency. However, the use of the web as a primary venue for discussion was not only mandated by necessity, but was also a matter of choice. Since its inception, the global jihad has relied very heavily on the internet and the nature of online communities to further its aims.

3.1. Ideological Lessons

The jihad movement's internet presence is most felt in ideological discussion. The web is essential to the movement's ideological development, as well as to the actual dissemination of this ideology to potential recruits and supporters. Such freely accessible discussions offer analysts a window into the jihad movement at both the "grass roots" level and the level of its top-level leaders. Senior ideologues, as well as mid-level operatives and up-and-coming scholars write on these sites. This is intelligence straight "from the horse's mouth."

Downing and Meese of the CTC point out that a fair amount of the jihad's key doctrinal literature is available online, in addition to documents that have been captured by the intelligence agencies of different countries.

> One of the best ways to learn about al-Qa'ida is to read the papers, manuals, and other documents which al-Qa'ida leaders have written to guide and discipline their own enterprise. Many of these documents have been captured by military and law enforcement forces and can provide insight into the way the organization works. Other key references are readily available on the World Wide Web.[2]

They emphasize that the more access is provided to these documents, the more benefits will accrue to the counter-terrorism community as a result. As archives of translated jihadi documents become more accessible to analysts, the understanding of the movement's key ideology, strategy, and motivation is growing. What is interesting is that such analysis is increasingly coming from the private and academic sectors.

An example of how analysis is leading to a greater understanding of the jihad movement's vulnerabilities comes out of the Combating Terrorism Center(already mentioned in full). In *Stealing al-Qaida's Playbook*, Jarrett Brachman and William

McCants demonstrated how "jihadi strategic studies" can be used to identify and exploit the weaknesses of the jihadi movement.[1] The authors point out that the key to defeating the global jihad from an ideological point of view is to understand its ideology from the inside out: who the main ideologues are, and the significant issues that unite and divide the movement. The authors note that jihadi leaders are remarkably open and blunt when discussing who their biggest competition is and what their public relations vulnerabilities are.

> In a sense, members of the jihadi movement have put their team's playbooks online. By mining these texts for their tactical and strategic insights, the United States will be able to craft effective tactics, techniques, and procedures to defeat followers of the movement."[1]

I'd like to give a few brief examples of the kind of intelligence that can be gleaned from examining some of the strategic dialogue of the global jihad movement. The following examples are from *Stealing al-Qaida's Playbook.*

3.1.1. Example 1: Abu Bakr Naji, The Management of Barbarism, 2004

Abu Bakr Naji is one of the more prolific Al-Qaida ideologues of the new generation. He is well-read and articulate, and his works are cited on numerous jihad websites, a testament to the high regard in which he is held by the movement. In *The Management of Barbarism*, Naji presents a thoroughgoing analysis of his own movement's strengths and weaknesses, as well as those of his opponents. Obviously, such an analysis is useful for us in that it allows us to see how our enemies see themselves. But no less important is the information they give us as to how they see us. This point is well illustrated by the examination of Naji's writings.[1]

Brachman and McCants studied *The Management of Barbarism; b*elow are some key points from Naji's work, as cited by Brachman and McCants.

- Naji urges fellow jihadis to study Western works on management, military principles, political theory, and sociology, in order to borrow strategies that have worked for Western governments and to discern their weaknesses.
- The jihadis cannot defeat the United States in a direct military confrontation. Rather, the clash with the United States is more important for propaganda victories in the short term, and the political defeat of the US is viewed as a long-term goal, as American society fractures and its economy is further strained.
- Naji also discloses weaknesses in the jihadi movement and the problems that predictably confront such an entity, such as difficulties in resolving chains of command, ferreting out spies within the organization, and reining in overzealous recruits.
- There is concern that the momentum of the movement may be slowed by clerics who challenge its legitimacy and siphon off its recruits among the youth.
- Naji observes that the jihadi movement has often split over theological differences.

3.1.2. Example 2: Abu Qatada, Between Two Methods, 1994

Another example of the vulnerability of the jihad movement given by Brachman and McCants comes from Abu Qatada's work, *Between Two Methods*. Abu Qatada is scathing in his criticism of a popular Saudi cleric, Rabi`al Madkhali, a serious rival of the salafi movement: "This man is content to claim that he is a Salafi so that he can be an imam for some inexperienced boys whom he feeds slogans and shimmering phrases." Jihadi ideologues are very open about which Muslim religious leaders they most fear.[1]

Brachman and McCants point out that although a specific enemy may no longer be a threat, by understanding why he was a threat in the past, we can look for—and perhaps exploit—similarly threatening enemies in the present.

Why was Madkahli a threat? To begin with, he was a quietest, and was supported by the Saudi government. More importantly, his popularity and outreach were such that he was able to draw off young recruits from the more radical movements.

3.1.3. Lessons Learned

What do we take away from these examples? Perhaps the most salient piece of information we gain is "what worries them." Naji openly discusses some of the vulnerabilities of the jihadi movement from an ideological perspective. From his discussion, we see that the movement is vulnerable to ideological splits and knowledgeable clerics who "call them" on their interpretation. What does Naji say should be done about it? His suggestion is to co-opt religious clerics to back the jihadi's interpretations. Where this is not possible, he suggests that rival clerics either be intimidated into silence or killed.

For his part, Abu Qatada provides us with some keys as to what sort of rival cleric is the greatest threat to the movement. The answer seems to be one who manages to appeal effectively to the same target audience, the youth, and yet who espouses a program directly at odds with that of the jihad movement.

3.2. Strategic Lessons

The term "jihadi strategic studies" was coined by Thomas Hegghammer and Brynjar Lia, of the Norwegian Defense Research Establishment in Oslo, to refer to books and articles on the strengths and weakness of the jihadi movement and those of its enemies. [3] Hegghammer noted that the Internet has become a vital venue for terrorist cells to organize and "brainstorm" about tactics in a decentralized way. This allows sleeper cells to operate virtually autonomously, deriving their inspiration and operational direction from texts published online by individuals on other continents.[3]

Hegghammer argues that it is online that "...you really get the early signs of the ideological developments, which are later going to affect us, or might affect us, physically."[4]

At the same time, the increasing use of the internet as the central brain of the movement has also made the movement more transparent to onlookers. Terrorist/Jihadi thinking is public and susceptible to infiltration. Hegghammer encouraged counter-terrorist agencies to create an "atmosphere of paranoia" on these websites, by posting fraudulent texts and subverting the readers' trust in the literature.

3.2.1. Example: "jihadi Iraq, Hopes and Dangers"

Hegghammer and Lia analyzed a document on the Internet, "jihadi Iraq, Hopes and Dangers", which detailed how terror attacks ahead of Spain's general election could drive Madrid to pull its troops out of Iraq and thus harm the US-led coalition. Brynjar Lia came across the document in December 2003 on a website called "Global Islamic Media." According to Hegghammer and Lia, "The main thesis proposed in the document is that America cannot be coerced to leave Iraq by military-political means alone, but the Islamist resistance can succeed if it makes the occupation of Iraq as costly as possible—in economic terms—for the United States."

Their analysis led them to see in the document a blueprint for driving a wedge between coalition members.

> The document therefore offers a number of specific "policy recommendations" in order to increase the economic impact of the insurgency and the jihadi campaign in Iraq. The most important of these recommendations consists of trying to limit the number of American allies present in Iraq, because America must not be allowed to share the cost of occupation with a wide coalition of countries. If the mujahidin can force US allies to withdraw from Iraq, then America will be left to cover the expenses on her own, which she cannot sustain for very long. The intermediary strategic goal is therefore to make one or two of the US allies leave the coalition, because this will cause others to follow suit, and the dominos will start falling [5].

The document's anonymous author emphasized that:

> ...It is necessary to make utmost use of the upcoming general election in Spain in March next year. We think that the Spanish government could not tolerate more than two, maximum three blows, after which it will have to withdraw as a result of popular pressure. If its troops still remain in Iraq after these blows, then the victory of the Socialist Party is almost secured, and the withdrawal of the Spanish forces will be on its electoral programme. [5]

A few months later, on 11 March—just prior to the elections—Madrid was rocked by a series of train bombings that killed 190 people. Partly as a result, Spain's conservative government, which supported the Iraq war, lost the vote to the opposition Socialists, who later pulled Spanish troops out of Iraq.

It would be hard to believe that the bombers were unaware of the "Hopes and Dangers" document, given its depth of detail and its widespread distribution. In fact, as pointed out by Lia and Hegghammer, the "nom de terror" chosen by an alleged Al-Qaida video spokesman after the attack—Abu Dujana, a warrior and contemporary of Mohammed—matches one mentioned in the "Hopes and Dangers" document.

Further evidence that the bombers had considered the political effect of their actions was provided by the behavior of an alleged ringleader of the Madrid attacks. A cell phone on an unexploded bomb led Spanish police to Jamal Zougam within a day of the bombings. As the *New York Times* reported: "When Mr. Zougam arrived in court after five days incommunicado, he reportedly asked the clerks, 'Who won the elections?'"[6]

3.2.2. Lessons learned

The "Hopes and Dangers" document is one example of the kind of basic intelligence that is freely available on the Internet. What can be learned from this example of jihadi strategic studies and the way it was put to practical use?

One of the most obvious messages is that the new generation of jihadi strategists place a greater emphasis on pragmatism, and less on ideology. The "Hopes and Dangers" document is written from the standpoint of cool analysis of political realities, with fewer references to historical battles and quotes from the Quran. Eschewing flowery rhetoric and exaggeration, it goes to the heart of current events and possible consequences. Clearly, Al-Qaida's future ideological leadership is evolving in the direction of greater professionalism.

3.3. Tactical Lessons

Just as the jihad uses the web for ideological discussion and dissemination of ideas, tactical discussions and training materials are also freely available online. Among the intelligence that can be gleaned from these documents is information on the tactics they see as effective, the weapons they favor and why, and perhaps even more importantly, their assumptions regarding the effectiveness of these weapons.

The following two examples are from Brachman and McCants's *Stealing al-Qaida's Playbook.*

3.3.1. Example 1: Abu Mus'ab al-Surri.

Abu Mus'ab al-Suri (aka Mustafa Setmarian Nasar) has made a study of failed jihads in the contemporary world. He identifies various reasons for such failures, including:

- The co-operation of local governments in countering jihads, and the failure of the jihadists to organize simultaneous attacks in neighboring countries. Suri cites as an example the co-operation of Syria with Jordan, Iraq and other neighbors during the 1960s to 1980s. He suggests that if the neighboring states had been struggling against their own jihads, they would not have been so quick to assist Syria.
- The failure to consider the influence of ethnic minorities and tribes, or the possibility that these populations may be co-opted by the state.
- The failure to provide jihad fighters with a sense of personal connection to their leaders, or with the vision that they too may become leaders.
- The failure to gain popular support from the Muslim majority; Suri identifies the role of propaganda as being crucial in this regard.
- The insufficient involvement of Muslim clerics. Suri argues that clerics' involvement is essential for developing new local jihad groups.

3.3.2. Example 2: Abu Bakr Naji

Abu Bakr Naji suggests various ways in which the jihad movement could be strengthened. He outlines three stages for establishing the Khaliphate, beginning with the bombing of crucial targets in order to draw the local security forces in around these centers. The ensuing chaos would allow jihad leaders to assume control of the more remote regions, and from there the jihad administrators could network towards establishing a Khaliphate in that country.

Naji also suggests that low-ranking jihadists should not be allowed to launch their own attacks, other than small to medium ones. He argues that major attacks such as 9/11 must be organized only by the High Command, since the wrong attack at the wrong time would undermine the movement.

Naji notes that mass-scale Muslim support is of prime importance, and that already the Muslim public is wary of the jihadists' use of violence, and may view the jihadists as trouble-makers. Without Muslim support, Naji cautions, new recruits cannot be found.

Finally, Naji argues that the education of young people is only complete when they participate in jihad. He views the education provided by Muslim religious leaders as being of lesser importance, while jihad involvement is seen as providing essential training. The involvement of youth in jihad is viewed as a further step towards establishing the "global Islamic resistance" and, ultimately, the global Khaliphat.

3.3.3. Lessons learned

One way to counter the jihadis' tactic would be by helping local surrogates establish their own enclaves in those regions left unprotected by security forces. Local ethnic groups can play a role in preventing security vacuums from forming.

The jihadis and the counter-terrorism community are competing for the same audience; however, public opinion is more important to the "irregular" side in a low intensity conflict. For this reason, a greater emphasis must be placed on psychological and information operations in the sphere of influence, in order to undermine the popular support on which the jihad movement is dependent.

4. Counter-Attacks

4.1. Terrorism Websites as a Key to Efficacy

As mentioned, the jihad relies heavily on the web for the dissemination of information from the upper echelons, as well as for discussion at all levels. Statements by the leaders of the movement should be considered a useful input for intelligence analysis. Content analysis of such statements may give some hint of intentions, although there is a significant amount of bluster, hype, and misdirection present in much of these statements. Context analysis may therefore be of greater utility. This kind of analysis can determine what the person making the statement *thinks that his followers want to hear.* Thus, context analysis can provide some measure of the "temperature" of the "street." This is a useful input to foundation intelligence, as well as an aid to determining the focus of tactical intelligence gathering.

Special attention should be given to the language, focus, and design of jihadi websites. The languages used can tell us who the jihadis see as their primary audience for recruits. In some cases, this points to perceived vulnerabilities among the targeted group. But it can also tell us who the jihadis *would like to have* as recruits—who can be of most use to them. For example, European Muslim converts are increasingly the focus of "narrow casting" on jihadi websites.

Another point of consideration is the imagery and design of the sites. These sites are frequently the work of some of the best and brightest of the new generation jihadi recruits. They know their audience, and they know what techniques are likely to be effective. With this in mind, we can say that imitation is the sincerest form of counter-terrorism!

But it isn't only the style of outreach that can be copied. Here too, a bottom-up approach can yield benefits that would be unthinkable for top-down institutionalized responses. Web designers from the jihadis' own target audience can be recruited by the counter-terrorism community to build a counter-offensive. The success of this kind of campaign depends on working at the "grass roots" community level. The main allies of the counter-terrorism community will be those singled out for recruitment by the jihadis themselves. Terrorists and counter-terrorists are competing for the same audience. The techniques used successfully by both sides will reflect this.

4.2. Psychological Operations

Analyst Stephen Ulph of the Jamestown Foundation has been monitoring jihadi forums, with a special emphasis on those dealing with the situation in Iraq, the new training ground for the jihadi movement. In July 2005, Ulph noted how news of ongoing discussions between the US military and the Iraqi insurgents was affecting traffic and commentary on jihadi forums [7]:

> The news evoked considerable notes of distress on the jihadi forums. On the al-Qal'a forum one signing himself al-Sharif al-Idrisi, noted, on June 28, the similarity of this potential development with the situation in Afghanistan, "when those fleeing the Tora Bora caves were met by the Pakistanis not intent on helping them but in selling them to the Americans. We pray God that this doesn't happen to our brothers in Iraq" [www.qal3ati.net].

This kind of commentary highlights a key weakness as perceived by the jihadi militants themselves—their vulnerability to betrayal by the wider society in which they operate. Such fears are, of course, easily played on. One obvious stratagem for exploiting this sense of paranoia would be to "feed" news of betrayals to local news media, then cast blame in the relevant forums upon elements within the organization, or in rival organizations.

Ulph also noted that the news of meetings between coalition leaders and insurgents was met with denial by many:

> At the same time strenuous denials were being posted on the internet forums that any such meeting took place, including from groups said to have participated in the talks. One posting on June 30 appeared on the al-Qal'a forum signed by The Islamic Army in Iraq, the Army of the Mujahideen and the Army of Ansar al-Sunna. It expressed exasperation at Ayham al-Samarrai's "lies and America's games" and swiftly pointed to the impending peril for the Islamist mujahideen in Iraq: "its intentions are to split the ranks of the mujahideen … to divide the Iraqis from non-Iraqis ... to pull the rug from under the mujahideen … How can a heroic mujahid Muslim brother in any country be a foreigner?" [7]

What is significant here is the perceived motive of these "American games," namely, "to split the ranks of the mujahideen … to divide the Iraqis from non-Iraqis ... to pull the rug from under the mujahideen." One can gain a good deal of insight into the state of insecurity of the opponent from this sort of posting. Obviously, one forum posting or internet statement does not give an indication of the psychology of a whole organization. However, if monitored on an ongoing basis, the total compendium of such statements can provide a real "feel" for the psychological state of the opponent.

And if "being sold out" is what they fear, then by all means, one should play up any and every possible case of such a sell-out, and milk it for all it's worth.

The potential of such suspicions to turn the opponents' forces against his own is indicated by the continuation of the same statement quoted by Ulph above:

> With America's designs being "to return the Baathists to power, in the name of the resistance" the statement accused Iyad Allawi of "giving orders to the Baathist Ayham al-Samarrai to intrigue against the mujahideen and the resistance … So we proscribe the life of Ayham al-Samarrai, and declare him to be a target of the mujahideen in general and in particular of all members of the three groups (The Islamic Army in Iraq, the Army of the Mujahideen and the Army of Ansar al-Sunna). … Anyone who allows himself to be seduced into doing what the fantasist Ayham is doing will share the same fate" [www.qal3ati.net]. [7]

In other words, whether correct or not, the suspicions served to drive a wedge between the organization and its perceived enemies. Ulph points out that if the same suspicion can be cast upon elements within the organization, instances of "red on red" firing between opponent groups will be seen to increase.

4.3. Countering Recruitment and Propaganda

The global jihad movement, like any other major social movement, depends on a broad base of support. Positive public opinion from within its constituencies is a must, not only for bringing new recruits into its cadres, but also for garnering support for its less obviously goal-driven activities.

As was pointed out by Yoav Mimran in his discussion of Arab society and culture earlier in this workshop, one of the key values of Muslim societies is social unity and harmony. Anyone seen as sowing discord or endangering public order is likely to be seen as an enemy of the public good, no matter how otherwise worthy are his goals. Brachman and McCants have noted that the movement "declines in popularity when it is perceived to be attacking fellow Muslims, causing public disorder, damaging critical national industries, or engaging in sectarianism." [1]. They point out that one effective point of counter-attack would be to "harness the power of the 'Shayma Effect' [referring to an incident where an Egyptian schoolgirl was killed in a jihadi attack], broadcasting images of jihadi attacks that have killed Muslim children. [1]

The authors point out, however, that any such campaign must be managed very much "from the rear" and by proxy. To this we may add that an institutionalized campaign can never have the same power or reach as a genuine "grass roots" campaign among the potential constituency of the opponent. Only when we are able to inspire local actors to join the fray will such a media counter-offensive really be effective. A government-sponsored campaign, no matter how skillfully managed, is no substitute for the participation of local bloggers, media people, and commentators.

4.4. Undermining Trust

From articles which are available online regarding the training of activists, we can learn of areas that the jihad movement itself sees as weaknesses, the better to exploit them. For example, a number of articles have been posted warning forum posters of possible digital interception, and suggesting ways of getting around the problem.

In other instances, internal debate can point to potential splits within the movement, or to a lack—or perceived lack—of leadership. In addition, these kinds of debate can show what issues are of greatest concern to the jihadis themselves.

4.4.1. Example 1: Divide and Conquer

In fact, the penetration of Islamist forums by counter-intelligence agencies has resulted in the arrests of several key figures. These may include the arrests of forty mujahideen in Saudi Arabia, and the arrests of the attackers of the Abqaiq (Buqayq) oil facility. Most certainly, the arrest of the infamous "internet jihad" Irhabi007 was due to a fairly common lack of attention to basic security procedures online.

A significant spin-off of such events is the mistrust and confusion which abounds on jihadist forums in their wake. During the first half of 2005, Stephen Ulph and other researchers at the Jamestown Institute noted a spate of warnings and debates which appeared on jihadist forums. The warnings were to the effect that forum participants should not enter certain websites—not even as a visitor—for fear of being identified by the "dogs in intelligence" (cited by Ulph, as posted on the Syrian site "Minbar Suriya al-Islami"). The debates concerned the authenticity of certain jihadist forums, or of site administrators or participants who were suspected of being counter-intelligence spies [8].

The end result is that jihadists' confidence in their ability to dodge state control via the use of Internet forums has dropped significantly in the past couple of years. Despite the jihadist forum administrators' best efforts to use proxies and to conceal participants' identities, this kind of confidence may not be all that easily recovered.

4.4.2. Example 2: Encourage Sectarianism

According to Ulph, "the pool of experienced, credible ideologues may be draining" amongst jihadist supporters. Ulph has noted that in the wake of the kidnapping and assassination of the Egyptian ambassador to Baghdad, al-Zarqawi's Al-Qaida group was criticized by the two main Egyptian militant Islamist organizations. [8] These detractors argued that Al-Qaida in Iraq was more focused on destroying the Shi'a and Kurd populations than it was on fighting the enemy occupier.

They also suggested that the organization was unable to learn from its past mistakes, and that it was causing "the average Muslim" to feel alienated from Islamist groups in general.[9] The latter criticism was echoed by Abu Muhammad al-Maqdisi of the al-Tawhid wal-jihad movement, who argued that the true nature of jihad was being distorted by al-Zarqawi and other mujahideen in Iraq. In addition, criticisms regarding the London bombings were posted on jihadi forums by Abu Baseer al-Tartusi [www.abubaseer.bizland.com]. These kinds of criticisms are particularly important since they have been publicly aired by members of the Muslim community.

In both instances cited above, internal debate and dissent can work to undermine the confidence of the jihadis in their own organizations. This drop in confidence can affect the organizations' infrastructure, as in the case of the perceived use of Internet forums by counter-terrorism operatives; alternatively, confidence in the leaders themselves may be affected, as in the case of the criticism of Zarqawi.

5. Conclusions and Recommendations

Terrorists' use of the Internet for command and control, propaganda, and intelligence gathering has been bemoaned for years by law enforcement officials and intelligence agencies. Too often, the field has been ceded to the terrorists, with the counter-terrorism community struggling to catch up—amid allegations of inefficiency, misplaced funding, and organizational ineptitude.

The examples presented above show the kind of conclusions that can be drawn from jihadi texts and inter-organizational dialogue, as well as examples of practical lessons learned. The fact that almost all of the raw material for this analysis is available online should not detract from its importance in the eyes of intelligence agencies. There is a tendency for intelligence agencies to get caught up in the "top secret implies top value" mentality, leading them to overlook some of the most crucial inputs to intelligence that are freely available to all takers.

5.1. Horizontal Dataflow versus Vertical Dataflow

In order to fight the type of terrorist entity that has evolved out of the communications revolution, intelligence agencies will need to learn the lessons taught them by the terrorists themselves. This means the establishment of a "ground-up" grassroots approach to counter-terrorism—a horizontally-structured counter-terrorism apparatus to replace, or at least complement, the older top-down hierarchical model. What is needed is not only a technological shift (which is now underway), but an organizational shift. In practical terms, this means that information flow, too, will need to be less vertical (from the top down) and more horizontal.

At the same time, the organizational structure of the counter-terrorist community as a whole will need to be reworked to take full advantage of the Internet, both as a tool for intelligence gathering and as a means for command and control.

It's a tall order. But the reality on the ground increasingly calls for it. What's more, the means now exist for doing it. If they can do it, why can't we? And in fact, it's already happening. Increasingly, we're seeing the initiative for intelligence gathering and analysis shifting from the hands of government agencies into private hands. This is as it should be. Terrorism endangers us all—citizens as well as military and law enforcement personnel. In fact, due to the nature of terrorism as psychological warfare, it is the private sector that is most at risk; and this is also the designated target. It is important that the "professional" intelligence community understand this shift from governmental to private inititiative, and in fact embrace it. They need to learn to make proper use of the ground-up efforts of academics, internet sleuths, and independent analysts.

Brachman and McCants note that at present few agencies have access to the kind of experts who have a deep enough background in the culture and language of the opponent to provide the kind of analysis essential for good foundation intelligence. Even when they do, such analysis tends to remain within the "orbit of the agency where it originated." They suggest that the United States government "might consider establishing a think tank staffed with highly trained experts on the Middle East and counterinsurgency whose sole purpose would be to identify the major jihadi thinkers and analyze their works." [1]

While such a think tank is a worthy goal, these problems will not be overcome by the establishment of yet another government-sponsored entity, no matter how well-funded or well-staffed. It is the nature of bureaucracies to become monolithic, self-perpetuating, and inward-looking. A better solution is for governments to establish links of mutual support with the grassroots organizations already doing these analyses. This means that government agencies need to know who these entities are, what they produce, and what they need in order to keep working—usually money. What is needed is a working relationship, though it isn't clear whether this should be based on a "consumer/supplier" relationship, or on cooperation.

In addition to fundamental intelligence gathering and analysis, information operations too are shifting from the government to the private sector. This is a positive step, regardless of the potential for "vigilantism". Governments will need to recognize the potential of the private sector, particularly in the sphere of interest, to manage their own information and media operations.

To be effective in low-intensity conflict, our definitions of tactical counter-terrorist operations will need to be expanded to include web-based operations. In particular, web-based PsyOps and counter-propaganda should be seen as key elements in the counter-terrorist arsenal. It is crucial that any information campaign make use of the same tactics and the same venue as those used by the jihad movement itself.

It is also crucial that the counter-terrorism community understand the implications of the change brought about by the communications revolution. Today's battles are being fought more and more in the sphere of public opinion rather than on the battlefield. The counter-terrorism community is competing for the same audience as the jihadis themselves. In this type of warfare, the Internet is both battlefield and weapon. For the jihadis, this is a two-edged sword; the greater their dependence on the Internet, the greater their reach and efficiency, but also the greater their vulnerability.

References

[1] Jarret M. Brachman, William F. Mccants. "Stealing al-Qa'ida's Playbook." CTC Report. February 2006.

[2] Wayne A. Downing and Michael J. Meese, "Harmony and Disharmony Exploiting al-Qa'ida's Organizational Vulnerabilities." Combating Terrorism Center Department of Social Sciences United States Military Academy 14 February 2006.

[3] Jihadi Strategic Studies: The Alleged Al Qaida Policy Study Preceding the Madrid Bombings *Studies in Conflict and Terrorism.* Routledge, part of the Taylor & Francis Group, Conflict, Security and Strategic Studies, Volume 27, Number 5/September-October 2004. p. 355-375. August 19, 2004.

[4] Australian Broadcasting Corporation. TV Program Transcript: Al Qaeda weaves web of terror. Broadcast: 18/03/2004. http://www.abc.net.au/lateline/content/2004/s1069029.htm

[5] Brynjar Lia and Thomas Hegghammer. "FFI explains al-Qaida document." Forsvarets forskningsinstitutt. 19 March 2004.

[6] *New York Times*. "As Europe Hunts for Terrorists, The Hunted Press Advantages." March 22, 2004, Section A, Page 1, Column 1.

[7] Stephen Ulph. "Islamist insurgents seek to contain PR disaster: notes of defeatism." Terrorism Focus. Volume 2, Issue 13 (July 13, 2005)

[8] Stephen Ulph. "Zarqawi's declining ideological support among Islamists". Terrorism Focus. Volume 2, Issue 14 (July 22, 2005)

[9] www.alsakifa.net, July 14, 2005; cited in Ulph [8].

Hypermedia Seduction for Terrorist Recruiting
B. Ganor et al. (Eds.)
IOS Press, 2007

The Shifting Sands of the Global Jihad Online

Aaron WEISBURD
Society for Internet Research, www.sofir.org

Abstract. This chapter provides a combatant's perspective on Internet use by the forces of the Global Jihad. The aim is to familiarize the reader with the terrain, the actors, the technologies, the benefits which accrue to jihadists by virtue of their Internet use, and the ways and means by which we may exploit various aspects of such use to defeat this global threat.

Keywords. Internet, jihad, counter-insurgency, terrorism, Al-Qaida

Introduction

The Internet has grown in importance for terrorists of all sorts over the last decade, and has become their preferred communications medium [1]. The role of social networks in developing jihadist identity and fueling the process whereby jihadists adopt progressively more extreme and fundamentalist attitudes has been well documented [2]. The Internet creates a bond between the individual and a virtual Muslim community ("ummah"), which approaches the ideal Muslim world in the salafist vision, devoid of national divisions, corruption, exploitation and persecution.

Computer-mediated communication promotes the development of in-group love—close bonds within the online community of jihadists—and out-group hate—a process whereby jihadists alienate themselves from the world around them, including (perhaps most importantly) the larger and relatively more moderate Muslim world. The absence of real-world constraints enables the expression of the most violent of goals and aspirations vis-à-vis the perceived enemies of the "virtual ummah" [3]. It also increases the jihadists' level of identification with the broad objectives of the global jihad, to the detriment of support for conflicts closer to home.

Thus the Internet is particularly suited to promoting a worldview of the global jihad, and delivering a minimalist version of Islam based on a literal reading of the Quran and the Hadiths (the oral version of the prophet's tradition). This stance summarily dismisses centuries of Islamic scholarship and jurisprudence.

The use of the Internet also requires, as a prerequisite, a degree of modern education. This fact encourages a selection process which favors those who have had less exposure to a more nuanced, complete, and thus less fundamentalist, Islamic education [2,4].

1. The Actors

1.1. Service Providers

There are roughly five categories of service providers [5] that jihadists may make use of in the course of operating their websites. We will examine each of these in turn.

1.1.1. Domain Name Registrars

The Nature of the Service
While it is possible to operate a website without a domain name, for promotional purposes it is desirable to have one. Domain names are not so much bought as leased, and the registrar holds the lease, collects payment for the service, and keeps a record of which individual or organization has registered any given name.

Benefit(s) to Jihadists
Unlike most companies who accept payment by credit card online, domain name registrars make little or no effort to match the name and address provided at registration with the name and address of the credit card used to pay for the service. As a consequence, the use of fictitious names is common, and identity theft or credit card fraud is frequently found to occur in cases of domain names registered for the use of promoting the global jihad. For the jihadist who registers a domain in this way, a short term benefit is some degree of anonymity.

A number of larger domain name registrars allow smaller businesses to resell their service. In such a case, the reseller is the one who actually engages in a financial transaction with the registrant of a domain name. It has happened that supporters of Islamist terrorism have become resellers of domain registration services, with the specific aim of providing services to those who operate jihadist websites and in order to mask the identity of the actual registrants of those domains.

Domain name registrars are resistant to efforts to hold them responsible for the consequences of the services they provide, where these services benefit the global jihad. They will, as a rule, ignore the nature of the domains they service, turn a deaf ear to public complaints, and wherever possible attempt to shift blame to other service providers. At the same time they will facilitate fraudulent behavior in support of both terrorism and more ordinary criminal endeavors.

Vulnerabilities
Domain name registrars are legitimate businesses that can be held accountable for the services they provide to jihadists. Even when the transactions involved in setting up a domain are carried out by a reseller, and even when that reseller is openly supportive of the goals and methods of the jihad, it is the registrar who remains responsible for the services provided. The revenue generated by registering jihadist domains is negligible, and is certainly dwarfed by the potential liabilities represented by negative publicity, not to mention civil or criminal liability.

When stolen credit cards are used to pay for domain name registration services, each such instance is a criminal offense, and constitutes a cause for investigating a particular jihadist site and the individuals operating it. This fact is especially valuable in the United States, where any investigation related to a website can be misrepresented as an assault on someone's free speech rights.

Finally, the use of fictitious information in a domain's registration is less problematic than it might at first appear. Particularly in cases where individuals register multiple domain names for a single site, or register domain names for multiple sites over a period of time, clear patterns emerge in the deceptive behavior. These patterns can, and frequently do, contribute valuable leads for investigations into the individuals who stand behind jihadist websites.

1.1.2. Network Access Providers

Nature of the Service

Websites have a physical address, known as an IP address. This address is assigned to a server, generally located at a datacenter. Sites can share an IP address, and a single web server can host multiple sites on multiple IP addresses. The network access provider assigns IP addresses to servers as needed, and provides the line out from the server to the Internet. Typically, it is the network access provider who has physical control over the servers used to host jihadist sites; the provider does not necessarily own either the servers themselves or the datacenter, though they often do.

Benefit(s) to Jihadists

As with domain name registrars, network access providers can be resistant to accepting responsibility for the websites that they allow to access the Internet, preferring to shift that responsibility onto the website hosting companies that contract with them for IP addresses and network access.

Vulnerabilities

Network access providers have very little vested interest in the websites that use their IP addresses and network connections. They are also more likely to have existing relationships with law enforcement agencies, since the latter will go through them to perform operations such as acquiring a mirror image of the hard disk drive of a suspect's server. As with domain name registrars, it is possible to determine exactly who the service provider is, with a high degree of accuracy.

1.1.3. Web Hosting Providers

Nature of the Service

Web hosts provide a website with space on a server, and typically also provide DNS service for their customers' domains. The web host will be engaging in financial transactions with the jihadist(s) who are operating a site, a fact that becomes quite significant when the site is connected to a specific terrorist organization. Free web hosting service providers provide hosting in exchange for advertising that is inserted into each page.

Benefit(s) to Jihadists

In the case of small web hosting enterprises, the income provided by hosting jihadist websites may actually be a significant part of their revenue, making such providers reluctant to discontinue service to their jihadist clientele. The process of operating a web hosting operation has been automated to such an extent that it is fairly easy for jihadists to operate their own hosting businesses, and we see this from time to time.

Vulnerabilities

The operator of a web hosting operation enjoys, as a rule, administrative control and access to all aspects of the operation of the sites they host. In the event that the host of a jihadist site should turn out to be unsympathetic to their cause, the host can become an import ally for agents and agencies investigating online terrorist activity.

More importantly, the ease with which a web hosting operation can be set up makes it a powerful tool for undercover operations. A server used for this purpose is referred to as a "kill box", and the scenario involves soliciting the business of jihadists, closely monitoring their activities as they pay for and set up their sites, identifying the site administrators and users, then closing down the operation and rounding up the suspects.

1.1.4. DNS Service Providers

Nature of the service
DNS service identifies the physical (IP) address assigned to a domain name. Generally this service is provided by the web host, but sometimes it is managed separately.

Benefit(s) to Jihadists
The benefit to jihadists of managing DNS service separately from web hosting is that it allows them to move their sites from one location to another quickly, as network access providers and web hosts decide to discontinue these services.

As with domain name registrars, DNS service providers are highly resistant to efforts to hold them accountable for the services they provide to jihadists and other terrorists. As one example, we note that during the recent Israel-Hezbollah war, multiple cease-and-desist orders and the very real prospect of criminal prosecution were required before the DNS providers and domain name registrars of Hezbollah began the process of discontinuing services to the sites of that organization. This was despite the fact that those providers were operating in the United States, where Hezbollah is a proscribed organization.

Vulnerabilities
It is possible, though difficult, to get DNS service providers to discontinue services to their jihadist customers. To the extent that they persist in providing services, they are in a position to log and report on every individual who seeks to visit the domains for which they provide services, regardless of where those domains are hosted. The investigative or intelligence value of such information is highly dependent on the nature of the site in question.

1.1.5. Whois Proxy Service Providers

Nature of the Service
Whois proxy servers replace the real information about the registrant of any given domain name with the name of the proxy service provider.

Benefit(s) to Jihadists
This service allows the registrants of jihadist sites to hide their identity from the most superficial of investigations, without having to resort to providing fictitious information.

Vulnerabilities
To the extent to which registrants actually provide the proxy service with real contact information, they expose themselves to investigators who are in a position to compel the service provider to turn over their customers' records.

1.2. The Jihadists

1.2.1. The Enablers

Central to the online activities of the proponents of the global jihad is a relatively small number of individuals who create, host, administer or otherwise support the operation of jihadist websites. The alleged proliferation of jihadist websites over the past four years can, in fact, be traced back to the activity of this group of jihadists. Operating websites in support of the global jihad is an inherently risky activity, and so it is not surprising that we find these same individuals consistently appearing in other areas of jihadist—that is, terrorist—activity. Two examples, one from the pre-September 11 period, and one from the very recent past, suffice to illustrate this phenomenon.

Ziyad Khalil moved to the United States in the latter half of the 1990s as a student; he lived and attended university in Florida, and later in Missouri. His online activities included registering the first two domain names used by Hamas, and serving as the web master for the Al-Qaida-linked charities known as the "Islamic American Relief Agency" and the "Islamic Scandinavian Relief Agency", among others. He also represented the jihadist Committee for the Defense of Legitimate Rights, a Saudi exile organization long associated with the leaders of Al-Qaida. Finally, Khalil purchased the satellite phone used by Osama bin Laden to coordinate the 1998 Embassy bombings in Africa [6].

Our second example, Irhabi007, best represents the current generation of jihadi enablers. He began his online activities shortly after arriving in the United Kingdom with his father, who had a position with the Moroccan government's office of tourism and trade. He rose rapidly through the ranks of online jihadists, became a moderator of two jihadist forums (bulletin boards), and was an active participant in Al-Qaida's virtual intelligence network. The latter was formally created on the "Montada al-Ansar" forum in 2004, and continues to operate through various other forums to this day.

As his technical skills improved, Irhabi007 distinguished himself as a problem-solver. He came up with a number of novel solutions to the problem of distributing large multi-media files, which had until then stymied the propaganda efforts of Zarqawi's Tawhed wal Jihad/Al-Qaida in Iraq operation. Irhabi007 pioneered the use of stolen identities and credit cards for the purpose of paying for domain name registration and web hosting services, and he was involved in setting up, designing, or administering a number of prominent jihadist websites. He was ultimately arrested for his real world terrorist activities, involved as he was in a loose-knit network of jihadist terrorists in Europe, Canada, and the United States [7].

1.2.2. The Leadership

The jihadist leadership, whether it is the Al-Qaida "home office" represented by bin Laden and Zawahiri, or the leaders of affiliate organizations such as the Algerian GSPC or Al-Qaida in Iraq, actually provide very little visible guidance or logistical support for online activities in support of the jihad. Their role is primarily as the producers of content, written statements, magazines, and videos. The distribution of this content is left largely in the hands of the enablers and of the organization's rank and file, who each in their own way do a more than adequate job of ensuring that the material produced by the leadership gets distributed widely.

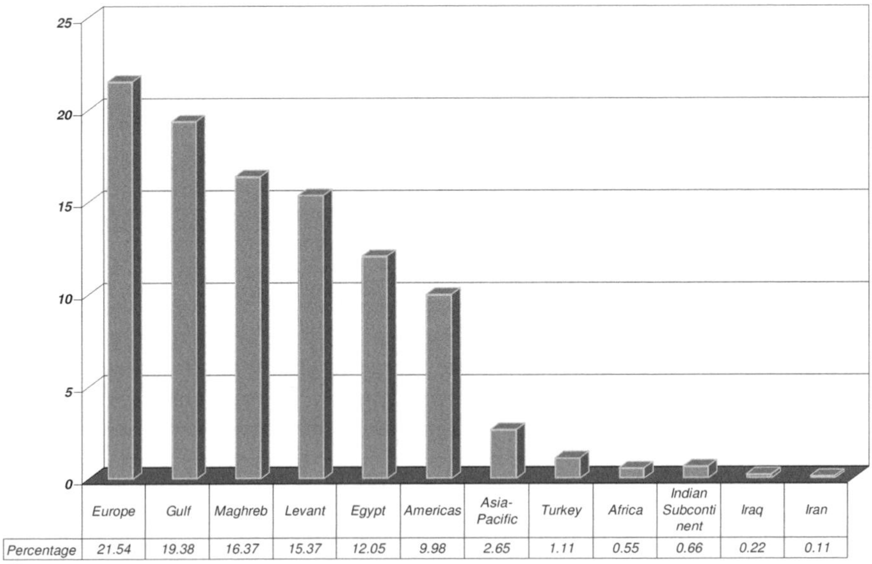

Figure 1. Geographic distribution (per country) of jihadists online.

1.2.3. The Rank and File

The jihadist rank and file, found mostly on the jihadi forums, are the most significant element of the global jihad as it manifests online. The ability of these foot soldiers of the jihad to interact with like-minded individuals globally serves to solidify their identification with the global jihad, and speeds the process of radicalization.

That jihadists gather online and are enabled to speak in their own voices, rather than acting merely as consumers of content provided by the leadership elements, provides investigators, intelligence operatives, and researchers with a wealth of opportunities. The situation allows us directly to observe and/or interact with the jihadists themselves. These opportunities remain largely unexploited, and if the reader comes away from this report with nothing else, it should be clear that it is imperative to take advantage of this opportunity while it still exists. The online landscape may change yet again, and not necessarily in a way that is to our benefit.

The author has been engaged for some time in actively monitoring jihadist forums, with the aim of identifying the approximate location of the participants and mapping out their geographic distribution.

2. Geographic Distribution of Arabic-Speaking Jihadis Online

Figure 1 illustrates the current geographic distribution of Arabic-speaking jihadists who are active online [8]. The predominance of European and Gulf Arabs is consistent with previous surveys we have conducted. The most significant trend is the rise in the number of jihadists in the Maghreb, Egypt and the Levantine countries. We are also seeing increasing numbers of jihadists in sub-Saharan Africa, though their numbers are relatively insignificant (the status of these individuals, in the greater scheme of things, is unknown to us).

Table 1. Rankings of regions based on IP addresses

Regional Distribution–All		Regional Distribution–Unique	
21.48	Europe	Europe	20.97
19.44	Gulf	Maghreb	20.14
16.37	Maghreb	Levant	16.16
15.37	Levant	Egypt	14.77
12.05	Egypt	Gulf	12.31
9.98	Americas	Americas	9.26
2.65	Asia-Pacific	Asia-Pacific	3.52
1.11	Turkey	Turkey	1.02
0.55	Africa	Africa	0.69
0.66	Indian Subcontinent	Indian Subcontinent	0.65
0.22	Iraq	Iraq	0.32
0.11	Iran	Iran	0.19

Before proceeding further in this discussion, we must take a moment to address the issue of unique IP addresses. When we acquired our data, roughly 30% of the IP addresses we gathered were not unique. There are two conflicting phenomena of which one needs to be aware. On the one hand, users of dial-up Internet connections will have a new IP address assigned to them each time they log in to the network. On the other hand, Internet service providers who require users to connect through proxy servers will have the effect of hiding many users behind a relatively small number of unique IP addresses, thus under-reporting the number of users in the country in question.

Table 1 compares the ranking of the regions or countries in our survey, based on all IP addresses versus only unique IP addresses. The reader will note that when we look only at unique IP addresses, the Gulf Arab states fall from second position to fifth. Table 2 provides an explanation for this, as we see Saudi Arabia drop from third place to eleventh. The reason is simple: the vast majority of Saudi Arabian web surfers, jihadi or otherwise, are funneled through a small number of proxy servers operated by the Kingdom.

In summary, counting unique IP addresses results in an under-reporting of Saudis in particular, while counting all IP addresses will account for particularly active users with relatively static IP addresses (e.g. broadband Internet users). We are trying to arrive at an understanding of the geographic distribution of current or future terrorists, and if we have to err, we would prefer to over- rather than under-count. Consequently, we use all the data, not just that gleaned from the unique IP addresses, in the remainder of this discussion.

Another question that arises when examining this data concerns how Internet availability in any given country relates to that country's ranking relative to the others. The answer—as far as we have been able to determine—is that if Internet access is available to any extent whatsoever, the jihadis will be found online. That is, they appear to be among the earliest adopters of Internet services in any country. Table 3, derived from our survey of the users of the "Mohajroon" forum, illustrates this point.

A more useful inquiry might be to compare the ranking of each country with the number of Muslims known to be living in that country. Figure 2 provides an example.

Table 2. Comparison of data – unique versus all IP addresses

Countries–All		**Countries–Unique**	
Percentage	Country	Country	Percentage
12.08	EG	MA	15.56
10.45	MA	EG	14.81
8.27	SA	PS	7.45
7.16	US	US	6.48
5.7	PS	DE	4.26
4.78	DZ	JO	3.33
3.76	AE	DZ	3.1
3.46	DE	UK	3.1
3.15	UK	YE	3.06
2.68	LB	IL	2.92
2.54	JO	SA	2.69
2.32	BE	KW	2.31
2.32	SY	FR	2.27
2.27	FR	BE	1.99
2.18	YE	CA	1.85
2.13	IL	AE	1.71
1.96	CA	OM	1.62
1.96	KW	ES	1.39
1.77	GR	IT	1.25
1.6	IT	LB	1.25

In 2006 Belgium accounted for 10.81% of the European jihadis in our survey, while France accounted for 10.55%. Belgium is believed to have a Muslim population of 450,000, while the Muslim population of France is estimated at between 4,000,000 and 6,000,000. Belgium's high concentration of jihadists among such a small population is likely to be a sign of future trouble, and we suspect that similar situations will be found in other countries.

Given the increasing difficulty that jihadists have in traveling internationally, it seems increasingly likely that jihadist terrorism will occur in countries with higher relative concentrations of jihadists, in light of the social nature of jihadist radicalization.

The predominance of Saudis among Gulf Arab jihadists (see Fig. 3) will not come as a surprise. However, the overwhelming number of Moroccans among Maghreb jihadis—outnumbering Algerians, for example, by more than two to one—should be cause for concern (see Fig. 4). Such a situation affects not only the Kingdom, but should also be of concern to authorities in European countries with large Moroccan populations.

From an investigative point of view, we will note here that the vast majority of the Moroccans in our survey were using static IP addresses on broadband (ADSL) lines. Thus it should be relatively easy to identify and investigate each of the individuals detected in our survey.

Table 3. Percentage (per country) of Internet users on the Mohajroon forum

Country	% of users	% of Internet Penetration
EG	13.22	7%
MA	9.06	12%
AE	5.62	36%
YE	5.07	1%
PS	4.89	5%
US	4.53	69%
DZ	3.99	3%
DE	3.8	59%
ES	3.62	39%
JO	3.62	11%
UK	3.26	63%
FR	3.08	43%
IL	3.08	45%
SY	2.72	4%
LB	2.54	13%
QA	2.54	21%
CA	2.36	68%
BH	1.99	21%
NL	1.99	66%
BE	1.81	49%

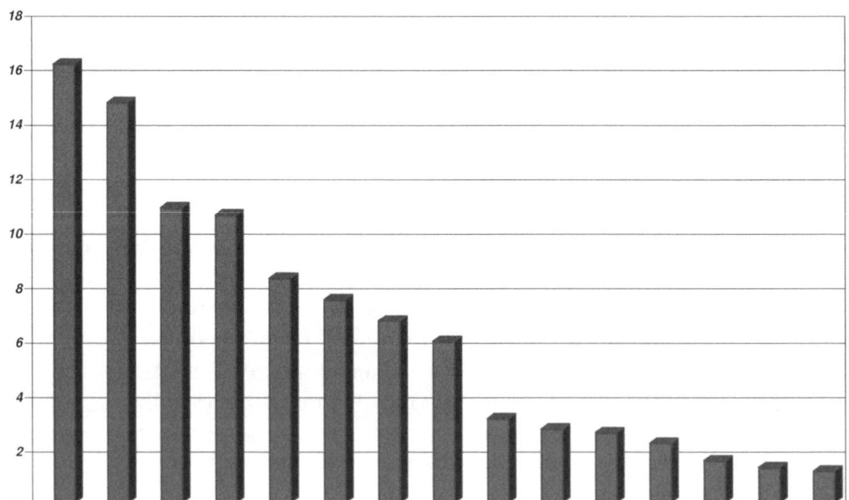

	DE	UK	BE	FR	GR	IT	NL	ES	SE	MC	DK	NO	CY	AT	CS
Percentage	16.09	14.67	10.81	10.55	8.24	7.46	6.69	5.92	3.09	2.7	2.57	2.19	1.54	1.29	1.16

Figure 2. Percentage per country (for top 15 European countries) of all Jihadis included in our survey.

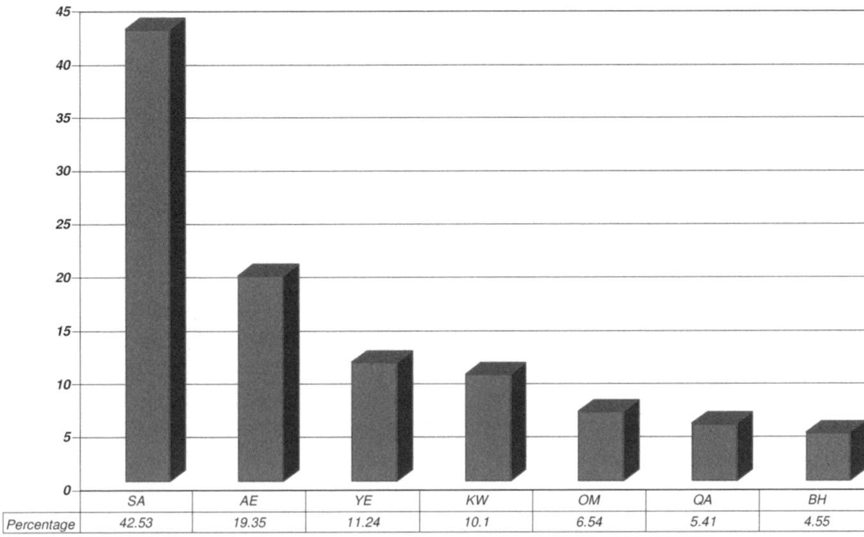

Figure 3. Gulf Arab states distribution of jihadis online.

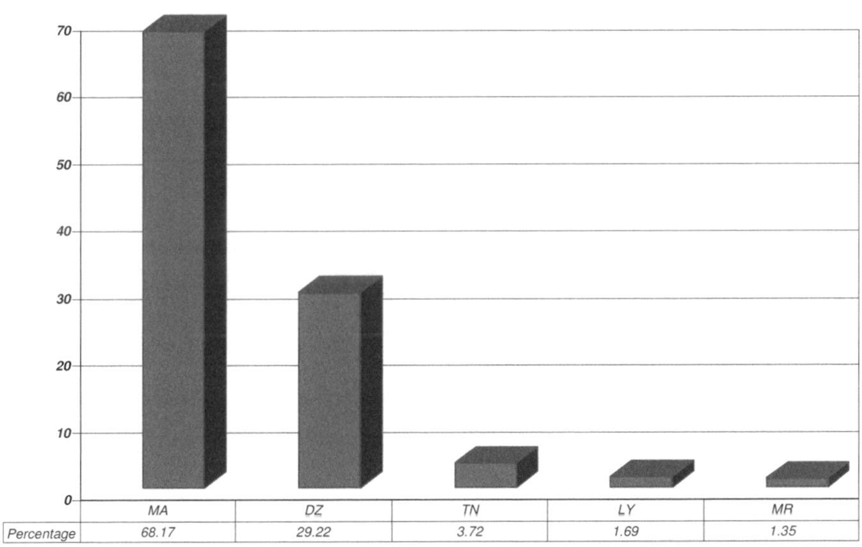

Figure 4. Maghreb distribution of jihadis online.

In the Levantine countries, we see primarily West Bank Palestinians being involved in jihadist online activity. However, during the course of individual website surveys conducted during 2006, we have observed a steady increase in the numbers of Syrians and Lebanese on the jihadist forums.

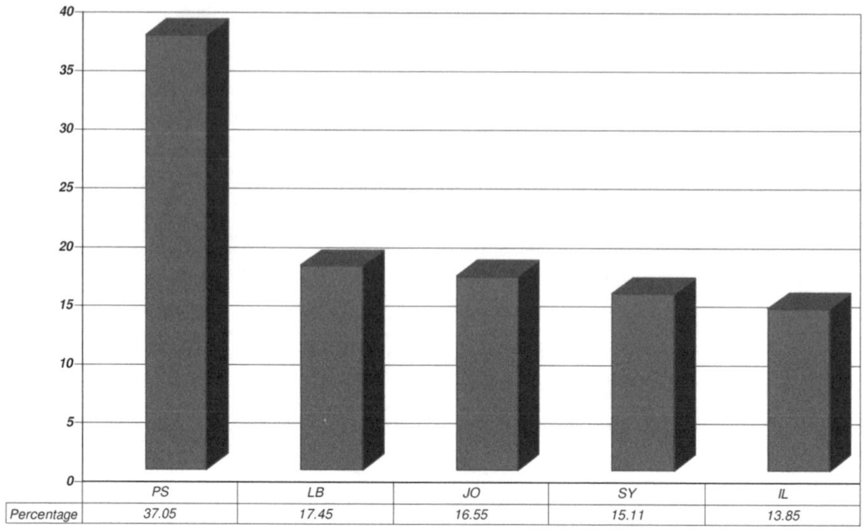

Figure 5. Levantine distribution of jihadis online.

3. Additional Opportunities for Counter-Measures and Data Acquisition

Our discussion of service providers and potential vulnerabilities noted a number of opportunities for acquiring data about jihadists by providing them with, or offering to provide them with, services they find necessary (e.g. web hosting). In addition, in our discussion of the jihadist rank and file, we noted how the jihadists' predilection for interactive bulletin-boards and forums makes it possible to observe, interact with, and infiltrate entire online communities of jihadists. Before concluding this chapter, we would like to point out a number of additional opportunities for counterterrorism.

3.1. File Sharing Services

As a result of the frequent removal of jihadist websites (as a consequence of their use in distributing controversial files), jihadists have taken to using online file sharing services. The now standard procedure for the distribution of jihadist multi-media files is to upload the files to multiple servers that allow anyone to do this. The links to the files are then posted on various forums, and perhaps also to static websites that are created for no purpose other than promoting the files in question.

While not quite as simple as setting up a web hosting operation, offering such a service to jihadists is still worthwhile, particularly when the files being uploaded are official communications from the jihadist leadership (e.g. bin Laden).

3.2. Open Proxy Servers

Proxy servers are often presented in the mainstream media as a "solution" to the security concerns of jihadists online. Through our considerable experience with investigating and successfully tracking down technologically savvy jihadists online, we note the following:

- Jihadists lack discipline. Even when they understand the risks, and present themselves to their cohorts as experts in computer security, they will not *consistently* make use of security measures such as proxy servers. There is an assortment of ways in which one may exploit such inconsistent behavior in order to discover the location of the subject.
- Open proxy servers allow any individual to relay their online actions through the proxy, masking the location of the user in the event that their destination (e.g. a jihadi forum) is under surveillance. If the point of observation is the proxy server itself, however, the user is exposed. Setting up and promoting an open proxy server is a relatively trivial matter.

4. Conclusions and Recommendations

The Internet is the glue that holds the global jihad together. Arguably, the global jihad only exists as such on the Internet itself. Internet use by jihadists provides ample opportunity for people working in counterterrorism—both to better understand the nature of the jihadists, and, more importantly, to locate them, identify them, win their confidence, or undermine their sense of security and commitment to their cause.

Jihadist websites can be allowed to stay online so that they can be monitored, or they may be shut down in order to interdict jihadists' communications and disrupt their efforts to build their community. Whether to do one or the other depends on a number of factors. First and foremost is the availability of resources to actively monitor any given website. The removal of a website from the Internet causes the current administrators to expose themselves as they move their site to a new location, and has the effect of encouraging other jihadists to expose themselves as well, through the effort to replace the removed site with a new one.

A dynamic and aggressive approach is best. The essence of counter-insurgency is the same, whether it is online or in the real world: it is not necessary, and rarely desirable, to try to exercise complete control over the contested territory. What *is* necessary, however, is to prevent the insurgents from enjoying free and unfettered use of that territory.

References

[1] Weimann, G. *Terror on the Internet: The New Arena, The New Challenges.* Washington DC: the United States Institute of Peace, 2006.
[2] Sageman, M. *Understanding Terror Networks.* Philadelphia: University of Pennsylvania Press, 2004.
[3] Roy, O. *Globalised Islam.* London: C. Hurst & Co, 2004.
[4] Paz, R. "Reading Their Lips: The Credibility of Jihadi Web Sites in Arabic as a Source for Information," ICT web site, 2006.
[5] Weisburd, A. "What you can do," Internet Haganah web site, 2004.
[6] Weisburd, A. "The Associations, Domains, and Web Sites of the Designated Terrorist Sa'ad Rashed Mohammed al-Fagih," Society for Internet Research web site, 2005.
[7] Weisburd, A. "The Decline and Fall of Irhabi007", Internet Haganah web site, 2006.
[8] Weisburd, A. & Kovler, A., "Global Distribution of Jihadis Online – 2nd Quarter Findings," pending publication, Society for Internet Research web site, 2006.

The Seductive Web: Technology as a Tool for Persuasion

The Seductive Web: Technology as a Tool for Persuasion

A Relativist Methodology for Website Design

Carlos A. M. DUARTE
Ph.D.; Associated Professor;
IADE - Escola Superior de Design and researcher member of UNIDCOM/IADE
Av. Dom Carlos I, nº 4, 1200-649 Lisbon, Portugal; cduarte@iade.pt

Abstract. This chapter is about design, here defined as a process that changes the interactions in a system so that the structure of the system itself is ultimately changed. Thus this chapter is about the capacity of design to bring about a new or an emergent behavior within a system. The present concept of design, and the difficulty arising from its methodology, do not allow for taking control of the various endogenous and exogenous components inherent to the working process in an objective way. Taking into account this drawback, the chapter presents a proposal for a methodology that will be an innovative concept of design. This methodology stems from the Gestalt approach, and is based on the method of measuring the amount of relativistic information in a system. Such an approach lends itself to implementation in the field of website design, with the aim being the development of persuasive tools.

Keywords. Design, seductive web, persuasive methodologies, perception, interaction, relativist information, emerging behaviors, accessibility.

Introduction

What is design? What is the "seductive web"? During the twentieth century, scientific interest in the design process grew rapidly, and it continues to do so. We have seen an amazing transformation of our society, such that the process of consumer goods being designed and created by hand, by skilled craftsmen, has been superseded by mass-production using new materials and technologies. Design is everywhere. Design history is not merely a chronology of objects and their forms, but is a record of the changing forms of life. The relation of human beings to the objects they produce and use reflects a large segment of cultural history, and this was especially true in the last century [1].

Design that gets people to act, perhaps to come to an event, give to a cause, buy a product or service, or ask for information, is an art in itself. Doing this successfully is part science, part intuition, and part experience. As Blum states, "You can't learn intuition or get experience from a magazine article, but I can share with you what I've learned about the science." [2]

In fact, designers and researchers must create tools for capturing and analyzing people's behaviors and perceptions, and illuminating what they know, in ways that transcend the barriers of time, place and language. The world of research compared with the universe of design is vast, and the depth of technique is deep and wide-ranging. The key is to choose wisely those techniques that will inform in ways that act as the catalyst for great design work.

For designers, visual techniques represent, first and foremost, an information-processing task; however, they are a lot more besides. For if both designers and common humanity are capable of knowing what is where in the world, the human brain must somehow be capable of representing this information, in all its profusion of color and form, beauty, motion, and detail. The study of vision must therefore include two aspects. The first concerns the way in which we extract from images those various aspects of the world that are useful to us. The second is an inquiry into the nature of the internal representations by which we capture this information, and thus make it available as a basis for decisions about our thoughts and actions [3].

Our dreams may sometimes might be realized in everyday life, and it is precisely through the management and manipulation of desire that we may come to obtain the objects of our desire. This possibility cultivates longing and imagining, as if these qualities were themselves a commodity. In design terms, the process is driven by the question, "What do the people want?"

In the field of design today, a seductive website is an essential way of understanding people's behavior in a rapidly changing marketplace. The relevant technology includes digital tools (such as digital photos and video, or digital audio), text messaging, e-mail, remote viewing of desktops, and blogging. Indeed, the present concept of design and the difficulties arising from its methodology do not foster control of the various endogenous and exogenous components inherent to the working process. Ideally, a more objective approach would provide useful principles for how to create persuasive designs for interactive websites that work, and that are built with high quality web design and functionality. In addition, such websites would need to be legally compliant and optimized for search engine registration.

1. Scope and Rationale of this Chapter

In line with the relativistic approach to studying the interaction between design and seductive web, this chapter is divided into three parts. In the first we begin with a brief look at some questions on the subject of seductive web, and examine concepts such as web design, Internet usability, persuasive design, graphic design, ergonomics or human factors, and socio-economics factors. The topic of website accessibility for users with disabilities is also addressed.

The second part is somewhat unorthodox, and consists of a set of points that are designed to help the reader understand the systems approach. This representational framework is proposed as a way of understanding the overall process of visual perception, as it is understood in the field of design. Systems theory (as applied in several diverse fields) provides an understanding of those processes which alter the interactions between the elements of a system, in such a manner that both a new behavior and a new structure emerge within the system.

The third part of this chapter describes informally, but in some detail, the theory of information. Information theory regards the communicative process as one of interactive changes in encoded information; such changes occur during the delivery of messages. It also explores the possibility of measuring the amount of information in a message, so as to enable a subsequent analysis of the meaning of the message. It includes the study of aspects related to the information involved and to respective decision-making.

The ways in which our senses interpret received messages provide relativist information. Designers must understand each system as an entity that resides in its elements and in the interactions of those elements; we know that a system can assume several different structures if the interactions are altered or if new interactions are introduced. It is also known that a variation in the quantity of information can result in modifications to the structure of a system. An illustration of this principle would be the emergence of new architectural or design styles whenever the quantity of relativist information undergoes appreciable alterations.

Each of the three parts of this chapter presents a brief review of the relevant aspects of how design and websites might be made more appealing and persuasive, simply by leveraging certain social rules and dynamics. But how is it possible to describe this process of altering the interactions between elements of a system as far as the creation of a seductive web design is concerned?

2. What is "The Seductive Web"?

While there are a number of methods and strategies for when, where and how to conduct web design, the primary tools have always been the eyes and ears of the designer. Normally, the designer is still limited largely to the use of his or her own senses when it comes to understanding how people interact with changing and prolific news tools.

In simplified terms, only recently have fields emerged which study concepts about seductive webs, and what factors make websites more appealing and persuasive. The next step in such studies is to examine if those factors also work in the human-computer domain. The focus is on the design, research, and analysis of interactive computing products that are created for the purpose of changing people's attitudes or behavior. This provides useful principles on how to think about creating persuasive design. Fogg pioneered the field's next frontier with his work on "captology", namely computers as persuasive technologies [4].

Fogg examines many principles of persuasion, which are useful as a systematic approach to thinking about different ways to make our own designs more persuasive. Mostly, these principles are not detailed enough to serve as prescriptive design guidelines. Nonetheless, Fogg lists six robust factors that lead to compliance:

- Liking: people tend to be more easily persuaded by people they like;
- Consensus: people tend to be influenced by group opinion;
- Reciprocation: people feel obligated to repay favors or gifts;
- Consistency: people want to appear consistent to others;

- Authority: people tend to believe authorities;
- Scarcity: people find scarce things more desirable.

Persuasion in itself is obviously not new. From Cicero's oratory to modern television commercials, communicators have long tried to persuade their audiences. What is different, these days, is that websites and other computerized designs are going beyond one-way rhetoric and have become interactive. Actively doing something is much more engaging, and thus potentially more compelling and persuasive, than passively receiving messages.

Fogg has researched captology for many years, and he presents many interesting examples of persuasive design, both from websites and from other forms of applied information [4].

Further frameworks for interpersonal persuasion exist, such as those described by Mbakwe and Cunliffe [5]. These authors have studied and developed models of hypermedia seduction so as to provide developers and researchers with a holistic perspective of this process. Their theory relates to the orchestration of specific actions and elements that facilitate the onset of desired changes within a user's behavior. Emphasis is placed on behavioral and emotional aspects of the user. The concepts presented suggest that seduction may enhance the formation of more meaningful user relationships.

Similar proposals come from the field of digital ethnography. This field provides an essential way of understanding people's behavior in a rapidly changing market. It involves the culture of design as far as it relates to the economic and ethical spheres; at the same time, digital ethnography considers technology as an instrument in the creative process, and not as the key of that process [6].

One type of persuasive technology is clearly ethical: when users ask to be persuaded. In many cases, you know something is good for you, but you may not have the willpower to meet your goals. Interactive systems could track your progress and motivate you to do better.

2.1 Human Factors: Website Access for People with or without Disabilities

When we examine several social groups in terms of the process of persuasion, we must consider that all human beings are, at some point in life, the carriers of some kind of disability. A design that takes into account multidisciplinary characteristics demands the integration of knowledge proceeding from diverse origins. Through the identification, coordination and association of those elements, the ergonomics of the web allow for the creation of a theoretical model. This model will make possible the systematic analysis of the seductive web, and the application of the design process to this [7].

Human factors comprise a very established field, with a well-documented foundation that remains the same decade after decade. Once you have learned the field's basic principles and practical skills, you can apply them to any new technology that comes along. Human factors are concerned with human behavior, and Homo Sapiens does not have a Moore's Law to change its capabilities every eighteen months.

Because of the robustness of human factors, many of the best books in the field were written in the late 1980s and early 1990s, a period in which usability researchers crystallized the formerly academic discipline and made it practical. The two main legacies were:

- 1975 to 1983: following an early emphasis on military design problems, the primary focus in human factors shifted to human-computer interaction.
- 1989 to 1994: usability engineering is victorious over the earlier cumbersome methodologies. Usability inspection methods become codified elements of the usability toolkit.

The growth of Web usability from 1994 to 2000 had a huge practical impact, but did not involve much new theory. We employed the methods we already knew, and tested websites to document the many stupidities that had dominated the early days of "killer design".

Web accessibility is about making websites accessible to all Internet users, both disabled and non-disabled, regardless of what browsing technology they are using. There are a number of myths and questions around accessibility [8], such as:

- "Do I need a text-only site?"
- "Should I have two sites, one with accessibility and one without?"
- "Accessibility is expensive, and good design and accessibility don't go together."
- "Blind or otherwise disabled people don't use the Internet."

Basically, one needs a single well-compiled site, developed by an accessibility specialist. There are substantial benefits to accessibility and to having the help of a professional. Some of these advantages are as follows:

- the website owner will not incur legal costs if taken to court.
- the website will be easier to manage via a style-sheet set.
- the website will be compatible with new technology as it evolves, such as PDA's and mobile phones.
- the website will appear higher up in the search engines (the more confident a search engine is of what a website is about, the higher it places that website in the search rankings.)
- the website's download time will significantly improve; accessibility improves download time.
- the usability of the website will be enhanced, and improved usability has been known to increase traffic by up to 100%. One thus gains good publicity from site promotion surrounding accessibility.

The European Union Council approved a Resolution entitled *"eAccessibility - Improving Access of People with Disabilities to the Knowledge Based Society"* [9]. In 1995, the Disability Discrimination Act was passed. In 1999, Part III of the Act, which refers to service providers, came into force. It refers to the provision of goods, facilities and services, and The Code of Practice specifically mentions websites. In 2002, the Code of Practice for Part III of the Act was published, and specifically mentions that accessible websites are now a legal requirement:

- "The Act makes it unlawful for a service provider to discriminate against a disabled person by refusing to provide any service which it provides to members of the public."
- "For people with visual impairments, the range of auxiliary aids or services which it might be reasonable to provide to ensure that services are accessible might include ... accessible websites."
- "For people with hearing disabilities, the range of auxiliary aids or services which it might be reasonable to provide to ensure that services are accessible might include ... accessible websites."

Since October 1st 2004, it has been a legal requirement that all websites adhere to certain minimum standards. As of that date, the majority of websites have been in breach of the law.

What do we need to do to comply? The W3C, The World Wide Web Consortium, is the Internet governing body, and its web accessibility guidelines can be found on its website. It is widely believed that whenever the first case makes it to court, the W3C accessibility guidelines will be used to assess the accessibility of the website in question, and thus will ultimately determine the outcome of the case. The W3C offers three different levels of compliance. Priority 1 guidelines (which *must* be satisfied, according to the W3C) will almost certainly have to be adhered to. Priority 2 guidelines (which *should* be satisfied, and are the EU recommended level of compliance), or some part thereof, will probably need to be adhered to as well.

In terms of accessibility for non-disabled people, for obvious reasons, the more people who can view a site, the more traffic the site is likely to get. In these early days of the twenty-first century, websites can already be viewed in a variety of different formats, each one facing different accessibility issues. Not everyone is using the latest version of Internet Explorer, with all the plug-ins and programs that may be required to for a specific website.

A good website must be able to function with all different browsing technologies. It must therefore provide alternatives to:

- Images - in the form of ALT text
- JavaScript - through the <noscript> tag
- Flash - with HTML equivalents
- Audio and video - by providing written transcripts.

By separating structure and presentation, a website can be flexible enough to face the future of the Internet, namely PDAs, mobile phones, in-car browsers, WebTV and ISOOpx screens. It should be noted that the *structure* of a document refers to how the document is organized; points of interest include navigational menu items, headings, sub-headings, paragraphs, lists, and links. The *presentation* of a document refers to the way in which the words and images are presented to the end user.

To return to the topic of website accessibility for people with disabilities, we might ask what types of issues the disabled face. The following section examines specific challenges faced by subgroups of the disabled population.

Visually impaired users: Internet users who have no sight at all utilize a screen reader, which reads the content of the web page (or rather the HTML code of the page) to them out loud. These machines sift through the HTML code, and the technology deciphers what needs to be read aloud and what should be ignored.

Users with partial or poor sight need to be able to enlarge the text on web pages if they are to take full advantage of the Internet. Website owners should check if their website allows such users to achieve this:

- Internet Explorer: View > Text size
- Netscape: Edit > Preferences > Appearance > Fonts
- Opera: File > Preferences > Fonts > Minimum font size (pixels)

If a site is accessible to this group of users, then the size of the text throughout the page will increase. Users with poor vision might also use a screen magnifier.

Color blindness: It is estimated that one in 12 men and one in 200 women have some form of color blindness. This should be kept in mind with certain graphics or text colors.

Deaf users: Deaf users are able to access the Internet in much the same way as able-bodied people, with one key exception, namely audio content. If it is a vital function of the website for people to be able to hear a message, then be sure to provide a written transcript.

Keyboard or voice-only users: Some website users do not have access to a mouse when browsing the Internet. Try putting yourself in their position by navigating a website using only the tab, shift-tab, and the return keys.

Epileptic users: Epileptic users must always be careful to avoid seeing flickering between 2 and 55 Hz.

Website accessibility checklist

		Yes	No
1	Check informational images for alternative text		
2	Check decorative images for alternative text		
3	Listen to video or audio content with volume turned off		
4	Check forms are accessible		
5	Check that text can be resized		
6	Check your website in a Lynx browser		
7	Check you can access all areas of your website without a mouse		
8	Check there's a site map		
9	Ensure link text makes sense out of context		
10	Check web pages with a specialist company		

2.2. Website Usability

On the web, usability is a necessary condition for survival. If a website is difficult to use, people leave. If the homepage fails to state clearly what a company offers and what users can do on the site, people leave. If users get lost on a website, they leave. If a website's information is hard to read, or doesn't answer users' key questions, they leave. The pattern here is clear. There is no such thing as a user reading a website manual, or otherwise spending much time in trying to figure out an interface. There are plenty of other websites available; leaving is the first line of defense when users encounter a difficulty.

Future Now's website [10] describes how is possible to create a seductive and persuasive website, in six phases. This reference is summarized here:

Phase One. The "uncovering" phase includes a needs analysis and persona development. Background information is collected about objectives, strategies, customers, and competitors and their environments. This provides the foundation of the project by documenting the particulars of the business, its customers and the competitive landscape. At the end of the uncovering process, personas are developed to provide insight into customers' buying and sales processes; this builds an understanding of customers' individual needs, wants and desires. On this basis, objectives can be mapped and strategies developed, and keywords and key phrases can be researched. We can begin the process of planning scenarios that will persuade our customers to take action. This first phase will also aid in answering specific and implicit questions from visitors.

Phase Two. Wireframe creation: wireframing methodology and tools are critical to planning persuasive scenarios, which to customers will be a click-though process. The wireframe is the culmination of the time spent on uncovering and persona development. The wireframe is a detailed description of the strategy which a website will take to persuade each persona to navigate in the website, according to carefully designed persuasion scenarios.

The wireframe helps to define the path or "click through pattern" each persona will take. This defines the objective of each page, and decides which web page each persona will see. The wireframe is an enhanced flow-chart, without graphical images, which demonstrates every click-through possibility for each of the different personas. It is about defining the entry points to, and exit points from, each page. It answers questions such as "What actions can be taken here?" and "What is the user's state of experience as she enters this page and when she leaves it?" It focuses on the click-through sequence and user-experience flow of the site. Nothing is more important in planning for conversions.

Phase Three. Copywriting may be carried out via in-house copywriting resources; alternatively, if one sub-contracts a copywriter, the wireframe will provide the tactical strategy on a page-by-page, element-by-element basis. The copywriter will know the responsibility of each piece of copy which personas will be reading, and will have specific directions for what to present to each persona, and how to present it in the copy. *Future Now's* website provides an online copywriting course for any copywriter wanting to learn more about writing for Persuasion Architecture [11].

Phase Four. Layout and design. This phase ensures that the website has a persuasive atmosphere and an appropriate look and feel, using scientific principles such as eye-tracking and color theory. A CSS template is also used in this phase, and website development is attended to in the following order:

- *Layout/Storyboard*: this rough sketch contains a mock-up of the visual aspects of the screen. Several elements will be considered: branding, navigation, page titles, header graphics, and footers, which include copyrights and privacy notices. After key elements are identified, each is assigned a priority so that developers know which elements need the most emphasis, and where these elements belong.
- *Design in grayscale*: this is the first step in designing for aesthetics. This mock-up will be in grayscale, so that one can assess how the composition of the piece works without the emotional influence of color. If it looks good in black and white, the design can probably come alive with the judicious use of color.
- *Design with color application*: at this stage the focus is finally on the mood and impact which color can create. Better use of color means a better website. A template can carry a color scheme from the landing page to the exit page, creating a cohesive look and feel for a site.
- *Master template (html) and style sheet*: once the color mock-up is completed, the image can be turned into an HTML document. Considerations addressed are download time, compatibility with multiple browsers, use of cascading style sheets for simplicity, laying out of tables with an understanding of how search engines spider a website, and choosing fonts with an eye toward easy screen reading.

Phase Five. Prototyping and implementation: based on the strategic plan created earlier, all of the phases are implemented and the website creation is completed. The site is launched, including all the individual pages, required back-end databases and middle-tier business logic. The correct functionality of the site is ensured.

Phase Six. Optimization: the wireframe lays the groundwork for tracking metrics and optimizing the site, based on those metrics which create an environment of accountability and a clear path to improving and optimizing results. Persuasion Architecture provides a foundational methodology for modeling customer behavior, and guides one to optimize proprietary changes and improvements on the site based on metrics, not on guesses or hunches.

Design or technology as a tool for persuasion is a distinctive mental activity, with a unique objective: to create a persuasive factor for inclusion in objects, artifacts or digital tools. Jakob Nielsen's website [12] includes a rank-ordered list of fifty-one design elements that help or harm a website's credibility. The four most harmful elements are:

- Links that don't work (-1.3 on a 7-point credibility scale)
- Content that is rarely updated (-1.7)
- Links to sites that lack credibility (-1.8)

- Advertisements that are indistinguishable from content (-2.1).

Note that poor linking accounts for two of the four most harmful design elements. Most web designers probably don't even have a formal linking strategy, but links are the Web's foundation, and they are well worth some consideration. Good links, both incoming and outgoing, boost credibility almost as much as poor links lower it.

Another main credibility guideline is the principle of real-world feel; a website's credibility increases when it shows the people or organization behind the site. This principle was also found in usability testing of e-commerce sites. For example, users were more likely to buy from a coffee site that included a photo of the company's coffee roaster. Seeing the equipment increased people's confidence that the company actually had coffee on hand, and would send a package if they placed an order [4]. Web persuasion is where the action is today, and persuasive design will have an even bigger impact when it becomes more tightly integrated with our environment. This is a clear trend as we abandon the personal computer as the center of interaction and the key view port on the Internet.

As we have seen, design is a highly complex and sophisticated skill and process. Many authors have tried to chart a route through the process, from beginning to end. The common idea behind all these models of the design process is that it consists of a sequence of distinct and identifiable activities which occur in some predictable and identifiably logical order. Design focuses on people's behaviors, capabilities and limitations in their interactions with the designed world, especially that portion of it represented by technology. It strives to understand humans as a function of their motor control, perceptual, cognitive and emotional components, as these factors influence people's use and abuse of things both product- and process-related.

The above approach seems, at first sight, to be quite a sensible way of analyzing design. Logically, it seems that the designer must do a number of things in order to progress from the first stages of getting a problem to the final stages of defining a solution. Unfortunately, these assumptions turn out to be rather rash [13].

3. The Systems Approach

This section illustrates an application of the systems approach to design, which gives us a method by which we can:

- check that what we want to do is worth doing;
- find out which things are feasible;
- get things arranged properly;
- use resources efficiently;
- learn from both success and failure.

The systems model provides vivid examples and fine details of the way the world works today. These give rise to new visions, which are embodied in planning frameworks and conceptual sketches of a future worth striving to attain. For example, we can articulate all the steps in a process, build decision-making models that inform how choices are made, and identify hierarchical models of consumer and user priorities. In a practical sense, descriptive models are only useful if they help people to see the world in a new way, or to manage complexity more effectively. Opportunities exist for improving the research process by allowing consumers to use the tools and technologies that already exist in our lives, so as to capture people's perceptions, processes, behaviors and interactions.

The difficulty of a task is driven by its complexity. Effective communication consists of information transfers that take place within a context that both parties understand. As already noted, design can be understood as a process that carries out an alteration of the interactions between the elements of a system. A system may exist with various structures; it may also pass from one structure to another when the interactions change, or when the total number of interactions changes relative to the number of elements.

In fact, for the same number of elements, when the number of interactions increases, the number of elements connected to each other also increases. It is thus possible to build a graph in which the ordinates show the size of the larger agglomerate on the Y axis, and on the X axis the ratio between the number of interactions and the number of elements (Figure 1). In the transition phase, one can establish from the vertical line indicated on the graph that a new behavior has emerged in the system.

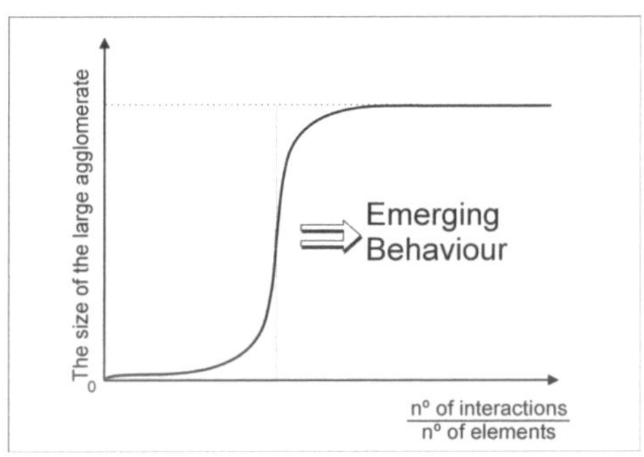

Figure 1. Analytic interpretation of emerging behavior

The above description is analytic. It outlines a system through its number of elements and interactions. Such interactions can also determine periodical patterns and repeated cycles, which could be called "rhythms"[1].

There is, however, another way of describing a system. This is the Gestalt approach, which assumes that examining the separate parts of a system can never provide a real understanding of how the system functions in its integrity. The Gestalt approach is global, does not analyze all details, and it studies the system's interactions.

---

[1] The world economic depression of the 1980s witnessed a growing confrontation between traditional functionalism and new design principles. The new principles of design rejected the old models of design-simplicity, reduction to pure function, and the definition of industrial mass production. Since the onset of the world economic depression there has also been a renewed worldwide interest in the study of long-term macroeconomic rhythms, most commonly referred to as long waves or Kondratieff waves [14]. In recent years, the theory of long waves has become a very useful tool to construct future scenarios, not only regarding the future of technology, but also the future of human reasoning (morals, ethics, and common values).

It is relativist, as the interpretation always depends on who is making the observations. It is also based on perception, in that it depends on the messages received by human senses.

There is some equivalence between the analytic description and the Gestalt description; they can, in fact, be seen as essentially the same thing. There is equivalence between, on the one hand, the structural description of a system via the number of its elements and their interactions, and, on the other hand, a description of the amount of relativist information contained in a system.

In one view, offered by the Gestalt approach, it is hypothesized that recognition of an entire scene is performed by a parallel, one-step process. To examine this hypothesis, a visualization of how a person views such an image is particularly helpful [15].

A number of diverse methodologies outlining techniques for human-computer interaction design have emerged since the rise of the field in the 1980s. Most design methodologies stem from a model of how users, designers, and technical systems interact. Early methodologies, for example, treated users' cognitive processes as predictable and quantifiable, and encouraged design practitioners to look to cognitive science in areas such as memory and attention when designing user interfaces. Modern models tend to focus on constant feedback and conversation between users, designers, and engineers; they also push for technical systems to be wrapped around the types of experiences users want to have, rather than wrapping user experience around a closed conceptual system.

4. Theory of relativistic information

Perception is one of the oldest fields within scientific psychology, and there are correspondingly many theories about its underlying processes. The oldest quantitative law in psychology is the Weber-Fechner law, which quantifies the relationship between the intensity of physical stimuli and their perceptual effects. It was the study of perception that gave rise to the Gestalt school of psychology, with its emphasis on holistic approaches.

Designers need to consider where and how to integrate the theory of information so as to create more appealing and persuasive interactive products. The theory of information values the communicative process as an interactive change of codes in the delivery of messages. It explores the possibility of measuring the amount of information in a message for a subsequent analysis of its meaning. It also comprises a study of aspects related to information, and to respective decision-making. The interpretation that our senses make of received messages is relativist information.

People have long attempted to understand the nature of information, or, more specifically, to understand what kinds of information propagate a specific structure. Simply put, we want to know what group of characteristics leads us to a desired result. Let us start by defining information, as follows: Information is the entity that measures the difference between knowing and not knowing; between perceiving and not perceiving; or even the difference between several possibilities of the occurrence of a future event, and knowing which specific occurrence will take place.

On the other hand, we also have to contend with the uncertainty of the actual information, which can be incomplete, inaccurate, fragmented, untrustworthy, vague, contradictory, or otherwise defective. Thus, the degree of knowledge that we have of any system is maximum if the uncertainty of it is zero. However, this leads us to another field of study, approaching the definition of uncertainty and the mathematical concepts of fuzzy measurement, and also to concepts related to complexity.

In this manner, we may analyze the possibility of quantifying the information received. Rodrigues [16] states that the measurement we make of the amount of information we receive via our senses is itself based on perception. He gives two examples, from the work of the Russian painter Kasimir Malevich[2], founder of Suprematism, who painted a black quadrilateral and another white one over white. Malevich had the necessary geniality to show in both these works the "supremacy of the pure sentiment or of the perception of the painting". In other words, the painting is merely a construction of colors in a bi-dimensional space.

When we ask what is present in both pictures, the immediate answer is: a black quadrilateral and another white one. However, we are only seeing a white texture in the first painting, and a texture with an almost unrecognizable stripe in the second. It is actually despite their textures that we attribute more information to the black and quadrilaterals, because these shapes induce in us certain feelings, which in turn generate action.

We also know that the rarer an event, the more disturbances it can cause within human structures [18, 19]. Thus we may conclude that the less probable an event, or the shorter its frequency, the more intense is the perception that we have concerning that event, and thus the larger will be the quantity of information which we store about it.

It is the supremacy of our perception over our feelings, and the supremacy of the amount of information that we measure in an event over its frequency, that influences our behavior, and that induces great changes within systems [18, 19]. It is not surprising that the structures created by man in the most diversified systems are fragile and precarious. They are structures which are unduly influenced by unpredictable events, thus becoming responsible for generating stronger and longer-lasting perceptions. We must remember that the structure of a system will only be compatible with an amount of information which is appropriate to it, because otherwise that structure would collapse, and that system would have to organize itself into another structure.

A system that contains a great amount of information is said to be a highly disorganized structure, because whoever observes the system brings a great amount of ignorance to the interpretation thereof. In contrast, a well-organized structure is said to have a small amount of information. There are, however, factors that contribute to decrease the amount of information in the structures. These are decisive factors for the progress of technology.

Marchetti [17] published a study in which he reveals the single most important factor behind great economic change, namely creative innovation. This is the decisive factor in terms of providing better structures for the development of science, art and human culture. As such, innovation incites a strong feeling, which enables the emergence of new structures, renovated by creativity and responsible for the cohesion

---

[2] Kasimir Malevich (1878-1935) affirmed that the reality of art depends completely on the effect that colour has on the scene. The painted picture is not subjected to the real world. It is in and of itself a real fact, as concrete as all the other objects that surround us. This theory signifies that the painted object does not imitate anything; it exists just as natural objects do.

of the system. These renewed structures also ensure individuality and enable alternative structures.

Creativity is thus responsible for the renewing of systems; it is a factor of reform and innovation, which brings about the emergence of new structures. Innovation is, as a result, quantifiable, having as a basis the parameter of the group of messages captured by our senses. This may result in the calculation of an algorithm of the amount of information received. This method was developed for calculating the amount of information which events provoke in us. It has already been tested in several fields, in studies related to the loss of cohesion of a society in the presence of effects related to widespread disease, or to casualties of an army in combat. Other studies relate to the detection of defects within the leather industry, and the determination of the distribution of filaments in the paper industry.

To calculate the amount of relativist information, we start by adding all the signals deriving from the system. According to Shannon, this amount of absolute information (S) is given as:

$$S = \sum_{i} -p_i \log p_i \quad (i = 1, ..., n) \tag{1}$$

in which p_i is the probability of the occurrence of message i.

In Eq. (1), all the contributions of all messages together account for the amount of absolute information. If, however, each observer j gives different weights to the various messages i, these weights being adjusted in the interval [0,1], and designated by G_{ji} (the weight given by each observer j to message i), the amount of information understood by each observer is given as H_j:

$$H_j = \sum_{i} -G_{i,j} p_i \log p_i \quad (i = 1, ..., n) \tag{2}$$

which represents an amount of relativist information. If observer j chooses messages k for which $G_{k,j} = 1$, then the amount of relativist information (previously H_j) can now be represented as H_k:

$$H_k = \sum_{k} -p_k \log p_k \tag{3}$$

From now on, the above formula will be represented as H, to simplify the notation. This symbol represents the amount of relativist information available for the system, and it takes into account the subjective perception of the observer. In fact, one could reach the same result if we bear in mind that an event with the probability p of occurring induces an amount of information as follows:

$$\log \frac{1}{p} \tag{4}$$

If there is σ of those events, the amount of information created is:

$$\sigma \log \frac{1}{p} \tag{5}$$

So that the amount of information does not enter the field of large numbers, one can normalize the previous operation dividing it by the total number of events N:

$$\frac{\sigma}{N}\log\frac{1}{p} \tag{6}$$

But $\frac{\sigma}{N}$ is the probability of occurrence of the chosen event, and it calculates the relevant amount of information. Thus Eq. (6) can be written as:

$$H = -p\log p \tag{7}$$

This equation yields the total amount of relativist information available (H), because it depends on the choice the observer made between messages. In addition, it is a special case of Eq. (3), which is deduced from the amount of information present, and which is stated as absolute, following Shannon's theory.

From Eq. (7) it is apparent that when a structure exists before a group of interactions of its elements takes place, with the probability of p occurring, the structure reaches the value of *l/e,* and a new structure emerges. A new behavior is said to have emerged (Figure 2).

Comparing Figure 2 with Figure 1, it is possible to affirm that there is equivalence in the descriptions of a system built on the theory of relativist information versus the analytical approach. The relativist explanation depends only on interactions, and hence it is global. But it is in line with the analytic description, which depends on the number of elements and the number of interactions; presented in analytic consequence (see Figure 1).

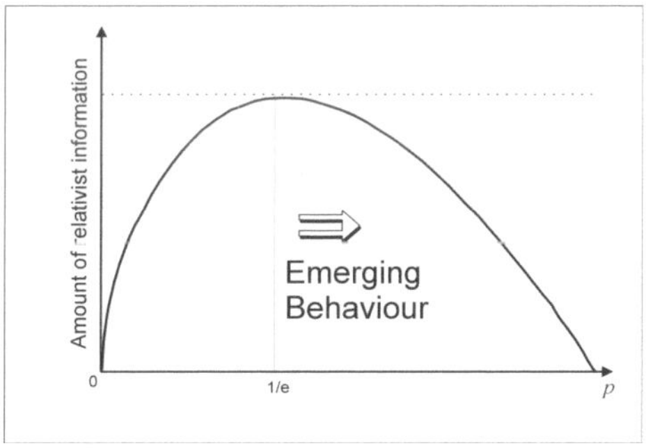

Figure 2. Emerging behavior in the description of the amount of relativist information

The use of analysis can serve as part of an overall effective research program, with methodology arising from the Gestalt approach, and based on the measurement of the amount of relativistic information. Most importantly, analysis is a complement, not a replacement, for the depth and insight of expert observation and participatory web design. Emerging research tools for measuring the amount of relativistic information may have a significant impact, not only on the accuracy and scale of user research, but directly on the design of seductive webs. Today, we believe it is possible to put this methodology of control in design into practice, via eye tracking applications which have been and are still being developed.

5. Conclusions and recommendations

Without wishing to draw any rash conclusions, we can interpret the above methodology as invariably being relativist and subjective. Nonetheless, this chapter is about design as a controlling influence, which is able to bring about a system's new or emergent behavior.

The present concept of design, and the difficulty arising from its methodology, do not allow for taking control of the various endogenous and exogenous components inherent to the working process of design in an objective way. We have stated that it is possible to define design as a process that introduces and changes the interactions in a system, so that the system's structure is ultimately also changed. Taking into account the methodological drawback, this chapter presents a proposal for a methodology that will be an innovative concept of design, and which can be used in the implementation and creation of persuasive webs. This methodology comes out of the Gestalt approach, and is based on the method of measuring the amount of relativistic information in a system.

Hence, we may state that the occurrence of changes in the structure of a system is synonymous with emerging behaviors or interactions within that system. These new behaviors determine the development of different rhythms, which in turn determine the evolution of design itself.

To conclude, it appears impossible to examine the idea of seductive webs purely according to an analytical methodology. We can affirm that persuasive design is a perceptual phenomenon that is governed by the coincidence of specific physical conditions; it is also dependent on relevant human factors, and on creative innovation. Persuasive design is, furthermore, a generator of change, both in quantity and in quality. As a field of formal enquiry, it is based on the relative frequencies of the occurrences of events in the various cases which have so far been analyzed.

This final section of the chapter describes and presents five proven techniques, which are recommended as powerful and effective tools for marketing design [2]. It is likely that in the future these techniques will routinely be adapted by designers and web designers to create seductive and persuasive websites.

1. Deconstruct the copy for "persuasion factors". The ability to persuade is built on ease of comprehension, which leads the user to interact with your message. In other words, once the reader is hooked, he or she goes on to absorb the rest of the design, and that indicates a willingness to be persuaded. Your job is to make the persuasive argument immediately understandable.

   To be persuasive, you must capture your readers' attention until they convince themselves of what you are trying to put across. Deconstructing the copy for persuasive factors means deciding how to emphasize the selling and action words through size, color, repetition, and the use of imagery. All of these factors should summarize the value and emotional appeal of the copywriter's intent. Incidentally, for websites, placing quote marks in a headline or subheader increases readership.

Alice Williams Taus, long-time creative director at Rodale Publications and now a freelance designer, says, "Repetition is important in persuasion. Clever design allows the message to be repeated in different ways, until it becomes almost subliminal and imprints on the mind. Think about where the person starts and stops, where the hand will touch the material and how the hands frame the message, and then what you want the person to feel and do - turn the page, tear out the coupon, fill out the order form" [2].

2. Find the emotion that sells (or "how to avoid puppies, kittens, babies, and babes"). In all effective creative work that's designed to persuade, emotion is employed, and is appealed to either subliminally or blatantly. Identify the emotions that do two things:

   - Get attention for the advertisement, mailing, or website
   - Keep driving the message home.

   Then select the images, colors, and typeface that immediately convey the emotional appeal. Next, identify the message which will convey to the reader how she or he will benefit. Find a way to lead the eye to this message of benefit, thus linking to the underlying emotion. The design formula is: Emotion to Benefit to Response. And, actually, it is acceptable to use puppies, kittens, babies, and babes, if all else fails.

3. Create "moments of decision" visually. To start with, describe to yourself in detail the act of responding. What makes the reader want to respond? How will he feel as he decides? What is the actual response mechanism? Does the reader pick up the phone? Go to the website? Send back a form? Find the copy that describes the act of response, and visually emphasize it. Make it stand out with symbols of value and text that emotionally sell.

   Text that directs the reader to respond, as well as information on how to respond, need to be featured in places where the hand and the eye come to rest. Sometimes those spots will be natural or instinctive, like the bottom-right corner of a page. But just as often your design will direct the viewer's eye to a specific place. Add visual elements that literally point out where or how to respond, leading from page to page or element to element. Arrows, rules, and swathes of color are all visual forms that support the response impulse. And counterintuitive tricks, like including a sticker that gets moved to a response card, help to lead the hand to the spot that triggers a response in the layout.

4. Practice weird science (*designus interruptus*). Design elements that cause the reader to go "huh?" momentarily can subconsciously refresh attention and arouse curiosity. These should be subtle, not like suddenly leading an elephant into the room. Doing something that is slightly visually uncomfortable, such as covering part of a face, printing a faux sticky note over a face, or putting a headline across the fold in a brochure, can be a powerful device to keep readers attentive without stopping the flow.

5. Use science. In designing for persuasion, don't let the desire to be original lead you to forego techniques that have proven to be effective. Adding these techniques to your repertoire can activate your talent in new ways, and allow you to create design that persuades, powerfully.

Finally, Blum [2] suggests using certain power words for persuasion and attention. These words are: now, today, immediately, free, money, love, safe, proven, results, discover, learn, know, understand, you, protect, trust, powerful, help, profit, easy, today, secret(s), health, guarantee, and exclusive.

6. References

[1] Duarte, Carlos A. M.; Devezas, Tessaleno; Carvalho Rodrigues, F., (2003). *The Paths of Design: Socio-Economic Pointers.* I Encontro Internacional de Ciência e Tecnologia do Design - *Senses and Sensibility in Technology*, Portugal, Lisbon.

[2] Blum, Sandra J., (1999). *Designing Direct Mail That Sells.* North Light Books.

[3] Marr, David, (1982). *Vision – a computational investigation into the human representation and processing of visual information.* W. H. Freeman and Company, New York.

[4] Fogg, Brian J., (1998). *Charismatic computers: creating more likable and persuasive interactive technologies by leveraging principles from social psychology.* Stanford University, USA.

[5] Mbakwe, Chanel; Cunliffe, Daniel, (2003). "Conceptualising the process of hypermedia seduction". I Encontro Internacional de Ciência e Tecnologia do Design - *Senses and Sensibility in Technology.* Portugal, Lisbon.

[6] Tattarini, Mirko, (2003). "Overmodern mermaids – digital design between ethics and local identities". I Encontro Internacional de Ciência e Tecnologia do Design - *Senses and Sensibility in Technology.* Portugal, Lisbon.

[7] Simões, Vitor, (2003). *A Ergonomia do impresso, uma via para o desenvolvimento: possibilidades e vantagens.* User's Design, International Congress, Lisbon, Portugal <*only in Portuguese* >.

[8] InfoDesign – Understanding by Design: *www.informationdesign.org/archives/cat_accessibility.php* Accessed February 2007.

[9] The European Union Council approved a Resolution of 6th February 2003 entitled *eAccessibility - Improving Access of People with Disabilities to the Knowledge Based Society.*

[10] Website of Future Now: *www.futurenowinc.com/designforconversion.htm.* Accessed February 2007.

[11] Website of Future Now: *www.futurenowinc.com/writingforweb.htm.* Accessed February 2007.

[12] Jakob Nielsen's Website: *www.useit.com.* Accessed February 2007.

[13] Lawson, B., (1997). *How Designers Think: the Design Process Demystified*, 3rd edition, Architectural press.

[14] Devezas, Tessaleno C., (2000). *Learning Dynamics of Technological Progress.* 3rd International Meeting of the Research Committee on Sociocybernetics (ISA/RC51), Panticosa.

[15] Duchowski, Andrew T., (2003). *Eye Tracking Methodology – Theory and Practice.* Springer-Verlag London.

[16] Carvalho Rodrigues, F., (1989). "A Proposed Entropy Measure for Assessing Combat Degradation." J. *Opl. Res. Soc. (UK), 40* (8):789-93.

[17] Marchetti, Cesare, (1986). "Fifty-Year Pulsation in Human Affairs". *Futures, 17* (3), pp. 376-378.

[18] Duarte, Carlos; Carvalho Rodrigues, F., (2005). "*Architecture Rhythms: as a perception phenomenon ruled by coincidence of specific physical conditions".* CUMULUS Spring Conference, II Encontro Internacional de Design, *Pride & Pre-Design – The Cultural* Heritage and Science of Design, Portugal, Lisbon.

[19] Duarte, Carlos A. M., (2001). *"Análise Relativista da Quantidade de Informação para Avaliação de Comportamentos Emergentes no Design"*. Covilhã, Universidade da Beira Interior *<only in Portuguese version>*.

7. Further recommended resources

Carvalho Rodrigues, F., Dockery, J. and Rodrigues, T., (1993). "Entropy of Plagues: A Measure Assessing the Loss of social Cohesion Due to Epidemics", *European J. of Operational Research* 71, 45-60.

Carvalho Rodrigues, F., Dockery, J., and Woodcock, A.E.R., (1992). "Models of combat with embedded command and control: Casualty based entropy calculations as a combat predictor", Chapter 2.5, in: *The Military Landscape, Mathematical Models of Combat*, Woodhead Press, Cambridge UK.

Carvalho Rodrigues, F., Carvalho, Fernando D., (1990). "Application of information entropy to defect characterization in leather", *Journal of the Society of Leather and Chemists*, **75**, p.10-14.

Carvalho Rodrigues, F., Carvalho, Fernando D., (1990). "Measurement of randomness in fiber distribution in paper using computer vision", *Tappi Journal*, **73**, 10-14.

Carvalho Rodrigues, F., Ramos, Luís, (1995). *"Ontem Um Anjo Disse-me - diálogos para o século XXI"*. Publicações Europa-América *<only in Portuguese version>*.

Carvalho Rodrigues, F., (1994). *"As Novas Tecnologias, o Futuro dos Impérios e os Quatro Cavaleiros do Apocalipse"*, Mem Martins, Publicações Europa-América *<only in Portuguese version>*.

Carvalho Rodrigues, F.; Duarte, Carlos A. M.; Ribeiro, André S.; Devezas, Tessaleno C.; Côrte-Real, Eduardo, (2001). *"Quantidade de Informação na Arquitectura Portuguesa, 1050-1950"*. Academia das Ciências de Lisboa *<only in Portuguese version>*.

Devezas, Tessaleno C. and Corredine, James T., (2001). "The Biological Determinants of Long Wave Behavior in Socioeconomic Growth and Development". *Technological Forecasting & Social Change*, **68**, pp. 1-57.

Devezas, Tessaleno C., (2001). *"Cycles Patterned Repetition or Chaotic Behavior"*. 21st International Symposium on Forecasting, Callaway Gardens, Atlanta, USA.

Devezas, Tessaleno C., (2001). *"The Causality of Long Waves"*. Presented at the 4th International Kondratieff Conference, Moscow, Russia.

Duarte, Carlos; Devezas, Tessaleno, (2005). *"Rhythms in Industrial Design: Kondratieff Wave Concepts"*. 6th International Conference of the European Academy of Design, University of the Arts, Germany, Bremen.

Duarte, Carlos; Carvalho Rodrigues, F. (2004). *Design Perception: Tempos in Portuguese Religious Architecture (1050-1950)*. 4th International Conference on Design History & Design Studies, Mexico, Guadalajara.

Fogg, Brian J., (1998). *"Charismatic computers: creating more likable and persuasive interactive technologies by leveraging principles from social psychology"*. Stanford University, USA.

Holland, John H., (1998). *Emergence from Chaos to Order*, Oxford University Press, Oxford.

Melo-Pinto, P., (1997). *"Aplicação da Teoria de Crenças ao Reconhecimento Visual"*, Universidade de Trás-os-Montes e Alto Douro, Vila Real, Portugal *<only in Portuguese version>*.

Pinto Peixoto, J., Carvalho Rodrigues, F., (1991). *"Sistemas Entropia e Coesão"*. Discórdia Editores, Lda, 3ª Edição, Portugal, Lisboa *<only in Portuguese version>*.

Schum, David A., (2000). "Teaching about Discovery and Invention in Engineering", New York, *Technological Forecasting and Social Change*, **64** (2) (3), 209-224.

Strzalecki, Andrzej, (2000). "Creativity in Design: General Model and Its Verification", New York, *Technological Forecasting and Social Change*, **64** (2) (3), 241-260.

Hypermedia Seduction for Terrorist Recruiting
B. Ganor et al. (Eds.)
IOS Press, 2007

Kamikaze Cyberpunk: Threats and Alternatives in the Age of Viral Power

Mirko TATTARINI
Communication Design Professor
Masters Program in Communication Design, ISIA – Florence, Italy

Abstract. The roots of contemporary, so-called Islamic terrorism are not uniquely religious; other historical influences are also at play: the crisis of modernity. The promise of modernity has resulted in a crisis of confidence. Globally, the hoped-for results of modernization have not been achieved, and in some instances this has led to extreme reactions. Today's terrorism should not be seen as a war between civilizations so much as a psychosocial reaction to the perceived untruths of modernity. This reaction takes on a sense of subcultural self-determination. Individuals who lack professional tools adopt a "bricolage" (do-it-yourself) approach in developing ideological "castles" and an Internet presence, relying on the promotional strategies of viral marketing. Resistance to "the" establishment is strengthened through recycling cultural artifacts which stemmed from the establishment itself. This recycling of what may be termed ideological waste parallels the social environment described by the cyberpunk literary movement of the late twentieth century. Thus, in terms of communications design, an answer to terrorism would require the development of alternative myths, which should also be given an Internet presence. Such myths need to be shaped by a new global consciousness, and animated by a new vision of the human journey.

Keywords. Modernity, aesthetics, terrorism, design, bricolage, mythology, Internet, warfare, viral marketing, branding, revenge, recycling, cyberpunk, resistance

Introduction

I am not an expert in terrorism. But I am a designer, and I can attempt to examine terrorism with the tools of design methodology. What I find is that there is no single interpretation, no simple answer, no monolithic solution. It's a matter of perspective. There are shades of understanding and symbolic exchange.

Sun Tzu, an ancient Chinese forerunner of today's military consultant, affirmed that the best success comes when one obtains victory without a battle [1].

1. The Crisis of Modernity

Modernity can be seen as a project that is concerned with human emancipation. Underlying this project is the yearning for positive knowledge, and the desire to functionally transform reality. The construction of an objective, absolute science is essential to this transformation, as is a universal and secular moral sense.

Modernity promised a smiling future to human beings, newly emancipated from their material needs through secularization, technology and welfare systems. It was the apotheosis of Western utopia, a vision which began with Aristotle and extended from Descartes to Einstein, all of whom increased our understanding of the world by reducing it to its basic elements and conceptually redesigning it in human terms. It was a project structured and a "design job" promoted via the simple promise of progress and growth.

Many cultures have believed in the promise of modernity. It first took root in Europe and the United States, and later in post-colonial Arabic and Asian populations. But the longed-for Western utopia has not come about, even though some nations have taken a good step towards it. In the Arabic countries, the Middle East, and Central Asia, it remained just a promise, while in Africa it was a mere mythological dream. Worst case scenarios include people getting a quick taste—a glimpse—of modernity, then nothing more. An example might be Kabul in the 1970s, as described by Khaled Hosseini [2], compared with the Kabul of today.

The crisis of modernity in Western countries is a matter of self-awareness. It is a psychological process of ever-increasing understanding, coupled with the progressive demolition of earlier mythological systems. For some Arab/Islamic populations, this has not been a process but a crash, followed, later on, by another awareness: the need to make some kind of break from the Western world before petrol and gas run out. Making this break requires powerful tools in the way of symbols and myths of self-determination: possibly a new project.

Such a socio-economic and cultural environment is likely, these days, to express itself through terrorism, if a break with the Western world is seen as the only hope for creating a reality that is rich and fertile in positive transformations. And the question is whether there is, in fact, any "world" other than the Western one, given the dominance of the Western paradigm.

However, members of such populations may become aware of not having the right tools for the job. So, they take a look back in time, searching through history and through the global cultural heritage. They make a rough inventory of cultural artifacts which might prove to be useful in their endeavor. Usually, they identify objects which originally carried vastly different meanings from those which they are now given; objects which were born out of completely different motivations, in completely different eras and so absolutely limited in their own nature. They identify and choose the Islamic religion, perhaps extracting—out of context—only a few aspects of it.

Thus, in the absence of sound economic, philosophical and technological foundations for building a competitive, future-oriented project, the focus defaults to the past, to the domain of the "savage mind" [3]. This is a term (also translatable as "wild and flowery thinking") coined by anthropologist Lévi-Strauss, and it is useful here. The "savage mind" mindset is intuitive, primitive and mythological, and it operates within all people. The savage mind has no really effective tools for building a competitive, future-oriented project. It can use ritual, and it can distribute and re-distribute symbols, or use and re-use cultural artifacts, but it can never really design the future.

But some of these tools can be seductive, as Ritu Menon mentions in Fatema Mernissi's book Karawan: "Where since millenniums the craftsmen weave symbols by spinning wool and plant fibers, you will find necessarily wizards of all kinds, from fakirs to snake-charmers, passing through storytellers, calligraphers and computer scientists" [4].

Figure 1. Karawan, cover page.

2. Jihad and Bricolage

Lévi-Strauss used the term "bricolage" to describe patterns of mythological thought, and the word has subsequently come to mean a range of things. In cultural studies, it refers to the process of building a new social identity by borrowing objects from various heritages and cultures. Bricolage is a particular feature of subcultures.

Bricolage is evident in the choice of the Islamic religion as a tool for supporting a reaction to "the establishment". The reference to superficial aspects of ritual, which do not take into account the centuries of brilliant philosophy and poetry which underpin such ritual, is bricolage. In the same way, the use of technology for disseminating awareness, promoting a project, or recruiting candidates for a cause, is approached by bricoleurs.

The advent of the Internet and digital technology has introduced the concepts of "easy to create" and "easy to access" for users. Just as one might fix a piece of furniture in one's garage, using specific tools from a toolbox, in the same way it is now possible to build up an Internet site, or to edit a video or create a graphic. All one needs is a home computer equipped with some simple software, which may be either free or pirated, essentially a do-it-yourself kit. Such computers and software are not usually of the same standard as those used by professionals, although they may approach that standard. More importantly, the users' know-how is not at a professional level.

The result is an Internet presence that offers mostly low-level artifacts, with an overcrowding of signs and symbols. The ability or possibility to create this kind of artifact may cause the user to feel that he or she is an authority, an expert of communication within a delirious professional democracy. The home computer user has become a bricoleur, one who uses bricolage.

The approach of the bricoleur doesn't exclude the possibility of attaining excellent results. Sometimes, the use of different tools instead of dedicated ones may bring about significant and unexpected visual innovations.

But the bricoleur is also an hacker, in the sense portrayed by P Himanen in his *Hacker Ethics*: someone with a passion for what he does; a lover of his task [5]. The hacker operates inside a community that is animated by a shared mythological interest, and which participates in something collective. This community is able to rebuild the archaic system of social ties, even in the digital age. Modernity had freed the individual from these social ties. In our contemporary, postmodern (hypermodern) world, the individual returns to recover these lost social contracts, sometimes as a consumer who is interested in specific products.

And the best place to rebuild these archaic ties and to recruit new members for the resistance movement is naturally the virtual domain of the Internet, because

- Virtuality is an ancient concept
- Paradise is virtual
- Umma already exists as a virtual nation
- The martyr's promise is virtual.
- In the same way, the Internet expresses the Islamic person's homeland language, because:
- It is not linear (nor is the original language)
- The characteristic of repetition is like a fractal iteration
- Both digital technology and the Arabic language appear to have magical qualities.

Regarding the final point above, we may note that the perception which users have of digital technology may be quite "magical", because incredible results can be obtained without a real awareness of the way the technology works. And the Arabic (Islamic) language is magical in that words and sentences—as well as the meanings they represent—are believed to have magical powers.

2.1. Viral Strategies

Contemporary Internet marketing strategies—including viral strategies—often take into account the increasing closeness of the consumer to the savage mind. This irrational tension is a strongly postmodern feature. Both tribal and viral marketing emerge via the meeting of the opportunities provided by information technology and the consumer's irrational tension, and seek to exploit pre-existing social networks so as to attain exponential increases in brand awareness.

Viral processes are similar to the spread of an epidemic. When a community of retro-fanatics meet in the digital domain, the viral process is primed, and we enter the era of "viral power" [6]. Kroker and Kroker state that "Viral Power is terroristic by nature precisely because it occupies the imaginary territory of symbolic exchange" [6].

In this view, the increase in Al-Qaida's "brand awareness" is viral. Beppe Grillo comments thus: "...a brand in franchising, a distribution model at low investment and big ROI [return on investments]. Who wants can use its brand and replicate its behaviors. A kamikaze is cheap: its life, a short training, some explosive." [7]

This seems to be nothing less than marketing and economics applied to terrorism, and it is highly effective. Terrorists release videos on the Web, and rock bands are releasing relevant new songs in an apparently random way (or they both release on YouTube). These videos are seductive, just as any terror or warfare aesthetics have always been seductive. This is a far cry from Aristotle's sense of beauty, charged with ethical values; instead, the implicit violence of these videos is guiltily seductive. (We are reminded here of Aristotle's theory of comedy, as in Umberto Eco's *The name of the Rose.*)

From Homer to the witch-hunts of the Middle Ages, from Robespierre to the Hollywood gloss of World War Two, we have been all seduced by the aesthetics of terror and warfare. And it's very disturbing to admit that the representation of war and of terrorism can be enjoyable, even for a pacifist.

Figure 2. The process of applying "warfare aesthetics" to consumer products is easily visible.

3. Kamikaze Cyberpunk

The current "survival attitude" in fanatical bricolage is also a good example of aesthetic and ideological recycling. There is recycling of ritual behaviors from the past, and of aesthetic seduction from the past as well as from the economic establishment of the enemy. There is also recycling of technology, recycling of weapons and recycling by using products and objects outside of their usual functional nature. We all use the car, the bus or the train to go to work, to the beach, or just to go back home.

The kamikaze uses it to go to paradise.

This recycling resistance movement bears an extraordinary similarity to the literary movement of cyberpunk. In the 1980s and 1990s, a group of science fiction writers, mainly North Americans and headed by William Gibson and Bruce Sterling, started switching the domain of science fiction from space trips to virtual reality. Their stories were animated by characters who fought against the establishment of multinationals, albeit with no great hope. In their battles these fictional characters used, interpreted and modified the products and technologies created by the establishment itself. These anti-heroes were very seductive in their survival efforts.

In the same way, the bricoleur army uses technologies, aesthetic codes and advertising strategies created by the enemy against the enemy itself. And the jihadist shares with the cyberpunk hero the perception of an unchangeable establishment, namely the Western paradigm. This perception is bolstered by the final non-tangible objective of attaining eternal grace after death.

The aesthetic of revenge outlined above is the same one that inspires the visuals of some American rappers. Moreover, the same motive underlies the revenge-through-sport concept of the "Just do it" slogan, famous worldwide. In fact, the starting point in each case is similar: a sense of marginalization, lack of power, lack of civil rights, and discrimination, together with the uncertainty of the future.

Figure 3. Martyrs' advert on the web.

Figure 4. Martyrs' advert on the web.

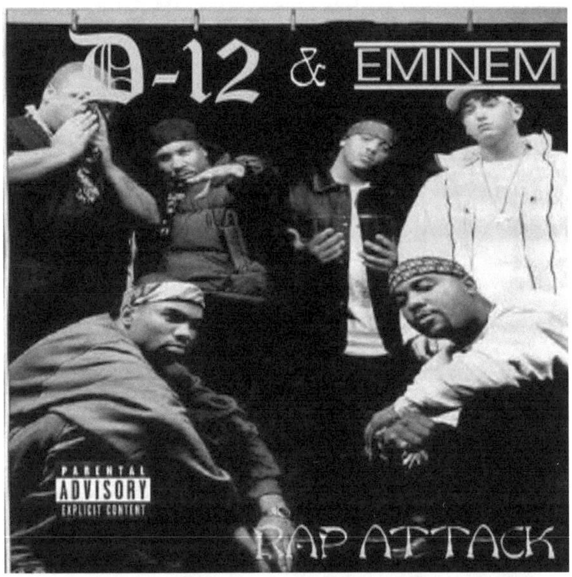

Figure 5. Rap graphics.

Figure 6. Counter-information.

Figure 7. Rap icon.

Figure 8. Just do it.

Figure 9. Just do it.

Figure 10. Just do it.

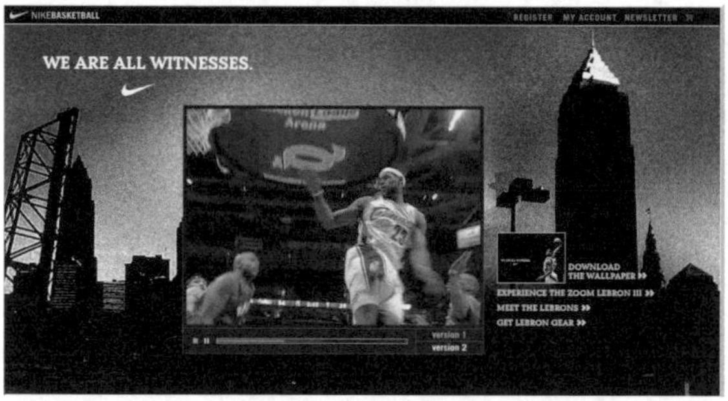

Figure 11. We are all witnesses.

Figure 12. Example of "sport revenge" advert.

Figure 13. Jihadists: A recycling policy with very explosive results.

4. Icons

As I said earlier, I'm not an expert in terrorism, I am a communication designer. After this brief discussion, it seems that the challenge for counterterrorism is not just a matter of stopping people from communicating, or arresting and capturing them. One cannot merely react with the same level of violence. There is, instead, the challenge of building up alternative myths, with equal powers of seduction to those currently in play, as a strategy in the battlefield of media aesthetics. A sustainable project with a credible promise is needed.

I think about the Islamic (rather than Islamist) Commander Massud, who was killed by Al-Qaida on 9 September 2001 as part of the strategy of the 11 September attack. Massud was a mysterious character, but with a strong potential to become a positive icon, able to withstand the cheap representation of lax values. And if I think about him, I cannot but be reminded of such another icon, whose aesthetic power is shared worldwide: Ernesto Guevara "Che".

In the hypermedia kingdom, there is still a lot of space for alternative myths.

It is appropriate, therefore, to help in designing an alternative mythology. Unicorns, mermaids or chimeras are able to multiply the epicenters of electronic emission, where the global village explodes in a liberating diaspora and forms geographic enclaves that are so diverse, and so attractive.

Human beings can regain control of geographical distance, and the conquest of language. They can experience the marvels of diversity within an exciting communicative milieu.

Finally, I would like to mention the "European Dream" described by Jeremy Rifkin. I'm European, so naturally I want to believe this dream exists—and that it is true, and it is European. Rifkin describes it as follows: "Although it is still in its adolescence, the European Dream is the first transnational vision, one far better suited to the next stage in the human journey. Europeans are beginning to adopt a new global consciousness that extends beyond, and below, the borders of their nation-states, deeply embedding them in an increasingly interconnected world." [8]

Values and myths are able to renew the world's enchantment with itself, in an instant shiver.

Figures 14/15. Massud and Che.

Figure 16. Ahmed Shah Massud.

References

[1] Sun Tzu and William Lidwell, *The Art of War*, Wowio: York, 2006.
[2] Khaled Hosseini, *The Kite Runner*, Riverhead Books. New York, 2003.
[3] Claude Levi-Strauss, *La Pensée Sauvage*, Plon: Paris 1962.
[4] Fatema Mernissi; translated from Italian, *Karawan*, Giunti. Firenze, 2005.
[5] Pekka Himanen, *Hacker Ethics*, Random House, New York, 2001.
[6] Arthur and Marielouise Kroker, *Terrorism of Viral Power*, Ctheory, www.ctheory.net, 2001.
[7] Beppe Grillo, *Economia del Kamikaze*, www.beppegrillo.it, 2006.
[8] Jeremy Rifkin, *The European Dream*, Tarcher/Penguin. New York, 2004.

Hypermedia Seduction for Terrorist Recruiting
B. Ganor et al. (Eds.)
IOS Press, 2007

Brand Terror: A Corporate Communication Perspective for Understanding Terrorism

By Michael MELNICK

Design Management Graduate Student
Pratt Institute, New York

Abstract. The concept of terror as a form of psychological warfare implies that there is more to terror than mere physical acts of violence. There is a non-physical component in action, which helps to carry and communicate the message of terror. Brands can be powerful tools for communicating ideas, building loyalty and driving choice. Brands have historically been associated with products and corporations, but the techniques of branding are applicable to every area of mass communications. The goal of this chapter is to examine the visual communication strategy of Hezbollah, using a brand identity and corporate communication framework. The underlying assumption is that any approach for countering terrorism should not only deal with the prevention of terror attacks, but should also enhance an understanding of the nature of the threat itself.

Keywords. Psychological warfare, visual communication, brands, Hezbollah

Introduction

Trying to understand terrorism and how organizations utilize it to achieve their strategic objectives is one of the biggest challenges of the twenty-first century. While there are many definitions for the term "terror", and even more interpretations for the causes and effects of terrorism, there is no dispute over the fact that terror is in essence a form of psychological warfare. In the article "Terror as a Strategy of Psychological Warfare", Dr. Boaz Ganor states:

> Terrorism works to undermine the sense of security and to disrupt everyday life so as to harm the target country's ability to function. The goal of this strategy is, in turn, to drive public opinion to pressure decision-makers to surrender to the terrorists' demands. Thus the target population becomes a tool in the hands of the terrorist in advancing the political agenda in the name of which the terrorism is perpetrated. [1]

The term "strategy of psychological warfare" indicates that there is more to terror than mere physical acts of violence. There is also a non-physical component in action, which helps to carry and communicate the message of terror. In simple terms, terrorism can be viewed as a methodical strategy by which to change the behavior of people by means of violence and fear. To be effective in a political sense, the fear generated by the transmission of terror must, in turn, produce a level of intimidation in the target population sufficient to elicit the response desired by the terrorists. As noted by Sam Sarkesian, terrorism "is the creation of fear in a population in order to force the existing system to respond to the terrorists' demands and/or objectives." [2]

One of the channels that terrorist organizations rely on to deliver their message is visual communication. A report of The Islamic Imagery Project states:

> We regard Jihadi imagery to be a primary vehicle for the communication and diffusion of Jihadi ideas, and an essential tool utilized by radical ideologues, terrorist organizations, and sympathetic propagandists, which plays to the particular religious and cultural experiences of their audience. Therefore, understanding how these images work, what ideas they convey, why they are employed, and what responses they may elicit, is vital to our struggle against the influence of Jihadi organizations and the violence they create. [3]

Given the broad definition of terror as a non-physical strategy to change people's behavior, one can see how visual communication—which ultimately serves also as a tool for moving people toward a desired behavior—plays a role in the strategy of terror. A simple example of the importance of visual communication in impacting behavior is a road sign (such as a stop sign), which not only informs the driver what he should expect up the road, but also directs him how to act in the given situation.

However, visual communication goes far beyond simply being a tool for informing and directing; it also has the capacity to contain and deliver highly complex systems of meaning, emotion and value. In this chapter I will show how organizations use visual communication (images, symbols, colors and video) to deliver their messages. I will offer another look at the phenomena of terrorism using a corporate communication framework and, specifically, brand identity terminology.

1. What is Corporate Communication?

In the business world, companies have realized that communications with their constituencies are a critical vehicle through which to form sustainable relationships with customers, partners, and employees, as well as shareholders. Corporate communication has become a critical aspect of management, marketing and public relations. Practically everything an organization does can be viewed as part of its communication.

The first part of an effective corporate communication strategy relates to the organization itself. The three subsets of an organizational strategy include (1) determining the *objectives* for a particular communication; (2) deciding what *resources* are available for achieving those objectives; and (3) diagnosing the organization's *reputation* [4].

In order to deliver a message appropriately, the organization must first analyze its constituencies, namely which target audiences need to receive the message and then understand and react to that message. Then, based on the audience and the nature of the message, the company needs to choose the communication channels it wishes to utilize, as well as how it should structure the message in order to maximize its effectiveness [4].

There are three basic stages to designing effective communication: the first stage has to do with defining, articulating and creating a *vision*. The second entails planning the *strategy* for delivering the message, and the third stage is *executing* the delivery of the message in a clear and valuable way.

Modern management and marketing models have mastered the utilization of communication strategies, developing sophisticated messages and using technology to effectively reach mass audiences. In order to maintain relationships with customers and build loyalty and trust amongst stakeholders, many (if not all) business models today recognize the importance of *identity*, *image* and *reputation*.

A company's identity is the visual manifestation of the company's reality, as conveyed through the organization's name, logo, motto, products, services, buildings, stationery, uniforms and all other tangible assets created by the organization and communicated to a variety of constituents [4]. Image is a reflection of an organization's identity, seen from the viewpoint of its constituencies. Managing the gap between identity and image—more specifically, narrowing that gap—is one of the primary objectives of corporate communications. One of the more advanced communication tools which companies use is called branding. A simple definition may be applied to all brands: A brand is a *promise* with resonant value that is defined by the *experience* between the individual and the organization *over time*.

A distinction should be made between the function (product, organization or service) and the emotional and intangible ideas that are associated with the function. Take Coca Cola, for example. The product of Coca Cola is a sweet carbonated brown liquid; the brand of Coca Cola is the energetic and vibrant taste of life, youthfulness and fun. The whole purpose of brands is to attach this bundle of emotional associations to an otherwise undifferentiated product. According to Howard Schultz of Starbucks, "Customers must recognize that you stand for something." [5]

Brands exist in the minds of people (consumers, employees, or other stakeholders). They are not something that a company or organization sells, but are instead made up of the perceptions of the users of that brand. Brands are the culmination of a user's total experience with the product or service (or company) over time. A clear and distinctive brand proposition can influence stakeholders' attitudes.

Three basic criteria for creating a viable and enduring brand are:

A. ***Differentiation:*** A strong brand always represents an alternative. One of the most important marketing and communication functions of a brand is to differentiate itself from the competition, and position itself as a better choice for the consumer. The brand's visual identity plays a critical role in helping audiences to identify this difference and believe in it. Highlighting the brand's benefits or advantages over the competition is not only a way in which branding creates a clear distinction in the marketplace, but is also the means by which to generate faith and loyalty amongst customers, thus creating brand zealots.

B. ***Credibility:*** In a world of unlimited choices and options, with so much visual stimuli and clutter competing for our attention, brands serve as icons of trust and credibility. The promise that the same experience will be delivered every time the customer interacts with the brand is critical for creating a sense of confidence in the product, service or organization. Consistency is a key factor in creating and delivering a memorable message. This is why brands maintain strict guidelines for the use and application of their visual identity elements.

C. ***Emotive impact:*** Every brand has a story to tell, a reason for being, and an inspiring vision. A strong brand wraps the product, service or organization it represents with a world of meaning and a set of values for people to identify with and believe in. The brand narrative helps to turn a complex reality into a simple and memorable idea that can be easily understood. Stories are also tools for creating an emotional response and tapping into values, which are extremely important for building not only a motivated audience but also a loyal one.

In the next section, I will use the above-mentioned brand criteria to illustrate how Hezbollah uses visual communication strategies to convey their message.

2. Hezbollah: "The Party of God": A Brief Overview

Hezbollah was formed in 1982 as a reaction to the Israeli invasion and occupation of southern Lebanon. The organization, whose name means "the party of God", originally sought to bring the Islamic Revolution to Lebanon, seeking the transformation of Lebanon's multi-confessional state into an Iranian-style Islamic state [6]. Hezbollah and Israel have been clashing since 1982, and Hezbollah has been responsible for multiple kidnappings, murders, hijackings, and bombings against Israel. These violent acts are characterized by some as terrorist attacks, while others regard them as legitimate resistance or even military jihad [6].

In 1992, Hezbollah participated in the Lebanese elections for the first time, gaining twelve out of 128 seats in parliament. It acquired ten seats in 1996, and eight seats in 2000. In the general election of 2005, it gained fourteen seats nationwide (of a total 128), and an Amal-Hezbollah alliance captured all 23 seats representing Southern Lebanon. The bloc it formed with others, namely the Resistance and Development Bloc, took 27.3% of the seats. Hezbollah is a minority partner in the current cabinet, holding two cabinet positions (and endorsing a third) in the Lebanese government [6].

Mohamed Fneish was appointed Energy and Water Minister in the cabinet, and has been quoted as saying, "We are a political force that took part in the polls under the banner of defending the resistance and protecting Lebanon and received among the highest level of popular backing... Hezbollah's resistance [against Israel] does not in any way contradict its political role. If joining the government and parliament is a national duty, then so is defending the country." [6]

Hezbollah's political success is regarded as a model for other Islamic parties in the Middle East, including Hamas and United Iraqi Alliance; its actions are thought to provide strong pointers as to how these other emerging Islamist forces might behave [6].

3. The Hezbollah Brand

"Successful branding requires a carefully crafted message delivered with consistency and discipline." –Allen Rosenshine, Chairman and CEO of BBDO Worldwide.

According to Wolff Olins, head of a firm of design and image consultants, brands have historically been associated with products and corporations, but the techniques of branding are applicable to every area of mass communications. Political leaders, in order to inspire people, need to become "brand managers" of their parties—and preferably of the entire nation [7].

In order to analyze Hezbollah from a brand identity perspective, we must examine how the identity and image of the organization are managed and carried out. We do so according to the criteria for enduring brands mentioned earlier.

3.1. Differentiation

Hezbollah is differentiated on many levels. Internally, it represents the Shiite minority in Lebanon, standing as an alternative to the Sunni government. This differentiation has defined Hezbollah from the very beginning, and it is a source of identification for its supporters. More importantly, Hezbollah has positioned itself as a counter-force to Israel, the regional super-power that is supported by the United States. This resistance has made the organization popular among non-Shiite populations as well.

Figure 1. Hezbollah's flag

Every brand has graphic and visual elements which help people firstly to identify the brand, and secondly to identify with the brand. Hezbollah's flag, shown in Figure 1, comprises a yellow background, with the organization's logo appearing in green and its main slogans in red. The logo serves as a powerful symbol representing the ideas that the organization stands for. The letters in green mean "party of God", and support an upward-lifted arm holding a rifle, which represents the military resistance of the Hezbollah. A report of The Islamic Imagery Project states that:

> The color green in Islamic tradition is always linked to the Prophet Muhammad, and is therefore synonymous with Islam itself. It is used ubiquitously throughout the Islamic world [...] While the act of "coloring" a state (or a flag, an individual, or a symbol) green, may seem simple and of little consequence, it should be remembered that the color green is deeply significant to all sectors of the Islamic world, and the act of "coloring" specific symbols green is an overtly political and explicitly Islamic statement. [8]

Beneath the rifle, there is a wheat-like crop or plant to represent growth and prosperity. In the background we find a circle with some coordinates, signifying the globe. Above the logo is the slogan of the Hezbollah, "We will be victorious". This represents the essence of the Hezbollah brand promise. The flag has become a powerful icon in the struggle against Israel.

3.2. Credibility

The high degree of credibility that Hezbollah enjoys is, in many ways, a by-product of the sophisticated and consistent use of mass media over time. Al-Manar (Arabic for "The Beacon") is a television station that was launched by Hezbollah in 1991. The station has an estimated daily viewership of ten to fifteen million people worldwide, and is the second most popular station among Palestinians in the West Bank and Gaza [9].

Al-Manar often serves to convey the official statements of Hezbollah and the speeches of the organization's leader, Sayyid Hassan Nasrallah. As of 2002, Al-Manar has regularly aired self-produced "patriotic music videos". These are composed of video footage from the West Bank, mixed with Hezbollah battle songs, and they are aired after news reports of suicide attacks in the West Bank and Gaza. According to the producer of one of these videos, they are intended to encourage Palestinian youth to become suicide bombers [9]. The frequency and consistency of these messages over time creates a powerful impact in viewers' minds, giving the organization a highly credible stance and creating a sense of legitimacy for their message and actions. At the same time, Hezbollah uses the media platform to communicate its message to their other foreign target audiences, whom they aim to terrorize.

Figure 2. Hassan Nasrallah on Al Manar

3.3. Emotive impact

According to O'Shaughnessy and O'Shaughnessy, "Emotion is the adhesive that, when mixed with trust, equals loyalty" [7]. When consumer trust in a brand is undermined, there is a corresponding loss in market power, as customers have less faith that the company will live up to their expectations [7]. The emotional appeal of an organization such as Hezbollah is immense.

Figure 3. Young Hezbollah protestors in a Washington DC rally

Hassan Nasrallah is considered by many to be a national hero, and an icon who represents resistance to Israel and the United States—not only in Lebanon but throughout the entire Muslim world. Incitement was always part and parcel of

Hezbollah's strategy. In an interview with Al-Safir, Nasrallah described his priorities and methods in the first months of Hezbollah's formation:

> The main effort at the time went into mustering and attracting young men and setting up military camps... . The second effort was spreading the word among the people, first, in a bid to raise their morale, and second to instill in them a sense of animosity towards the enemy, coupled with a spirit of resistance... this required us to use the language of indoctrination rather than real-politic. People then were not in need of political analysis, they were in need of being incited and goaded. [10]

4. Conclusion

The idea behind this chapter was to examine the communication strategy of a terrorist organization such as Hezbollah, using a brand identity and corporate communication framework. The underlying assumption is that any approach for countering terrorism should not only deal with the prevention of terror attacks, but should also enhance an understanding of the nature of the threat itself.

In the recent conflict between Hezbollah and Israel, communications, propaganda and psychological warfare have played a critical role. The fact that both Israel and Hezbollah have claimed a victory shows that much is in the eye of the beholder, and that perception is as important as reality, if not more so. Having said this, following the recent war, Hezbollah is now at a crossroads. On the one hand, it has shown a great deal of courage and an unprecedented resistance to the mighty Israeli army; on the other hand, it is facing internal pressures and having to take responsibility for creating complete chaos in Lebanon.

From a brand management perspective, any organization must be flexible enough to respond to changing conditions, yet constant enough to maintain a distinct identity. Facing this new political reality, Hezbollah's "brand identity" is now at stake. Only time will tell what direction the organization will take, and if it will be able to maintain public support in the future.

The premise of this chapter rests on the notion that terror is a war fought in the mind of people. Therefore, regardless of the actions of Hezbollah, counterterrorism policies should consider the long-term psychological impact of "terrorist brands". Counterterrorism policies should implement strategies (inspired, perhaps, by the world of marketing) to dilute and undermine these brands' differentiation styles, their credibility, and the effectiveness of their use of emotion to reach and sustain a following.

References

[1] Ganor, Boaz (2002), "Terror as a Strategy of Psychological Warfare", http://www.ict.org.il
[2] in Richard W. Schaffert (ed.) *Media Coverage and Political Terrorists: A Quantitative Analysis.* Praeger Publishers. New York. 1992.
[3] The Islamic Imagery Project: *Visual Motifs in Jihadi Internet Propaganda,* Combating Terrorism Center, Department of Social Sciences, United States Military Academy, March 2006
[4] Argenti, A. Paul, (1994) *Corporate communication*, 3rd edition, McGraw Hill, NY
[5] http://www.allaboutbranding.com/index.lasso?article=22, retrieved August, 2006
[6] http://en.wikipedia.org/wiki/Hezbollah, Retrieved August 2006
[7] John O'Shaughnessy, Nicholas Jackson O'Shaughnessy, *The Marketing Power of Emotion.* Oxford University Press., New York. 2003.
[8] *The Islamic Imagery Project:, Visual Motifs in Jihadi Internet Propaganda*, Combating Terrorism Center, Department of Social Sciences, United States Military Academy, March 2006 http://www.ctc.usma.edu/imagery_other.asp
[9] http://en.wikipedia.org/wiki/Al-Manar, Retrieved August 2006
[10] Hala Jaber, *Hezbollah Born With a Vengeance*, Columbia University Press, 1997

Hypermedia Seduction for Terrorist Recruiting
B. Ganor et al. (Eds.)
IOS Press, 2007

Hypermedia Seduction: Further Exploration of the Process of "Seductive" Online User Interactions

Chanel MBAKWE and Daniel CUNLIFFE
Hypermedia Research Unit,
Department of Computing and Mathematical Sciences,
Faculty of Advanced Technology, University of Glamorgan,
Pontypridd, United Kingdom
E-mail: cmbakwe1@glam.ac.uk; djcunlif@glam.ac.uk

Abstract. This chapter aims to extend the examination of seductive online user interactions, specifically when divergent hypermedia domains are presented for evaluation. In our earlier work, we recognized the need for researchers in the field of Human Computer Interaction (HCI) to consider seduction beyond what is popularly assumed of this phenomenon. Accordingly, the benefits we believe attainable when the concept of seduction is applied were highlighted. In this chapter we expand our initial ideas, by giving developers and researchers a practical insight into how the mechanics of hypermedia seduction currently may enhance and sustain the appeal of more engaging hypermedia products. A demonstration of our approach allows us to reveal the seductive strategies characteristic of an e-commerce website which delivers a more engaging user experience. In the second part of this chapter, we consider how our conceptual model may be utilized to evaluate more "extremist" hypermedia forms, specifically websites that are created and hosted on behalf of terrorist organizations. In our method, we consider the negative aspects of hypermedia seduction. We focus specifically on how certain factors may play an active role in influencing user behavior, by exploring the seductive strategies apparent in "extremist hypermedia". At this stage, we are able provide some indications of the seductive means used by the developers and creators of websites in this genre. From an interdisciplinary perspective, we also provide researchers with a useful starting point towards understanding the mechanics of "extremist hypermedia seduction". We anticipate that as knowledge about the nature of the interactions and communications of both the users and the creators of extremist websites improves, more suitable and effective methods for countering online extremism or terrorism may emerge.

Keywords. Hypermedia seduction, conceptual model, human computer interaction (HCI), framework, user interaction, e-commerce, evaluation, extremist websites, extremist seduction

Introduction

A fair deal of consideration has been given to the concept of seduction in the field of Human Computer Interaction (HCI) recently. Despite this, it appears that the idea that aspects of seduction may significantly influence user attitudes or behavior during system usage is still not widely accepted. In our earlier work [1], we argued that despite the apparent negative ideological manifestations surrounding the concept of seduction, seduction should be considered beyond what is popularly assumed of this process. Accordingly, we focused on highlighting the benefits which we believe are attainable as a result of further consideration of this significant phenomenon. Our beliefs to date remain unchanged, since there is still little evidence to suggest that seduction has been, or is currently being, utilized to its full potential to influence the development, creation, and evaluation of more engaging hypermedia products.

With respect to the goals and themes of the NATO advanced research workshop "Hypermedia Seduction for Terrorist Recruiting", we feel it is necessary to consider the concept of seduction in a new light. In this case our intention is to shift our focus towards consideration of the *negative aspects*[1] of hypermedia seduction, and reflect on how certain characteristics of seduction may play an active role in influencing user behavior, particularly in cases identified as supporting more extremist[2] agendas.

Our research recognizes that seduction is potentially an unpredictable and subjective phenomenon, which at worst may be used to influence human behavior, actions, and decisions in a negative and manipulative manner. And at best, it may be used positively to enhance levels of pleasure, fulfilment, and immersion during system usage.

The aim of this chapter is to explore how effective our existing conceptual model may be when applied to a situation requiring greater understanding of how seduction may enhance the potency of extremist websites. In such cases, the website will have been created for the purpose of both promoting and supporting the ethos of the extremist organization. We begin with a recap of the ideas presented in the paper entitled "Conceptualizing the process of hypermedia seduction" [1][3], in order to provide a context in which to place the work which follows. In this first section we briefly highlight the current dilemma in HCI research surrounding the phenomenon of seduction. We present a brief reminder of our definition of "hypermedia seduction" from the perspectives of hypermedia development, evaluation and usage. Finally, we present our conceptual model of the process of hypermedia seduction, pinpointing the factors we believe assist progression within the hypermedia seduction lifecycle.

Presentation of our model is followed by a brief e-commerce case study for the purpose of demonstrating how we are able to pinpoint the seductive strategies characteristic of a sample e-commerce website. In this instance we are able to provide developers and researchers with a practical insight into how the mechanics of seduction are currently utilized (or feature) in mainstream hypermedia.

In the second section of this chapter, we focus more specifically on the goals of this workshop. We do so by examining how the principles outlined in our approach may translate when utilized in the evaluation of more extremist hypermedia, specifically "terrorist news networks" and "terrorist websites"[4]. This chapter by no means

---

[1] The terms "negative aspects" and "positive aspects", in the context of this paper, are technical ones utilized within our conceptual framework of hypermedia seduction. They are used to define the characteristics of specific interactions or behaviors, effects, or events believed to occur during the process of hypermedia seduction, rather than to define an activity, a person, or any organization in particular.

[2] The term "extremist", in the context of this paper, defines the activities or the ideologies of groups (or organizations) perceived as existing outside of the conventional political, ethical, and ideological centre of the society in which these groups exist.

[3] The paper entitled "Conceptualizing The Process of Hypermedia Seduction", originally presented at the conference on *Senses and Sensibility in Design and Technology* (Lisbon, 2003) focused on the development of a conceptual model of hypermedia seduction. The purpose was to provide a means for developers and researchers to better understand the mechanics of hypermedia seduction.

[4] According to Reid et al. (2005), "terrorist websites" are websites produced or maintained by extremist groups or networks of people, or their sympathisers. They are believed to be utilized for the purpose of acquiring sensitive intelligence or information, controlling and organizing "operations", or as a broadcast platform for the dissemination of a particular group's political beliefs and ideologies.

aims to define the nature, conduct, or the occurrence of extremist organizations[5] or extremist groups of people. An abundance of expertise and research exists in relevant fields that specifically address the above phenomena. Our intention here is simply to provide a better understanding of how the process of seduction may feature and be used within the hypermedia domain recognized as the "Dark Web".[6]

We pinpoint the methods most likely be utilized as a means of attracting, alluring, influencing, captivating, and persuading users. We believe it is necessary to consider both the negative and positive aspects of hypermedia seduction. We thus specifically concentrate on how certain characteristics may play an active role in influencing user behavior, in an attempt to better understand some of the reasons why negative aspects of seduction may perhaps also enhance the potency of these particular websites.

The goal of this chapter to is to provide researchers with a useful starting point towards understanding the mechanics of "extremist hypermedia seduction", from a human factors research perspective. It is anticipated that through improved knowledge of both the means of online communications and the nature of interactions between the users and creators of extremist websites, more suitable and effective methods for countering online extremism may emerge. This chapter concludes with thoughts and suggestions for future research.

1. Seduction in the Field of HCI

Previous research in the field of HCI has addressed the concept of seduction in relation to a) software development [2–4]; b) the idea of more seductive interfaces [5,6]; c) guidelines for the development and evaluation of more seductive websites and products [7,8]; and d) methods supporting more seductive design practices [9,10]. Seduction is frequently defined as a strategic process, not dissimilar to the procedures identified in human to human seduction theories, as evident in the works of Skelly et al. [5,6], Hofmeester et al. [9], Mantovani [11], and Dormann [12].

Themes which have featured predominantly in existing approaches include attention seeking, the establishment of a relationship and proposal of promises, fulfilment of these promises, the receipt of rewards, and parting on good terms. Current hypotheses borrow heavily from classical theories of seduction, as evident in the work of Khaslavsky and Shedroff [7], Papsadore [8] and Mantovani; these do little more than echo the theories present in the literary and psychological fields.

Seduction in the field of HCI is frequently conceptualized as a sequence of interactions or events, characterised by different "steps" or "phases". It is most often considered only in terms of what it is assumed may "happen" to the user, emotionally, cognitively and psychologically as a result of being seduced, rather than how the system itself performs as a seducer. Only in the works of Scanlon [13], Papsadore, Khaslavsky and Shedroff, and Spool [14] is the need for additional consideration of the behavior of the software or system recognized. Despite this, these approaches remain predominantly user-centric.

---

[5] As hypermedia researchers, in no way do we intend to influence or to promote a particular view or political stance on the subject of extremism or terrorism; neither do we intend to single out any existing organization(s) or networks of people who identify themselves as terrorists, or feel they fit this particular category or group.

[6] According to Chen, H., et al. (2003), the term "Dark Web" defines the alternate side of the Internet, believed to be used by terrorist and extremist groups to promote their ideology.

To date, we believe the work of Papsadore best describes seduction as a strategic approach, and succeeds in communicating the idea that specific elements or factors may be utilized in the creation of more seductive experiences. A decade or so ago, Nielsen [2] acclaimed the work of Hofmeester et al. [9] as "one of the few case studies with a systematic approach to seductive design". However, it may be argued that in fact Nielsen's assumption is slightly incorrect, since the work of Hofmeester et al. focuses on creating a sensual product—a product with "sex appeal", rather than a product with strategic seductive qualities.

1.1. Problems with Existing Approaches

Like many areas of research which struggle in the attempt to better understand a subjective phenomenon, the field of HCI faces an abundance of problems concerning the concept of seduction. No single approach to date has succeeded in providing a systematic method for the creation and evaluation of more seductive hypermedia products. Existing hypotheses and theories are limited, since although a myriad valuable suggestions and considerations regarding the phenomenon of seduction have been provided, there is a notable absence of sufficient methods, guidelines, and solutions to enable developers to actually achieve, or increase, the "seductive potency" of hypermedia products.

Existing hypotheses are often domain-specific. Thus, guidelines have a tendency to place the onus on the designer to interpret, assess, or judge whether the goals defined are fulfilled adequately within the relevant field. In addition, much of the existing work fails to offer any detailed justification of the significance of seduction. The process of seduction is seldom considered as a reciprocal or interactive relationship which may occur between the user and the system. Hence, we feel that there is still insufficient focus on what may be required of the interface or the system to increase levels of seduction.

1.1.1. Seduction and HCI: The Current Dilemma

One of the major dilemmas remaining in the field of HCI is that little is known about how to make hypermedia seductive. Progress is apparently inhibited because "developers simply don't understand the mechanics of seduction in the same way designers and engineers from other disciplines do" [7]. As a result, limited effort has been expended in making hypermedia more seductive, since it is believed that "seduction is highly subjective, and not measurable in the same ways other forms of software development are" [7].

Research to date is criticized for being "...limited by methodological problems"; it is argued that exhaustive descriptions of the seductive process have been provided, but without any means for studying the process in more depth, and without any indications of how such study might be planned [11].

2. Conceptualizing the Process of Hypermedia Seduction: An Overview

One of the wider goals of our research is to address the need for better methods to assist in identifying which elements make hypermedia more seductive. Coupled with this is the need for developers and researchers to better understand the mechanics of hypermedia seduction. On this occasion we will refrain from an in-depth discussion about our work undertaken 2003; in summary, this work enabled us to reach our initial definition of hypermedia seduction, to model the process of hypermedia seduction, and then to identify the elements we believe occur in each phase of the seduction life cycle. A detailed account is available in Mbakwe and Cunliffe (2003) [1].

However, because our original work provides an important backdrop for what is presented in the current chapter, we shall provide a brief explanation of what we mean by the term "hypermedia seduction" within the context of hypermedia development, evaluation, and usage. This explanation is followed by the presentation of our conceptual model of hypermedia seduction, with summaries of each of the phases believed to occur within the seduction life cycle.

2.1. Hypermedia Seduction: A Definition

The concept of hypermedia seduction is two-fold: firstly, seduction relates metaphorically to the orchestration of specific actions or elements that facilitate the onset of desirable changes within a user's behavior. Achievement of desirable behavioral changes relates to idea of the system as the seducer, where behavioral changes are believed to occur due to the system's ability to sustain its appeal.

Secondly, the act of seduction defines the particular path (or course) of a relationship formed as a consequence of interaction between the user and system. Seduction, in this instance, is believed to enhance the user's perception of the nature of the communication; we believe this results in the formation of a more meaningful "bond" between the user and system.

2.2. Conceptualizing Hypermedia Seduction

Our goal is to provide developers and researchers with practical insight into how the mechanics of hypermedia seduction may enhance and sustain the appeal of more engaging hypermedia products. Our conceptual model of hypermedia seduction emerged from the belief that desirable (positive) emotional responses may influence both the user's attitude towards the system as well as his or her behavior, and that such responses relate to the depth and duration of system usage.

Our model represents the relationships between the factors which we belive occur at each phase within the hypermedia seduction lifecycle. An illustration of this model is presented in Fig. 1, and is accompanied by descriptions of the phases we believe occur during the process of hypermedia seduction.

2.2.1. Seduction Phase 0 (PSP0: RS): Inducement

Inducement occurs instantaneously (or momentarily) prior to interaction, and its role is to deliver instant allure. Inducement provides the necessary "eye-candy" to attract or distract, by utilizing qualities that either please or evoke the idea of further fulfilment. Developers often mistake the power of inducement, believing it to be sufficient to achieve the overall seductive appeal of the product. Preliminary research findings reveal that inducement functions only to influence the onset of seduction, and alone is insufficient to determine, evoke, or sustain user seduction. However, despite these findings we recognize that there can be no seduction without inducement.

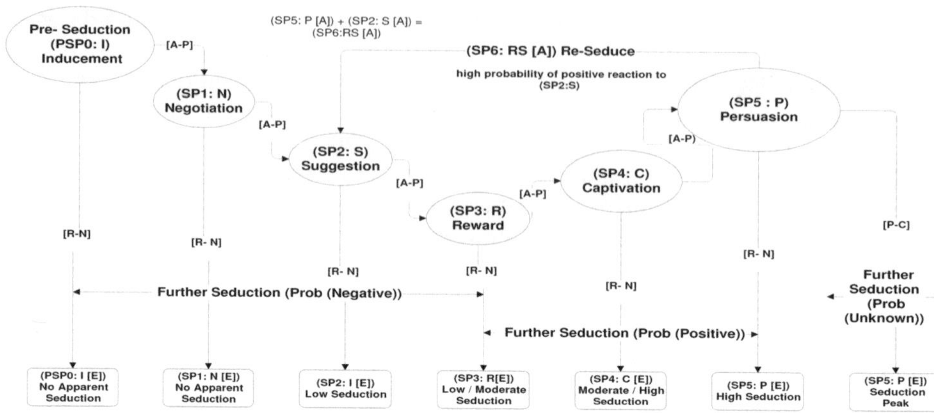

N=Negative Emotion State; P=Positive Emotion State; R=Rejected; A=Accepted; C=Completed; E=End; (SP1)=Seduction Phase One Identifier; (PSPO)=Pre-Seduction Phase Identifier.

Figure 1. Hypermedia seduction: a conceptual model.

2.2.2. Seduction Phase 1 (SP1: N): Negotiation

Negotiation may arise either prior to or (in most cases) during interaction, and procures the delivery of an engaging sensorial experience. Negotiation occurs subtly and dynamically as a form of gesture, and triggers the onset of dynamic changes to the look and feel of the interface environment. Negotiation focuses specifically on the arousal of the user's senses, for the purpose of influencing desirable changes within the user's emotion state or mood. In some respects, the "personality" of the software negotiates its compatibility with that of the users. This process might be regarded as the start of an assessment (or gauging) of the probabilities of successful influence.

2.2.3. Seduction Phase 2 (SP2: S): Suggestion

Suggestion arises before negotiation ends, subtly easing its way into negotiation and quietly phasing out on its completion. Suggestion functions to support the influencing of user needs. It is a critical phase, since we believe that at this point in system usage the probability of abandonment may increase due to the user experiencing a state of overexposure to the aesthetic qualities of the user interface.

Suggestion is best conceptualized as a form of "subliminal transition" which functions both to influence and to motivate the user by inducing "desired outcomes". A desired outcome in this context refers to the ability of the system to assist the generation of desirable (positive) user perceptions of their own needs. Importantly, these perceived needs are compatible with the actual capabilities of the system. Suggestion simply acts to inspire and guide the user to make "system-compatible" decisions and choices.

2.2.4. Seduction Phase 3 (SP3: R): Reward

Reward occurs when the user navigates the system. Reward functions to incite further interaction, or to operate as "a motive for the performance of actions" [15]. Reward involves the delivery of an immediate user need rather than the entire fulfilment of the user's goal. Fulfilment is delayed at this point, since it is believed that the probability of the user abandoning the system may increase once goals have been fulfilled, since there would be no reason to go on using the system.

Thus, we believe that if a reasonable amount of delay in goal fulfilment occurs, a greater sense of fulfilment may be felt with the eventual achievement of the goal. This would be true so long as goal accomplishment is not perceived as being threatened or hindered.

2.2.5. Seduction Phase 4 (SP4: C): Captivation

Captivation occurs on fulfilment of a goal, and its function is to facilitate the formation of a relational bond between the user and the system. Successful captivation instils the need for further exploration, since its role is to connect the user with the possibility of the attainment of higher goals. Captivation is communicative rather than interactive, and is delivered in order to convey meaning(s) that actually echo the user's momentary values. Captivation may be enhanced by carefully considering variables such as surprise, shock, oddity, cuteness, playfulness, and novelty.

2.2.6. Seduction Phase 5 (SP5: P): Persuasion

Persuasion begins during captivation, and its function is to induce behavioral and attitude changes without coercion or deception. Persuasion is generally perceived as a way of achieving improved levels of influence. It is also seen as a means of inducing behavioral and attitude changes, and improving the user's overall receptivity, acceptance and engagement with technologies.

However, literature in the field of HCI commonly considers persuasion as an isolated occurrence. This view is problematic, since it implies that the act of persuasion occurs without the support of a framework in which the process manifests. Our research argues that a framework is required to guide the user through a series of phases, to the point where they may receive and respond readily to persuasion. The framework we propose is "hypermedia seduction", since we believe persuasion is inhibited unless it is accompanied or supported by the vital interplay and pre-established rapport between user and system.

2.2.7. Seduction Phase 6 (SP6: RS [A]): Re-Seduction

Re-seduction occurs during persuasion, and its function is to fuel the persuasive process to a point where further suggestion is more readily accepted. The probability of the user reaching this phase is determined by two significant factors:

1. Preliminary acceptance in the persuasion phase
2. Acceptance of further suggestion.

Re-seduction occurs where persuasion and suggestion meet, since for successful re-seduction it is necessary for the user to accept further influence. The probability of the user being re-seduced is high if persuasion is successful and the user accepts further suggestion. We hypothesise that beyond the persuasion phase, the user may have entered a "state of flow"[7] [16].

[7] The concept of Flow refers to "the state in which people are so involved in an activity that nothing else seems to matter; the experience itself is so enjoyable that people will do it even at great cost for the sheer sake of doing it." Mihaly Csikszentmihalyi (1975) [16].

2.2.8. Seduction Phase 7 (PSP7: PS): Post-Seduction

At this stage in research, the probabilities of future seduction following re-seduction are unknown. We therefore recognize that empirical research is required to investigate whether users at this phase will either be positively or negatively influenced toward re-using the system. Variables to consider for measurement include user experience ratings, attitude assessments (including consideration of perceivable future needs), goals, and the user's perception of the system's ability to fulfil these goals during future usage.

2.3. Time and Duration

The goal of hypermedia seduction is the formation of meaningful relationships maintained over duration of time. The advent of those relationships may be affected considerably by the constraints of the delivery medium. Our research acknowledges that time invested at each phase in human-to-human seduction often determines the success rate of the overall seduction process; time investment may vary from days to weeks to months, or even to years, within each phase of the cycle.

In HCI, traditional human factor guidelines indicate "10 seconds as the maximum response time before users lose interest"; on the Internet this figure increases slightly to approximately 15 seconds [2b]. Table 1 indicates that the scope for concern may be limited to the time-frame that occurs before the user's interest is captured, namely between the seduction phases of inducement and suggestion. Beyond this point, the probability of the system failing to arouse interest is lower, according to predictins.

Table 1. Estimated time duration for progression within the seduction life-cycle

Seduction Phase	Minimum Estimated Time Duration (ETD)*	Maximum Estimated Time Duration (ETD)*
0. Inducement	2 sec	3 Sec
1. Negotiation	3 Sec	5 Sec
2. Suggestion	10 Sec	20 Sec
	Interest is Captured	
3. Captivation	6 Sec	10 Sec
4. Persuasion	10 Sec	20 Sec
5. Re-seduce	30 Sec	Flow**

ETD* – Estimated Time Duration: time estimates refer to time after initial download of visible assets.
Flow** – timing is limitless, according to Flow theory, since it is recognized that people generally lose track of time once immersed in a state of flow.

2.4. Summary

Our conceptual model of hypermedia seduction enables a better understanding of the significance of seduction, and may be utilized as a framework for the development and evaluation of more engaging hypermedia products. This model offers a potentially useful conceptual tool for understanding the process of hypermedia seduction. Specifically, it supports the idea that hypermedia seduction relates to the orchestration of specific actions and elements that may facilitate the onset of desired changes within a user's behavior.

This model also provides a framework for describing the process of hypermedia seduction, and the precise nature of seductive interactions. It provides a detailed explanation of the way in which seduction may enhance the probability of the formation of more meaningful relationships between the user and the system.

However, it is recognized that this model at present serves as a conceptual tool only. Empirical testing and further decomposition are required to test the hypotheses. In addition, this preliminary version did not account for the possibilities which have recently emerged, which suggest that it may be equally viable to fuel and to sustain progress within the seduction life-cycle by eliciting both negative and positive emotions at each of the phases identified.

3. Conceptual Model: An Application and Evaluation

We now expand further on the ideas raised in the above hypothesis, and also provide practical insight into how the mechanics of hypermedia seduction currently have a role in enhancing and sustaining the appeal of mainstream engaging hypermedia products. An example of a brief evaluation using our framework is presented. This example is used to explain and highlight the seductive strategies characteristic of an e-commerce website capable of delivering a more engaging user experience. For this evaluation, an increasingly popular e-commerce website, namely ASOS.com, has been selected.

ASOS has existed in the domain of e-commerce since the year 2000, and is now considered one of the top two most visited on-line clothing stores in the UK [17]. According to the Interactive Media in Retail Group (IMRG), ASOS attained a sales growth rate of 86 percent, equivalent to £8.1million in the six months prior to end September 2005 [18]. Our intention here is to pinpoint more precisely what it is about this website that has enabled its continued success, despite there currently being no affiliated ASOS real world retail outlets.

Figure 2. Screen dump of the ASOS.com website interface.

3.1. ASOS – Inducement

Drawing on the popularity of "celebrity culture", ASOS targets Internet-savvy 18 to 34 year-old male and female users. It offers affordable replicas of celebrity-style womenswear, menswear, jewellery, beauty, accessories and footwear.

Key attraction factors of this website are the products and services on offer, rather than the overall look and feel; on user exposure to the website, the site appears more functional rather than novel.

Key perceivable benefits apparent prior to interaction is the availability of celebrity-style clothing, and the ability to purchase this clothing online.

Key motivational factors are supported by the essence of celebrity endorsement this website provides. We use the term essence here rather than endorsement itself, since it is stated (in small print) that the celebrities named or featured on this website have not endorsed, recommended or approved any of the items offered for sale.

3.2. ASOS – Negotiation

Negotiation, in this environment, occurs in the form of slight dynamic changes to the look and feel of the interface. These are apparent in the notable alternation of the display of three sub-nodes entitled "Trends", "So I heard", and "Today I am", in contrast with the remainder of the interface, which once loaded remains static.

The language utilized on this website echoes the perceived philosophy of the user. This is particularly evident in the navigational links, which use language such as "New in Today", "Most desired", "In the Press", and "Today I am". Emotional connections may be formed at the negotiation phase as a result of these suggestive hyperlinks. For example, the "Today I am" link provides an emotional connection for users who perhaps yearn to be like, dress like, or even look like the celebrities they admire.

Emotional connection is strengthened by the website featuring press coverage of celebrities wearing outfits similar to the clothing available for purchase; exposure to such material may help to increase feelings of desire, want and need. The psychology behind the "Today I am" link is interesting, since this idea is central to our theory which addresses the importance of the system's ability to recognize users' emotions and moods. In this instance, the system poses an indirect question, along the lines of "Which celebrity appeals to your own ideas of how you would like to dress or look today?"

3.3. ASOS – Suggestion

This particular style of "open communication" undoubtedly functions to pose a similar subliminal question within the thoughts of the user through exposure to the "Today I am" link. These types of questions, raised during negotiation (as illustrated above), incite greater thought from the users when compared with a typical "click here" navigational link or a direct decision-making prompt.

In some respects, the above subliminal question may be considered as "mood-centric", since the word "today" suggests a notion, thought, or feeling that is not static. In other words, the link assumes that the same question posed tomorrow or another day may perhaps generate an entirely different answer.

Selection of the "Today I am" link suggests that with further navigation a user may "steal" the style of the featured celebrities. This suggestion thus presents a two-way need-influencing mechanism. It may suggest to existing users that there is now an opportunity to recreate a particular style or "look" they may seek to achieve. On the other hand, for new users, this suggestion may perhaps influence an increase in the desire (if they don't already have it) to mimic "celebrity style".

Further acts of suggestion are notable on selection of a product, where thumbnail images of complementary products are also presented to the user. In addition, most product descriptions are accompanied by the statement "in the style of celebrity X", which may further the supposed "celebrity-endorsed" ethos of this website, perhaps increasing desirability. Additional justification for the credibility of the products is given at the "In the Press" node. Here, articles featured in reputable fashion magazines are highlighted, further heightening the credibility and/or desirability of the products on offer.

Communication via newsletters is also consistent with the ethos of the website. A tabloid layout is utilized, containing paparazzi-style shots of celebrities and models wearing clothing similar to the latest arrivals in ASOS stock. This site also supports the sharing of "finds" with friends, cleverly capitalizing on the social value of peer assessment or peer-to-peer approval.

3.4. ASOS – Reward

Users may experience reward even before purchasing on this website, through the ability to access style, trend, and other celebrity-related news articles. Users may also subscribe to the site's mailing list and enter competitions, without any purchase being necessary. They can also browse as many sections of the site as they wish without the obligation of registering or purchasing.

3.5. ASOS – Captivation

The ASOS "Catwalk" function is very much in line with the idea of supporting captivation, as described in our conceptual model. Captivation appears, in this case, at a stage where the user may perhaps have fulfilled his or her primary goal of finding an appealing item. Hence, they may have reached a point where they now require further and more detailed information before they make a decision about purchasing.

The "Catwalk" function enables users to launch and view video footage of select items of clothing modelled on the catwalk, as a complement to the visible still images presented. This method of captivation functions on two counts. Firstly, on the practical level it enables the demonstration of the cut, length and potential fit of a garment, and how this garment might move when worn. This fulfils a need that naturally arises whilst shopping, the need to "try something on"—or in this case, have a model do that for you.

Secondly, this form of captivation supports the notion of conveying a meaning which may echo a user's momentary values. In this instance, the "catwalk" may assist in instilling the idea that as a customer you are able to purchase an outfit "hot off the catwalk", an idea perhaps desired by most fashion-conscious people.

In addition, zoom features are available for images of all products, and when different colourways are an option, a variety of views are presented. Again, this ability to draw the user temporarily closer to an object supports the goal of captivation, since the possibility of a bond forming is heightened by the display of greater intricacies.

3.6. ASOS – Persuasion

The price of the clothing featured on this website is relatively inexpensive when compared with the actual price of the designer clothing worn by featured celebrities. Catwalk images and those of celebrities wearing similar styles function to increase the desirability of the inexpensive look-alike products available. Purchasing from ASOS presents a smarter option. Hence, this notion of having the means to obtain stylish, desirable clothing at a fraction of the price of designer clothing is persuasive.

The actual purchasing and transaction process is secure and easy to use; the Verisign certificate featured on this website offers an assurance of safe online shopping. The shopping basket allows the user to select and review each item before purchase, and all items added to the basket are reserved for two hours to allow users to leave and return to the site later to purchase if they wish. Purchases can be transacted in six difference currencies, and eight different types of credit card are accepted. However, language support is limited to English only.

3.7. ASOS – Re-Seduction

As a mark of completing the persuasion process and beginning the re-seduction phase, a user who has purchased from this website will receive an immediate e-mail notification containing transaction status information. The re-seduction process begins when further e-mail notification is provided on dispatch of items to be delivered. Communication at this point serves as a means of maintaining the relationship formed with ASOS, along with providing the customer with meaningful information.

The second area where re-seduction occurs is in the maintenance of regular communication with existing and potential customers. ASOS stock is sometimes updated daily. This form of communication supports the re-seduction phase, since it may assist in drawing the users back to the suggestion phase of the seduction cycle, where it is perhaps possible to incite further progress in the persuasion-seduction cycle. In addition, the website regularly features competitions where tickets to desirable social events may be won. This provides another reason for users to return and as a result perhaps reenter into the seduction cycle.

3.8. Summary

The ASOS website appears to comply with many of the principles outlined by our conceptual framework, and therefore provides some hints as to which attributes should be considered in the development of more seductive hypermedia products. It is also apparent that there is a significant focus on aspects which are supportive of the suggestion phase. Again, this complies with our model, since we believe that it is at this phase that the risk of abandonment increases; hence, the creators of the ASOS website have focused on taking the necessary steps to counter this risk.

However, with regard to the role of the "website as seducer", the ASOS site is in fact considerably passive. It is the marketing and presentation of the content that is seductive, rather than the behavior of the medium in which it is presented. Similarly, the look and feel of the website interface, in comparison with others within the same genre, is not outstandingly novel.

Nonetheless, the site's content, its products, its mode of communication, and most significantly its "ethos" (and particularly how this ethos is utilized), are elements that make this website more seductive than other e-commerce websites in the same genre.

Suggestions for improving the ASOS website would be that further consideration may be required in the area of customer purchasing styles. Customers are not only selecting a product they desire, but also the style of a particular celebrity who appeals to them most when the purchase is made. This presents an opportunity for the collection and the collation of data relating to a user's personal preferences or purchasing style. Therafter, the purchasing style of an individual may be predicted through the analysis of the customer's purchasing patterns via his or her order history. Findings from an analysis of this nature may then be used to influence future user interactions by ensuring existing customers are offered a more personalised or tailored user experiences.

4. Seduction, Extremism, and Terrorism

Seduction features strategically in many aspects of our lives—socially, culturally, and politically. Arguably, at times it may provide an overwhelming influence on much of our decision-making, motivations, and behavior. Each of us is susceptible to seduction by someone or something during the course of our lives. Yet the means to seduce are both conditional, and variable from person to person.

Much of our work to date has focused on highlighting the benefits we believe to be attainable if more consideration is paid to the concept of seduction in the domain of mainstream hypermedia development, evaluation, and usage. We believe that seduction should be considered beyond what is popularly assumed of this phenomenon, and should not be limited to the idea of one person seducing another (or many) so as to cause wrong, inflict harm, or manipulate. However, we also accept that seduction may be utilized in a more extremist manner, and perhaps for the purpose of inducing conduct of a more *divergent* nature.

In this section, we discuss how the principles raised within the framework of our conceptual model translate when applied to the genre of hypermedia emerging from the "Dark Web". Our intentions are to investigate whether our current conceptual model of hypermedia seduction may suitably support the analysis and evaluation of websites that employ more negative aspects of seduction. We aim to provide a better understanding of how the process of seduction may both feature and be utilized within this domain. Specifically, we focus on unveiling the methods most likely be used to attract, allure, influence, captivate, and persuade users.

Moreover, our aim is to examine the seductive strategies apparent in hypermedia forms which may relate to the success or effectiveness of the deployment of extremist strategies. However, on this occasion our approach is slightly different, since we shall not present an example of an evaluation of any single website from this genre. As hypermedia researchers, in no way do we intend to single out any existing organization(s) or networks of people who identify themselves as part of this particular genre. Neither do we intend to "promote" any particular organization or group by featuring screen shots of its website in this chapter. References to examples of websites from this genre can be viewed on the *Site Institute* website maintained by Katz and Devon [19].

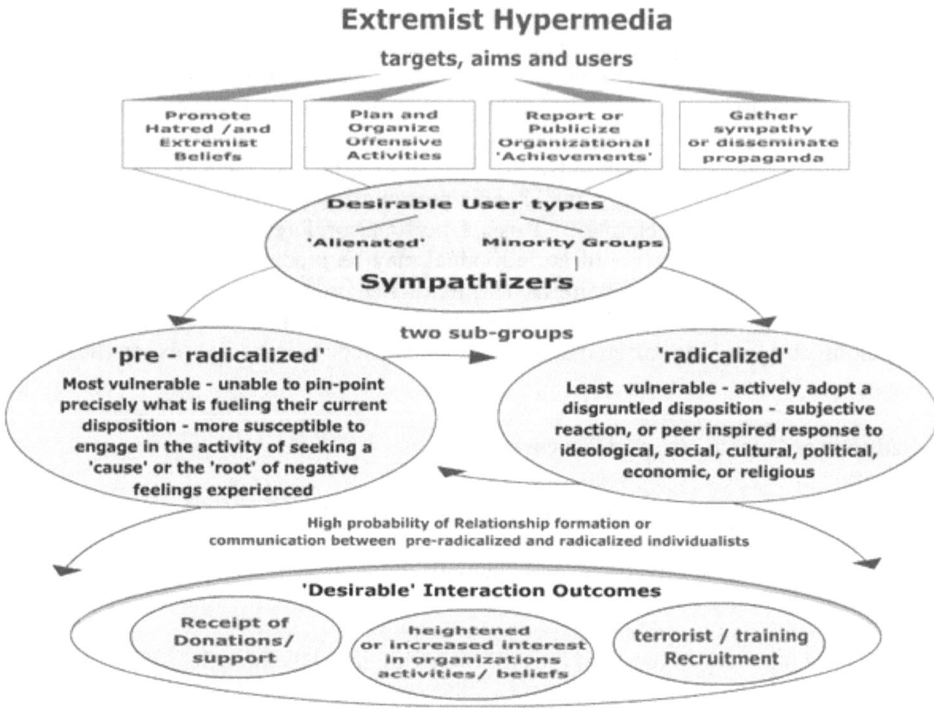

Figure 3. Diagram of extremist hypermedia: targets, aims, and users.

4.1. Extremist Hypermedia: Targets, Aims, and Users

On reviewing the relevant literature in the field, it was decided that before attempting to examine or evaluate samples of "extremist" websites, a more suitable approach to this investigation would require consideration of the following factors:

1. who the target audiences of these websites might be;
2. the aims and intentions of the creators of these websites;
3. the effects on the user of either accessing or using websites of this nature.

According to research, websites hosted by extremist organizations strategize to target Internet-savvy individuals, identified as either "sympathizers", "international audiences", or "the enemy public" [20]. On further examination of the above user categories identified by Weimann [20], and after a review of the literature focusing on recent strategies for countering international terrorism, it was decided that it was perhaps necessary to add two sub-division to the group identified as sympathizers; see Fig. 3. We believe that within this user group, in particular, there is potential for the existence of two types of individuals, whom we define as "pre-radicalized" and "radicalized" individuals respectively [21].

In the context of this chapter, the term "radicalized individual" defines one who actively adopts a disgruntled disposition, as a result of a subjective reaction or peer-inspired response to the recognition that the ideological, social, cultural, political, economic, or religious beliefs held by such individuals do not match those of the society in

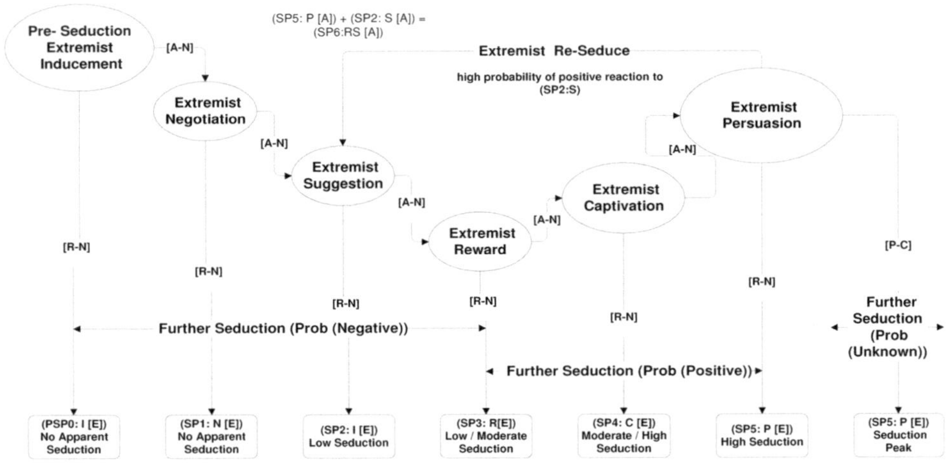

N = Negative Emotion State; P = Positive Emotion State; R = Rejected; A = Accepted; C = Completed; E = End; (SP1) = Seduction Phase One Identifier; (PSPO) = Pre-Seduction Phase Identifier.

Figure 4. Extremist hypermedia seduction: a conceptual model.

which these individual exist. Hence a radicalized individual may possess a notion of exclusion, or at worst the feeling of social, cultural, political, or economic alienation.

"Pre-radicalized" individuals are similar to radicalized individuals; however, they have not yet recognized, or are unable to realize, the root of their disgruntled disposition. Like radicalized individuals, pre-radicalized individuals may possess a notion of exclusion, disadvantage, or a feeling of alienation within their social milieu. However, a pre-radicalized individual is unable to pin-point precisely what it is that may be fuelling their discontented outlook, and is therefore more susceptible to outwardly seeking a "cause" or the "root" of their current disposition.

Despite the above definitions, we feel that it is important to state that only a very small minority of either radicalized or pre-radicalized individuals actually do cross over to become terrorists [21].

4.2. Extremist Hypermedia Seduction

In an attempt to contribute to the wider goals of this workshop, our aim is now to provide developers and researchers with practical insight into how the mechanics of "extremist hypermedia seduction" may be utilized to enhance and sustain the appeal of terrorist websites.

The conceptual model in this instance has been modified slightly to represent the relationships between factors which we believe occur at each phase of a more "extremist seduction lifecycle". Modifications to our original model emerged from the belief that "desirable" (although possibly negative) emotional responses may influence both the user's attitude towards the system, as well as his or her behavior in terms of the depth and duration of system usage. An illustration of this model is presented in Fig. 4, and is accompanied by descriptions of the phases we believe occur during the process of "extremist hypermedia seduction".

In addition, it was decided that for the purpose of this chapter our focus will rest on examining the effects which extremist websites may have on both pre-radicalized and radicalized users. We examine how both positive and negative characteristics may play an active role in influencing user behavior, particularly in cases recognized for their purpose of inciting support for more extremist agendas.

4.2.1. Extremist – Inducement

Instantaneous allure or attraction may occur simply through the existence of an extremist website, even before consideration of its content takes place. This is because in some cases the presence of these websites may shock, enchant, exhilarate, threaten, or intimidate a user even prior to interaction.

Unlike Greene [22], we believe that the "anti-seducer" possesses an equal amount of power to seduce as that of the traditional seducer. Not everyone is seducible by the same means. Hence, the role of the anti-seducer is to generate and maintain a level of allure or attraction via inciting or arousing extreme (perhaps negative) emotional responses. For instance, if negative emotional responses are incited on preliminary exposure to a website of this nature, immediate rejection should result; yet, in some cases, we believe further intrigue may occur via the influence of "anti-seduction" tactics. Careful and strategic use of negativity is an effective means of sustaining allure.

At the phase of inducement, a radicalized individual may find the existence of a website promoting similar shared values exhilarating, since websites of this nature often act as idyllic forums for promoting, celebrating, and encouraging radical thought. It is not surprising that concerns are raised by governments, security agencies, or the media in response to knowledge of the existence of websites of this nature. Consequently, in most instances it appears these websites are held partially responsible for provoking and instilling radicals to action.

However, the above notion appears questionable when the actual "added value" or "benefits" to be gained by radical individuals who access these websites are considered. In retrospect, these benefits are marginal, and may consist of basic offerings such as assurance, confirmation, restoration, and some fulfilment. We believe that it is less likely, in such cases, that the benefits will provide an impact or create a tangible impression in the lives of these individuals.

By comparison, when the possible impressions or impact of extremist websites on a pre-radicalized individual are considered, they may be far greater, due to the way in which these websites communicate, deliver, and present their ideas. A brief analysis of the content of a small sample of extremist websites showed that notable interface elements and characteristics consisted of the following: audible rhetoric (military or religious compositions); provocative graphics and imagery; the display of political or controversial leaders or "muses"; provocative stories and/or animations; and religious or nationalist iconography.

Arguably, as well as perhaps creating the notion of access to a particular ideological community or a network of like-minded people, such content may also assist in instilling a deeper sense of difference, isolation, anger, fear, or insecurity in those who do not already possess such feelings. Such emotional valence would later function to suggest a route by which a sense of purpose, direction, or belonging may be restored to a potentially vulnerable or disesteemed individual.

For example, as a result of visiting an extremist website, a feeling of empowerment may be restored to a person experiencing anomie[8] or feelings of social, cultural, political, and economic alienation.

The key perceivable benefit apparent prior to the user interacting with the system is the ability to gain access to a particular ideological community, or a network of like-minded people. Effective, thorough presentations of ideological "likeness" appeared to be an essential factor in facilitating the necessary emotional elicitation required to support negotiation. More than 70 percent of the websites in the sample of extremist websites viewed at the time of writing were responsible for hosting either a forum or a blog, where there appeared to be evidence of the "monitoring" of all correspondence—perhaps to ensure the publication of only "ideologically sound" messages. Facilities to vote, obtain news, and request copies of the organization's newsletters also featured.

4.2.2. Extremist – Negotiation

Extremist negotiation may occur in the form of minor dynamic changes to the look and feel of the interface environment. Interface elements featured on websites of this nature operate beyond general functional purposes, and assist in promoting the ethos of an extremist organization. In some instances, they may echo (or oppose) the perceived ideologyy of the user. Examination of a sample of extremist websites showed that the imagery presented was either for the purpose of advertizing organizational activities and their ethos, or to shock, evoke sympathy, intimidate, or stimulate the need for further enquiry.

One of the most prominent negotiation features in the genre of these websites was the inclusion of animations or movies which denoted "desirable" acts of war, destruction, or murder on either target countries or on figures of hatred. The occurrence of extremist negotiation is less subtle, and is perhaps more aggressive, urgent, or drastic in nature, when compared with the delivery of mainstream negotiation. It is believed that this approach serves to heighten a feeling of tension, thereby helping to instil a form of unease within the user. In the case where either a pre-radicalized or radicalized individual visits an extremist website, negotiation will be concerned with the delivery of materials which either condone or support existing radical thought, or materials that may incite or increase further radical thought.

The "system" (or, more precisely, the organization) is able to negotiate its compatibility with the user's mood, and may in turn elicit a desirable emotional state. Hyperlinks featured on websites of this nature echo the ethos of the organization, and also either echo or challenge the perceived ideology of the user. Links feature language which further promotes the philosophy of the organization and the nature of the website's content.

The use of indicative hyperlinks may also function to provide an emotional connection with users, who perhaps yearn to "be like", admire, or support the achievements or activities of the muses and leaders featured. On the other hand, it is anticipated that a negative emotional connection will be formed between users who oppose

---

[8] French sociologist David Émile Durkheim (1897) used the term "anomie" to describe a condition of malaise in individuals. It is characterised by an absence or diminution of standards or values (referred to as normlessness), with an associated feeling of alienation and purposelessness. He argued that anomie is common when the surrounding society has undergone significant economic change, and when there is a significant discrepancy between the ideological theories and values commonly professed and what is actually achievable in everyday life.

the ethos of such organizations; it is believed feelings such as fear or anger may be the only connection fostered in these instances. Yet despite the elicitation of negative feelings, a user may in some cases continue with site usage.

4.2.3. Extremist – Suggestion

Utilization of indicative navigational links, coupled with provocative website content and images, provides a means for extremist websites to facilitate the process of suggestion, via a form of grooming. The term "grooming" in this context defines a normalising process, which when successful functions to lower a user's resistance to the acceptance of further suggestion.

Suggestion may occur in such instances as a result of the initial "over-exposure" to provocative language, content, and images, extremist ideology or violent behavior. This, in turn, may succeed in lowering a user's resistance to the receipt and acceptance of the often unethical values, ideas, beliefs and practices which many extremist organizations support, practice, or condone. However, we believe that if this form of suggestion fails to deliver the desired outcome, the adverse affect may result in inciting emotional aggression.

Emotional aggression[9] [23] in this context is believed to occur as a result of the sustained arousal of negative feelings. It is possible that provocative and indicative features may be utilized on these websites precisely as a strategy to incite negative affects (such as fear, anger, disgust, hatred). This could represent an attempt to increase the user's level of motivation towards challenging such suggestions, since the incitement of a challenge against the ideology being promoted may present the possibility of further user interaction with these websites.

The above scenario presents an interesting notion and an alternative way of considering how negative emotions may be used to fuel or sustain user-system interaction. However, without empirical testing of whether such negative feelings (specifically aggression) may be utilized to facilitate further acceptance to suggestion, it is difficult to determine how successful a strategy of this nature might be. Nonetheless, we believe that it may be possible to sustain seduction in such instances, since the arousal of strong negative emotions may prove to be a powerful facilitator in fuelling a user relationship with these websites. We recognize that the depth, duration, and success of this approach to suggestion is questionable, and is perhaps an area of research which requires further investigation.

4.2.4. Extremist – Reward

Extremist reward functions to incite further interaction between the user and the system. It occurs via the inclusion of interface elements, which facilitate communications, enable user involvement, or present opportunities for users to acquire relevant knowledge (such as access to updated information or news).

Within extremist domains, accessible rewards appear in abundance; delivery is not delayed as it is in more mainstream hypermedia, where fulfilment delay is specifically used as a way of decreasing the probability of system abandonment. The absence of this delay tactic may be said to relate to the "personality of the system", which in most

---

[9] Emotional aggression is believed to be the product of a predisposition to negative affect or arousal, resulting in an inability to regulate that affect/arousal. This effect will be most pronounced in individuals experiencing stressors or adverse social environments, where negative affect and arousal is increased [23].

occasions appears to be more aggressive and displaying greater urgency. The "persona" of the site is thus such that it may perceive any form of delay as an unnecessary obstacle, rather than as a tactical device for further enticing a user.

The perceivable benefits of using this type of website are the opportunity to communicate or network with other like-minded people; in some cases , the chance to apply for organizational membership may be presented too. Perhaps the most valued reward is that the user is provided with a sense of belonging, inclusion, purpose, and community.

4.2.5. Extremist – Captivation

Captivation is most likely to occur in this particular genre at a point where the user has perhaps fulfilled the initial goal of finding an item, or information which appeals most. Hence, a user may have reached the point where further and more detailed information is now required about an item before a decision of approval is reached.

Key additional captivating factors are perhaps the presence of further radicalizing elements, which may include the following: video footage containing the doctrine or rhetoric of organization leaders; shocking imagery revealing persecutions and scenes of suffering; audio containing the doctrine or rhetoric of a religious leader; radical literature including news, legends, religious, and historic content; images portraying the treatment of prisoners and civilians as a result of ideological struggle; exposure to forceful and inspiring figures; and media coverage of isolated incidents.

Captivation occurs through furthering the possibility of forming an ideological bond. In some cases, this is achieved by featuring coverage of the achievements of organizational members, muses, or ideological leaders. Exposure to such material may assist in facilitating the incitement of emotions such as desire, want, or need. Alternatively, the negative poles of these emotions may be induced, depending on the user. Desire, want, and need are complex notions, since it is not anticipated that all users may desire, want, or feel the need to mimic the activities of other members, idols, or muses. However, as a result of viewing this type of material, notions of desire, want, or need may increase.

For example, the desire to provide support, or the need to indoctrinate others in the physical world about events or items featured on these websites, or the need to achieve wider dissemination of extremist ideology via communications on the web might occur.

At this point it is not possible to provide an exhaustive list of additional captivation strategies, due to our limited understanding of the languages in which some of existing websites of interest are published. Hence the above serves only to highlight one of the possible strategies (of perhaps many) that may be utilized at this phase in the seduction lifecycle.

4.2.6. Extremist – Persuasion

Traditionally, persuasion occurs during captivation, and its function is to induce behavioral and attitude changes without coercion or deception. However, within extremist domains, the likeliness of the use of coercion or deception increases, since the act of persuasion presents a foreseeable vulnerability, namely the risk of rejection. Hence extremist domains may tactically attempt to beguile a user rather than persuade them in the traditional sense.

Online fund-raising is one of the reasons or motivations for extremist groups and terrorist organizations to host websites. In some cases, extremist websites present a means of directly appealing to or prompting the user to support a specific cause during the captivation process. However, in other cases, extremist websites will not directly promote or support fundraising activities; instead, they may provide links to other domains and websites hosted by external "supporters" or "fund raisers".

One purpose for this strategy may be to provide a gateway to a domain where more extremist activities take place. In doing so, the onus with regard to the content of these linked domains, or to any direct or adverse effects experienced by the user, is detached from the original host organization. Whatever experiences a user encounters thereafter will occur on navigation away from the initial host domain, and the likeliness of the original host being seen as responsible for any adverse actions (such as a viral or network attack) are lowered. Hence, such adverse experiences would not affect the likeliness of the user returning to the original host website in the future, despite the fact that the host may in fact be affiliated with these linked domains.

On examination of some of these linked domains, it was evident that these websites do not support a stable web presence, and were often hosted for a limited time period before they vanished. Sites of this nature were often untraceable back to a specific owner or location, since the majority are hosted by international domain hosting companies. Such companies may hold no direct information concerning the exact location or the owners of the websites they host, since often times these domains are acquired via a fraudulent means.

In such instances, linked websites may attempt to beguile users via alluring navigational links, which in fact when selected deliver a false promise. For example, the user may click a particular hyperlink, but by doing so will not receive access to the information they thought they would obtain, but instead may activate a malicious script, perhaps capable of disabling a firewall (if present). This would allow the downloading of spyware or viruses to assist assailants in accessing bank details and/or other personal information, which can the be used to support the purchase of fresh domain space to host future websites.

The above is intended to provide an example of the potential activity which may take place as a result of interaction via websites within this particular genre. However, this description is not an exhaustive account of the means in which these particular websites may operate; there may, in fact, be numerous and more sophisticated persuasive tactics in operation at present.

4.2.7. Extremist – Re-Seduction

Re-seduction may occur as a result of further communication which takes place beyond the boundaries of the initial domains of extremist organization websites. Currently, the likelihood of an extremist organization's website openly advertizing and providing facilities for users to access terrorist training and/or related recruitment directly are much lower nowadays than in the past, when material of this nature was more readily available.

To date it has been impossible to find websites that directly offer or support features of this nature. Hence it is believed that activities of this nature are most likely to occur as a result of a contact or an acquired communications link featured within another domain, such as a chat room, user group, or in the physical world. Users subscribed to an extremist organization's newsletter or who contribute to blogs may receive information or maintain regular communication with a particular network of people. This form of contact or involvement supports the re-seduction phase, since such activities assist in drawing users back to the suggestion phase of the seduction lifecycle.

4.3. Summary

With further consideration of the structure and the interactive features of the websites we viewed from the "Dark Web" genre, it was evident that the sites themselves were not the seducer. In these cases, it was the content which was seductive, rather than the behavior of the medium itself, or the means by which the content was delivered.

Moreover, from this single application a greater understanding of the approaches and the "personalities" of websites typical of this genre emerged. Extremist websites were identified as being more aggressive by nature, and as having a tendency to incite a larger proportion of extreme negative emotion states, when compared with mainstream hypermedia.

However, this is not to say that mainstream hypermedia do not utilize strategies to incite negative emotions, such as envy, specifically where the aim is to incite a purchase. But where extremist websites differ is that they have a tendency to go beyond inciting acceptable levels of emotional arousal to the arousal of extreme levels of hatred, shock, anger, disgust, or fear. This is in contrast with the aims of mainstream websites, which in general aim only to incite perhaps a dash of envy to motivate enough desire to incite a purchase. In general, mainstream hypermedia have a tendency to focus on the arousal of pleasurable feelings instead, such as fulfilment, happiness, excitement, and joy. They focus on inciting more positive emotions, as a means of sustaining the duration of user interaction and the process of seduction.

Although there has been much amplification of the negative connotations and manifestations surrounding the concept of seduction, we accept that seduction may be utilized in an extremist manner to influence human behavior. However, despite what is popularly assumed of this phenomenon, the act of seduction may be perceived differently when it is considered that seduction is potentially an overwhelming and all-consuming process. The risk of fruitless investment is high, since it is necessary for the seducer to be willingly seduced themselves. Moreover, it is seldom recognized that a genuine seducer, who devotes full attention to the seduction, is effectively enslaved by the seduced subject until the seduction is accomplished [24].

4.4. Limitations

At this stage it may not be correct to state that our conceptual model is capable of supporting a precise evaluation of extremist hypermedia forms. We acknowledge that it is impossible with the use of any single tool to determine exactly what the intentions of the creators of a website are, or what the actual effects are on individuals who access and utilize such websites. Such an assessment would only be made possible by interviewing or asking the people involved directly. It is difficult to provide an extensive, accurate evaluation of something when you are unable to use, test or directly experience the consequences or benefits of that thing yourself, both subjectively and objectively.

In addition, due to our limited understanding of the content or the languages in which some extremist websites are published, and also through having a limited understanding of the some of the topics, issues, and events featured, it is difficult for us to determine whether the nature of the strategies we have identified are correct. Doing so would require several in-depth evaluations using a varied sample of real users, who would accurately be able to report on how using these websites may have influenced or affected them personally.

In addition, we recognize that when an ideology or principle is underpinned by not only a culture but also by a set of specific religious beliefs and values, then an in-depth understanding of the pure, original (non-extremist) religious scriptures is necessary. This kind of knowledge would enable the detection of any potential deviations or extremist distortions in that faith.

5. Conclusions

The above findings raise some thought-provoking points, and demonstrate the usefulness of the conceptual model of hypermedia seduction. This approach can fruitfully be used to increase the current understanding of the mechanics of seduction, as found within the domain of extremist hypermedia. Our model enables a clearer definition of extremist seduction, and a clearer understanding of the strategies utilized. The model also provides a valuable dimension to supplement the current knowledge in the field of counterterrorism, regarding what to expect in terms of the content of extremist websites. Specifically, it supports a more in-depth understanding of the function and purpose of website interface elements.

Moreover, the model presented in this chapter offers a potentially useful hypothesis for understanding the overall process of hypermedia seduction. It supports the idea that hypermedia seduction relates to the orchestration of specific actions and elements that facilitate the onset of desired changes within a user's behavior. Furthermore, in both case studies discussed here, it was possible to detect the presence of some (if not all) of the specific factors and elements believed to occur at each of the phases identified within the seduction cycle.

Application of this model has enabled us to examine how these featured elements correspond to the goals identified in the original model. Utilization of the model has also enabled the detection of similarities and differences between the two case studies. Interestingly, in both cases the role of the system as seducer was considerably passive. It was apparent that in each case it was the website *content* that was the most seductive factor, rather than the behavior of the medium in which that content was presented. This finding alone suggests that greater consideration of the concept of hypermedia seduction is urgently required.

Furthermore, the results of the e-commerce case study provide additional support for the suitability of this model as an evaluation tool. At this stage, it is perhaps fair to assume that our conceptual model sufficiently supports the evaluation of extremist hypermedia forms as well. Finally, it also provides an interesting starting point to enable a better understanding of more divergent hypermedia.

6. Future Work

At this stage, our conceptual model provides a representation only of the phases and the relationships we believe occur during the process of hypermedia seduction. We therefore recognize that further decomposition of this conceptual model will be required to provide more precise definitions.

This model is currently in the process of being tested via specific tools developed for use within our virtual experiment lab (www.mbakwe.com). We anticipate that the results which emerge will enable us to to report more specifically about the reliability of the model as a means for conceptualizing and evaluating seductive hypermedia.

References

[1] Mbakwe, C. & Cunliffe, D. (2003). "Conceptualizing the process of hypermedia seduction". In *Proceedings of the 1st International Meeting of Science and Technology of Design,* E. Côrte-Real, C. Duarte & F. Rhodrigues (Eds.) Lisbon, 2003: 20–27.

[2] Nielsen, J. (1996). "Seductive User Interfaces". Online: http://www.useit.com/papers/seductiveui.html Access Date: 31/07/06.

[2b] Nielsen, J. (1997). "User Interface Design for the WWW". Proceedings of CHI 97, Los Angeles, USA. Access Date: 31/07/06.

[3] Bruck, J.S. (1998). "Seductive Computing: an exploration of the senses of persuasion." Online: http://www.jonbruck.com/thoughts/papers/Summary_of_Seductive_Computing_11.29.1998.pdf Access Date: 31/07/06.

[4] Fogg, B.J. (2003). *Persuasive Technology: using computers to change what we think and do*. London, Morgan Kraufmann.

[5] Skelly, T., Fries, K., Linnett, B., Nass, C., Reeves, B. (1994). "Seductive Interfaces: Satisfying a Mass Audience." *Communications of CHI 1994*, Boston Massachusetts. Pages: 359–360.

[6] Skelly, T. (1995). "Seductive Interfaces – Engaging, Not Enraging the User", Microsoft Interactive Media Conference, July 20, 1995. Online: http://www.designhappy.com/sedint/TheMaze.htm Access Date: 31/07/06.

[7] Khaslavsky, J. & N. Shedroff (1999). "Understanding the Seductive Experience". *Communications of the ACM; 42*(5): 45–49.

[8] Papsadore, F. (1999). "Getting the order: the secrets of web site seduction", *Netgain*. Online: http://netgain.co.nz/library/basics_seduction.htm Access Date: 31/07/06.

[9] Hofmeester, H., Kemp, M., Blankendaal, M. (1996). "Sensuality in Product Design: A Structured Approach." *Proceedings of CHI 1996*, Vancouver, BC, Canada, ACM.

[10] Agostini, A., Michelis, G.D., Susani, M. (2000). "From user participation to user seduction in the design of innovative user-centred systems." Proceedings of the Fourth International Conference on the Design of Co operative Systems, Sophia Antipolis France.

[11] Mantovani, F. (2001). "Networked Seduction: A Test-Bed for the Study of Strategic Communications on the Internet." *CyberPsychology & Behavior (4)* 1.

[12] Dormann, C. (2001). "Seducing Consumers, evaluating emotions." *People and Computers XV*, Lille, Joint Proceedings of IHM-HCI 2001.

[13] Scanlon, T. (1998). "Seductive Design for Web Sites", *User interface Engineering*. Online: http://www.uie.com/articles/seductive_design/ Access date 31/07/06.

[14] Spool, J. (2002). "The Search for Seducible Moments," *User Interface Engineering*. Online: http://www.uie.com/Articles/seducible_moments.htm Access Date: 31/07/06.

[15] Bentham, J. (1843). "The Rationale of Reward," Bowring edition of *Bentham*, Volume II. Online: http://www.la.utexas.edu/research/poltheory/bentham/rr/ Access Date: 31/07/06.

[16] Csikszentmihalyi, M. (1975), *Beyond Boredom and Anxiety*, Jossey-Bass, San Francisco, CA.

[17] Wallop, H. (2005). "Business Profile: 'As seen on Sienna? I'm a Prada man myself'." *Telegraph* – UK Online: http://www.telegraph.co.uk/money/main.jhtml?xml=/money/2005/01/22/ccprof22.xml&menuId=242&sSheet=/money/2005/01/22/ixcoms.html Access Date: 31/07/06.

[18] IMRG, (2006). "Sales boom at ASOS.com." *Interactive Media in Retail Group* Online: http://www.imrg.org/ItemDetail.aspx?clg=News&cid=nws&pid=News_asos&language=en-GB Access Date: 31/07/06.

[19] Katz, R. & Devon, J. (2006). *Site Institute*. Online: http://www.siteinstitute.org/websites.html Access Date: 08/08/06.

[20] Weimann, G. (2004). "www.terror.net: How Modern Terrorism Uses the Internet." United states Institute of Peace. *Special Report No. 116* Online: http://www.usip.org/pubs/specialreports/sr116.html.

[21] HM Government. (2006). "Countering International Terrorism: The United Kingdom's Strategy" July 2006. *HM Government. TSO*. Online: http://www.mi5.gov.uk/files/pdf/ct_strategy.pdf.

[22] Greene, R. (2001). *The Art of Seduction*. Profile, London.

[23] Scarpa, A. & Raine, A. (1997). "Psychophysiology of anger and violent behavior." In *Anger, Aggression, and Violence; Psychiatric Clinics Of North America. 20*(2). June 1997. Elsevier, USA. Online: http://www-rcf.usc.edu/~raine/PsychoPhysiology.pdf. Access Date: 31/07/06.

[24] Baudrillard, J. (1990). *Seduction*. New York, St Martin's Press.

Further References

Boeree, C.G. (2000). "The Ultimate Theory of Personality." Online: http://www.ship.edu/~cgboeree/conclusions.html Access Date: 31/07/06.

Carotenuto, A. (2002). *Rites and Myths of Seduction*, Chiron, Rome.

Chen, H., Chung, W., Larson, C., Reid, E., DLS., Xi, W., Bonillas, A., Yin, C., Su, Y., Lai, G. (2003). "Advanced Methodologies for Collecting and Analyzing Information from the 'Dark Web' and Terrorism Research Resources." *The Terrorism Knowledge Portal.* Online: http://66.249.93.104/search?q=cache:wuhrKEhifWIJ:ai.bpa.arizona.edu/hchen/docs/DHS-Dark-Web-2003.ppt+Dark+web&hl=en&gl=uk&ct=clnk&cd=6.

Ciceri, R. (2001). "Seductive Communication: Paradoxical Exhibition, Obliquity and NonVerbal Synchronization. in Communication – Say not to Say": *New perspectives on Miscommunication* (3): 106–118.

Frijda, N.H. & Mesquita, B. (1994). "The Social Roles and Functions of Emotions." In S. Kitayama & H.R. Markus (Eds.), *Emotion and Culture: Empirical Studies of Mutual influence*. APA, Washington DC.

Lisetti, C. (1999). "Modeling Cognition-Emotion of Users for Improved Interaction with Software Systems." *Proceedings of UM.* 307–309

Reid, E., Qin, J., Zhou, Y., Lai, G., Sageman, M., Weimann, G., Chen, H. (2005). "Collecting and Analyzing the Presence of Terrorists on the Web: A Case Study of Jihad Websites." *Intelligence and Security Informatics*: IEEE International Conference on Intelligence and Security Informatics, ISI 2005, Atlanta, GA, USA, May 19–20, 2005. Proceedings. Lecture Notes in Computer Science. Volume 3495 / 2005 Springer Berlin / Heidelberg. Online: http://ai.eller.arizona.edu/research/terror/publications/ISI_AILab_submission_final.pdf. Access date 31/07/06.

Whine, M. (2004). "Online propaganda and the commission of hate crime." OSCE meeting on the relationship between racist, xenophobic and antisemitic propaganda on the Internet, and hate crimes. Online: http://64.233.183.104/search?q=cache:l8MTEy-jCfQJ:www.inach.net/content/whine.pdf+why+do+extremist+groups+publish+websites&hl=en&gl=uk&ct=clnk&cd=5 Access Date: 31/07/06.

Hypermedia Seduction for Terrorist Recruiting
B. Ganor et al. (Eds.)
IOS Press, 2007

Use of Media and Challenges in Countering Terrorist Rhetoric

Thalia TZANETTI

International Analyst
International Security, Terrorism/Counterterrorism issues

Abstract. The reality of globalized media critically influences the terrorist threat by increasing the "inequality of arms" [7] between terrorism and counterterrorism. Such concerns can never overshadow the innumerable benefits of global communication networks in most areas of human activity, but, as long as counterterrorism is considered a priority, these concerns do need to be taken into consideration. They have a bearing on any decisions which may have a direct or indirect impact.

Keywords. Global media, asymmetry, radicalization, virtual networks, linguistics, interpretation, ideals, policies, consistency.

Introduction

The suggestion that terrorists are using modern media and global communication techniques to their benefit is certainly not new. Nor is new the fact that terrorist networks have proven themselves to be very successful and sophisticated in exploiting counterterrorism messages to their own advantage. Global terrorism is currently enjoying unprecedented worldwide visibility; terrorist sympathizers are using modern communication technologies widely, and they seem to be very successful in employing the same global media both to communicate their own messages and to deflect the "propaganda by the kufr" (infidels).

This interplay between the terrorist message and that of their adversaries has not gone unnoticed. Some of its repercussions and various dynamics—and, more importantly, its implications for counterterrorism—were, however, until recently to a great extent inconceivable. By and large, it was unimaginable that the time-honored "Western media" apparatus could be challenged, or its impact paralleled. Thanks to the significant expertise in using what was perceived as a "Western" tool, the odds of winning the "media war" seemed more than favorable.

Undoubtedly, recent experiences have shattered such assumptions, and have revealed new, multidimensional issues. Rather unexpectedly, recent experience has sufficiently proven that the penetrability, appeal and impact of a message do not depend as heavily as previously thought upon cutting-edge systems. And most notably, any relevant technological sophistication and expertise is hardly exclusive to specific countries. In a globalized world, knowledge quickly becomes universal, and so does the ability to apply it.

1. Global Media and New Levels of Asymmetry

In addition to our preconceived notions regarding the effectiveness of Western media, prevalent perceptions about the "enemy" have only reinforced our lack of caution. Not unreasonably, the enemy's assumed "rejection of modernity" was not imagined as being compatible with technological savvy and media mastery. Apparently, however, denouncing the "infidels' propaganda" and its "corruptive influence" does not necessarily imply rejecting the means *per se*, or modernity in all its forms. This is especially the case when the primary target audience is the youth, for whom global networks are an integral part of the world as they know it, rather than an element of modernity.

The use of globalized media for the propagation of information, either directly by terrorists or by their supporters, has clearly become central to their operations, as well as to radicalization and terrorist recruitment.

Carelessness? Lack of imagination? Excessive confidence? Perhaps; but whatever the reason for our having failed, thus far, to appreciate the extent, pace and significance of the change, it should not allow for further negligence. We need to consider the implications of terrorists' current operational capabilities, and the interaction of competing media messages. Given the nature of real-time communication and of global media, the capacity to gather and transmit information globally and continuously significantly influences the future evolution of the threat posed by terrorism. More generally, this is a threat posed by ideologies advocating violence.

Therefore, examining in depth the interplay between such factors and the actual threat environment is imperative if we are to gain a clearer understanding of the evolving threat, and, subsequently, to formulate any relevant response. Recent experience has revealed new factors and important lessons to be considered, on multiple levels.

The extensive use of networks, and especially the Internet, in ways that facilitate terrorists, is a widely documented phenomenon. Websites linked to—or sympathetic to—terrorists burgeon on the World Wide Web, and efforts to effectively monitor the emergence of new such websites and shut them down swiftly have proven insufficient. This ongoing "hide-and-seek", although necessary due to the friction which the situation creates, has admittedly failed to deprive terrorists of a pulpit which they seem to use more extensively and more diversely than in the past. The Web, now more than ever, is utilized for several different purposes, from public relations, indoctrination and recruitment to instruction and weapons' training.

This unrestricted dissemination of information undoubtedly intensifies the asymmetry of the current terrorist threat, by increasing the terrorists' capacities and, at the same time, inhibiting counterterrorism. Any aspirant Siddique Khan (the infamous London suicide bomber) or simply a curious but harmless "surfer"' can easily access material which, in the past, would have been accessible only through elaborate and often clandestine human networks.

Religious guidance is no longer the exclusive privilege of the local imam; self-proclaimed religious authorities willingly offer their services, while various and wide-ranging interpretations of holy texts, religious duties, history and so on are easily accessible, and may be scrutinized by any individual.

Nor is terrorist operational knowledge, such as that regarding tactics, weapons-making and training, necessarily channeled through training camps or a few experts.

Detailed manuals on everything from bomb-making for beginners to operating complex weapons are now easily accessible.

The significance of the emergence of virtual networks needs to be examined, not only as a factor which increases the terrorists' capabilities to conduct attacks, but also as an important element in determining the overall appeal of existing terrorist groups. This includes, more generally, the appeal of ideologies which justify or incite terrorism.

2. Virtual Networks

The significance of virtual networks and spheres is definitely not easy to grasp, especially when we don't fully realize the extent of recent changes. E-mails, chat rooms, web logs and intranets all create looser or tighter virtual communities, which are often no less real or appealing than physical ones.

Co-workers will often choose e-mail instead of face-to-face communication, and, especially amongst the younger generation, they use virtual networks extensively and widely for socializing. As David Morley points out, "physical and virtual public spheres can function either as complements or as substitutes for each other" [1]. So virtual networks are effectively "divorcing experience from physical location" [2] and are assuming, at least in part, the role that physical, social networks were hitherto believed to play.

From a terrorism-related perspective, social networks are important not only for planning attacks and sustaining terrorist support, but also, and most alarmingly, for indoctrination and recruitment. According to Florence Passy, the functions which social networks perform include influencing individuals' frameworks of interpretation, building and strengthening individual identities, indicating "opportunities for mobilization", recruitment, and shaping actors' decisions [3]. Thus, one's social circle can be expected to influence how one interprets information, how one defines oneself, what entities one may choose to relate to, and how (and with whom) one may decide to act.

If virtual networks have indeed begun to take on such functions, going so far as to affect individual affiliations and mobilization, then their use within the terrorist milieu is not limited to a wide but barren propagation of ideas. Such networks would lead to the creation of new allegiances and the shifting of existing ones, and to the creation and maintenance of motivational influences. They would even define the limits of acceptable or desirable action—in short, they would do everything which "real" social networks have done in the past.

Even if the strength of a virtual message cannot be compared to the influence of face-to-face communication, the sheer volume of such messages—and the fact that they are becoming so predominant in our lives—probably guarantees the long-term significance of globalized networks. These networks and messages can be used for the purposes of radicalization, terrorist support and recruitment.

The level of individual affiliation to terrorists' ideologies, "causes" and groups, the level of terrorist support or tolerance, and the distinction between justifiable and unacceptable means, are all substantially influenced by individuals' virtual explorations and encounters. In other words, it's not just that terrorist acts are becoming easier to prepare, thanks to the modern communication networks and resultant availability of technical expertise. Terrorist affiliation, in general, can become an easily available and appealing option for an individual.

This unexpected aspect of the universalization of information, apart from making violence-advocating ideologies and terrorist mobilization a more reachable option than before, also undermines traditional counterterrorism. If an individual or a small group can find convincing motivations, instruction, tactics, guidelines for target selection, and other essential elements for conducting an attack—with limited or no outside influence, then the taken-for-granted indispensability of human networks is challenged. At the same time, the signs which would traditionally sound the alarm are increasing exponentially, thus becoming redundant.

While the numbered "graduates" of terrorist training camps might be a legitimate concern and achievable target for surveillance, how credible a threat—and, more importantly, how worthy of the limited counterterrorism resources—is any one of the numerous visitors to suspicious websites?

With countless IP addresses of questionable investigative value to keep track of, numerous websites to monitor, and with inevitably limited resources (especially given the technical and language skills often required), the aspiration of effectively monitoring the World Wide Web resembles, more and more, wishful thinking. And although human networks remain extremely important, they can no longer be considered a *sine qua non* for individuals and groups to become operational. As a result, the value of monitoring them is declining. This situation not only undermines the usefulness of a trusted intelligence tool, but, more importantly, creates new, pressing challenges for counterterrorism.

More specific aspects of certain networks and their use can be particularly pertinent to terrorist radicalization. The Internet, in particular, can be characterized as a "pull-system" (as opposed to a "push-system"); each user actively has to seek information, and it is at his or her discretion to choose the source of information, as well as how much and what kind of information is obtained. Information is to a lesser extent "pushed" to the user by its source.

As with other types of networks, once inside a particular web of information sources, it's less likely that the user will coincidentally come across—let alone seek—contradicting or diverse information. It becomes a lot easier to reaffirm currently-held opinions than to challenge them. Following the same pattern, the viewer of a specific television news channel is more likely to read a newspaper with a similar outlook rather than a contradicting one. Even social groups may deter their members, either implicitly or explicitly, from what may be considered as unacceptable association with a differing group.

What differentiates Internet-based communities is their capacity to, subtly but effectively, exclude nonconforming views. Websites are connected via links, and visitors are retained because they can easily draw on the resources of several different websites to satisfy their curiosity. This creates wide communities of like-minded "surfers", and these communities may diffuse the inclinations of their visitors to explore other perspectives. Of course, that is not to say that conscious efforts to compare and contrast viewpoints will be thwarted. Nevertheless, expediency is often a sufficient incentive for a large number of surfers.

As already mentioned, such loyalty to virtual networks, whether chosen or induced, influences people's allegiances, motivations and, eventually, actions. Clearly, this situation becomes problematic if violence is encouraged in any way. Advocating for violence, justifying acts of violence, providing resources for executing attacks, dehumanizing the victims, promoting ideologies and fuelling motivations to which members can relate, are all recognizable phenomena to those familiar with terrorist-sympathizing websites. The existence of multiple such websites, and the continuous emergence of new ones, is far from a revelation. What remains to be realized and analyzed further is their radicalization potential.

If such networks-based communities are—as indications suggest—indeed successful in enabling terrorist support and attacks, influencing affiliation with terrorist groups and their aims, and resisting effective monitoring, then they drastically change the threat environment. This change needs to be sufficiently taken into account when designing new policies and amending existing ones.

Networks have come to replace hierarchies, and virtual communities increasingly complement or replace physical ones. These sweeping changes have not been sufficiently acknowledged. Until they are, counterterrorism strategies will not be able to effectively target the latest mutation of the threat. Essentially hierarchical government structures need to develop forms of organization that are capable of comprehending and targeting networks, in all aspects of their operations, instead of trying to adjust their image of the threat to the existing combating structures.

As Arquilla, Ronfeldt and Zanini state, "Fighting networks [...] involves the ability not only to neutralize critical nodes when these exist and are identifiable (for example, key physical targets or a group of operatives shortly after a strike), but also to disrupt the information." [4]

3. Terrorism, the Media, and Linguistics

In the current era of globalized media, it seems that terrorist sympathizers find in their enemies' communication, almost exclusively, valuable ammunition for their own communication campaigns, and useful promotion of their efforts. They thus come away from the communication battlefield significantly empowered, rather than undermined. The high-priority, worldwide coverage of anything related to terrorism undeniably offers a promotion mechanism that is unique in its reach, and which simultaneously portrays terrorists as successful, boosts their appeal, and multiplies their terrorizing effect.

Terrorists' demonstrated destructiveness, along with the credibility of the threat they represent, guarantee high public demand for any relevant information, and, consequently, its immediate dissemination. This sets up a terrorizing vicious cycle. Terrorist attacks generate insecurity, which, in turn, generates high public demand for information. Satisfying this demand, however, entails recycling the relevant images and stories, thereby keeping the threat ever-present in the public mind. This reinforces the terrorizing effect initiated by the terrorists themselves.

The fact that the threat of further attacks remains a credible one, despite the governments' stated focus on counterterrorism, suffices to portray terrorists as successful—or governments as unsuccessful. At the same time, terrorism can be viewed as a vehicle for social activism, if the terrorists' arguments sound convincing enough.

Once again, this scenario is not only beneficial for the terrorists, but poses further hurdles for long-term counterterrorism. The media cover topics from different parts of the world, thus making those events part of anyone's own experience; coverage is widely accessible, and frequently itself becomes the subject of reporting. Media correspondents have become an important part of the scenes which they try to describe, and thus they also become informal representatives of the "side" for which they are supposedly seeking information. So, willingly or not, reporters provide access to each side's broad constituency.

Particularly pertinent to counterterrorism is the terrorist practice of using material from media sources associated with their adversaries. Footage designed to address audiences in the US, Britain or France is used (edited or not) by terrorist sympathizers as evidence for differing interpretations. The current emphasis on anything related to terrorism ensures that there is ample material for anybody to use.

Statements which can be interpreted in a way that is favorable for the terrorists, inevitable inaccuracies or serious mistakes, decontextualized coverage: these are all being frequently used as evidence of the "crusaders'" true intentions and their hypocrisy, as well as of the perceptiveness or truth of the terrorists' own outlook.

Long-term counterterrorism, with its focus on public perceptions and ideology, cannot disregard the importance of the media's impact. The one-sided visibility of terrorists and their religious perspectives critically inhibits any efforts at marginalizing terrorists, and almost precludes any debate within Islam. As the oft-stated truism suggests, media are, above all, businesses, and need to comply with the rules of the marketplace. They cover what the public demands. So, because the public in affected countries demands reportage on terrorism, the media delivers it.

And most importantly, since the public does not demand any coverage on the diversity of the Muslim world, the media cannot afford to pursue this topic. Given this state of affairs, there is nothing to prevent terrorists from becoming the only visible Muslim interlocutors with "the West", at least in the eyes of potential sympathizers. They are the only Muslim interlocutors that Western countries seem to recognize and respect enough to engage with.

The main role of the media lies in this function of empowerment—of who is perceived as a real interlocutor. What is worrisome is that such implicit legitimization reinforces the terrorist rhetoric, and also deprives counterterrorism from valuable support.

The main repercussion derives from the essentially confrontational relationship between the terrorists and governments. With each being a sworn enemy of the other, in a zero-sum game, the messages they exchange are, naturally, limited to promises of mutual elimination. However, in the absence of any other equally empowered interlocutor from the Muslim world, attacking terrorists cannot easily be distinguished from attacking Muslims *per se*. Such an inference clearly reinforces the terrorists' credibility and prestige. The idea that Muslims as a whole are threatened is central to their rhetoric, and it acts as an incentive for essentially non-violent parts of Muslim societies to support—or at least tolerate—actions which they would not necessarily otherwise endorse.

More importantly, though, this situation makes it hard for more mainstream Muslim religious interpretations to gain visibility and openly challenge the terrorists' beliefs, without compromising their own credibility. At a time when it's not self-evident and widely accepted that the attack is not, in fact, against Islam or the *Ummah*, attacking the self-proclaimed—and fairly unchallenged—defenders of Islam can be interpreted and dismissed as betrayal.

It can definitely be argued that the media are not primarily responsible for the "side-effects" mentioned above. After all, they only report what others create. Reporting, though, is more than a copy-and-paste function which lifts events from reality and places them on the screen or in the newspaper. Reporting is also an act of interpretation which is, in its own right, a "form of making" [5].

Undoubtedly, politicians, academics and other key individuals, who provide the raw material and who may be called upon to inform the media and the public, are also responsible for the explicit and implicit messages they choose to send. However, the media's multiplying effect, which derives from their global nature, their function as a vehicle for second-hand experience, and their "rich symbolizing powers" [5], makes their role crucial. Especially for people with no personal experience related to the covered topic, reporting does not merely influence personal understanding; it determines it. In other words, the story, with all its emphases and its omissions, becomes second-hand experience; it becomes "reality".

So if terrorists are the only Muslim element presented, one's mental images of "jihad" or "nuclear mujahedin" will probably overshadow the fact that this group does not represent all Muslims. And the consequent generalized messages of destruction are bound to reach radio listeners in Mosul as well as in New York, either to offend or to reassure them.

Choosing terms, defining them (or not) and using them carefully (or not), as well as deciding on the newsworthiness of specific parts of a story versus others, are the prerogative of reporters and editors. And in making such choices, any indirect impact, especially of such a sensitive nature, needs to be taken seriously into account.

The framework of the public debate on terrorism and the terms used in that debate, which are primarily defined by governments, are also critical. The choice of terms, through their definitions and connotations as well the associations they trigger, is a substantive one. As the endless disputes over terms such as "terrorist", "freedom fighter" or "insurgent" suggest, terminology has important implications for counter-terrorism. Due to the terrorists' claims that they are fighting in the name of a religion which is, nonetheless, shared by millions of people who do not in fact embrace terrorist logic, semantic choices become extremely political and particularly relevant to counter-terrorism.

Creating labels for terrorism, such as "Muslim", "Islamic", "Islamist", "Salafist", "jihadi" or anything else linked to Islam and Muslims, or to specific strains of Islam which are not followed exclusively by terrorists, provides a necessary working term. However, it also leads to counterproductive inferences. Entire religious schools of thought, and even Islam in general, become associated with terrorism, which is essentially malicious and violent.

Not all Islamists are terrorists, and jihad is not necessarily violent. The term "Islamist terrorist" implicitly equates Islamism with terrorism, while the use of "jihad" as synonymous with "terrorism"—especially because of jihad's centrality in Islam—portrays the entire religion as prone to violence.

Edward W Said warned, in 1996, that "the mere use of the label 'Islam', either to explain or indiscriminately condemn 'Islam', actually ends up becoming a form of attack, which in turn provokes more hostility between self-appointed Muslim and Western spokespersons" [5]. So unless terms are carefully selected and used, they can become overgeneralizations. When these terms are linked to terrorism, they may well offend. This offence may, in turn, "convert the war against terrorism into precisely what it claims it is not, a war against Islam", and thus, "make a gift of the defense of Islam to the extreme" [6].

What defines terrorists is no longer terrorism, but their religious interpretations. Numerous people relate to such interpretations, although they do not necessarily relate to terrorism. The terrorists' rhetoric is significantly strengthened by these religious views, while counterterrorism is critically undermined, or even seen as a threat to

Islam. The dissident voices among Muslims cannot easily challenge the terrorists' logic.

4. Interpretation, Ideals and Consistency

The overall framework of the terrorism debate comprises an even more powerful form of interpretation. Thus, it may, apparently, lead to even broader generalizations. Governments rightly seek to separate themselves from terrorists and appear unrelenting. However, this effort to isolate and de-legitimize terrorism currently seems to be working to the terrorists' advantage. The Manichean terms which have, by and large, prevailed in the public discourse, stress the contrast between two extremes, with each claiming moral superiority over the other.

The ensuing polarization can be beneficial for domestic mobilization and unity within countries aiming at countering terrorists, but it is not without important repercussions. Especially when combined with terms which allow, even inadvertently, for generalizations, polarization reinforces the terrorists' longstanding argument about "belligerent Westerners" attacking Islam.

Consequently, it is not just terrorists who become alienated, but also Muslims in general, who may disagree with the terrorists' logic and practices, but nonetheless refuse—or cannot afford—to distance themselves from their coreligionists and thus become associated with "Islam's enemies". When the choice is framed in this or similar ways, Muslims are forced to choose the lesser evil. With black-and-white terms dominating the terrorism debate, the cohesion among terrorist networks and their sympathizers increases, and middle ground is subtly appropriated by the two extremes.

Especially in today's era of globalized communication, the polarizing effect of generalizing terms and antagonistic frameworks is not limited to the intended recipients. It is also felt by numerous members of a global audience. Terrorist networks clearly benefit from this impact, since their membership and appeal increase and become international. This in turn ensures the terrorists' survival, international relevance and ability to conduct attacks.

While terrorists have only benefits to reap from this expansion and increased cohesion of their circles, counterterrorism strategies need to find the un-polarized middle ground if they are to undermine the terrorists and neutralize the long-term threat they pose. Terrorists' survival depends on the non-terrorist majority of Muslim societies tolerating their existence and activities. As a result, non-terrorist Muslims become critical in relation to counterterrorism. Broad polarization, however, precludes certain alliances, and counterterrorism is deprived of critical partners in an area where initiatives by non-Muslims have little chance of achieving significant impacts.

The discrepancies among various messages have become yet another fertile ground for terrorists to exploit. The long-argued need for a carefully selected counter-terrorism message is undoubtedly crucial, but the need for broad consistency cannot be overemphasized. Even the most culturally sensitive, nuanced and non-confrontational approach will be critically undermined if it is not conveyed consistently.

Given the openness of modern communication networks, and the demonstrated focus of terrorist (or terrorist-supporting) expertise on those networks, it can safely be assumed that global communication networks are closely monitored. Any discrepancy in media messages can quickly be used to the terrorists' benefit. Even common inconsistencies, which the traditional audiences of government messages have been accustomed to and have learnt to rationalize, can easily turn into major vulnerabilities. For example, the typical Western tolerance of discrepancies between domestic and international messages, between declarations and results, or between messages delivered by different agencies or levels of government, is rather unlikely to be mirrored by more suspicious audiences—let alone terrorists themselves.

It is probably not unrealistic to presume that anything associated with the "infidels", and capable of conveying any kind of symbolism, be that in words, actions, or policies and so on, is under close scrutiny. Implicit and explicit messages, as well as opportunities for distortion, will be seized on. Lack of consistency can become the Achilles' heel of strategic communication in counterterrorism, since it limits the effectiveness of such communication. More importantly, it critically undermines the credibility of the communication. Without a single, clear message, the interpretation of communications is left to the discretion of the audience or any self-proclaimed interpreters.

Apparently, this leeway in interpretability is already being successfully exploited by terrorists to demonstrate the hypocrisy of the "infidels", and to further reinforce their own rhetoric. And, ironically, they do so by employing the very same global networks, and with remarkable consistency around the core elements of their rhetoric. Anything with potential for validating the terrorists' claims can, and most likely will, be used to this end.

Unfortunately, there is little to be done to prevent the use of images from the hurricane Katrina disaster or the Columbia tragedy as evidence of God's support for the terrorists' "cause". However, preventing the backfire of communication initiated for counterterrorism purposes is not impossible. But this can only be done by ensuring broad consistency, with a carefully selected and clearly expressed message.

Achieving even partial consistency is not only a function of successful coordination within large, bureaucratic and diverse government apparatuses. It also depends on choosing messages that are conducive to such coordination. Unfortunately, the current emphasis on ideals and values seems to be inhibiting rather than facilitating the communication of a coherent message. When ideals and values are presented as motivations for action, and as the driving force behind policies, discrepancies become unavoidable. Asking for consistent interpretation and representation of an ideal, however noble it may be, by entire government structures cannot possibly be viable. This is especially true when government representatives are no longer restricted to a few official spokespersons.

More important, however, is the fact that unrealistically high expectations are created when it is claimed that policies are aiming at ideals—even if they genuinely are. Values and ideals can only be defined in absolute terms, while both the implementation and the outcomes of any policy can never live up to such standards. When policies' goals are defined in such terms, they are doomed to be seen as unsuccessful.

Or, even worse, imperfect policies can end up actually defining so-called ideals, in people's minds. When the only point of reference to an ideal is the policy of a country that claims to be guided by that ideal, those policies seem to embody its very essence. For people who do not associate their own experience with, for example, democracy, policies which claim to establish democracy, are, logically, believed to illustrate what democracy actually is. So the unavoidable flaws of policies, instead of signifying aberrations from an ideal, may be perceived as the very definition of that ideal.

In this way, powerful ammunition is offered to the terrorists' arguments. All the terrorists need to do is to point out such inconsistencies, and capitalize on them by portraying them as examples of "Western" insincerity, incapacity or insufficiency. In other words, instead of creating additional friction for terrorists' arguments and building a strong alternative to terrorist rhetoric, the excessive emphasis on ideals and values has weakened the counterterrorism message by making success impossible. This scenario has undermined the values which counterterrorism is supposed to embody, and has cast doubts on the sincerity of the pursued goals, making the accusation of hypocrisy inevitable.

5. Conclusion

The globalization, openness and transparency of modern media, combined with the active engagement of all parties in this "virtual battlefield", allow for repercussions of great significance. The present communication environment offers important opportunities which are exploited by terrorists. The same environment impacts the character of the terrorist threat and its evolution, as well as the terrorists' rhetoric; it also affects strategic communication efforts aiming at countering the terrorist message.

The potential benefits for terrorists are numerous, and more are likely to surface. Some of the recently evident benefits include greater and longer-lasting visibility and appeal for the terrorists' message; wider propagation of their ideas and "know-how"; unprecedented capacity to access and neutralize counterterrorism messages (or even to use them for furthering their own rhetoric); and increased potential for radicalization and recruitment. At the same time, this communication environment—along with the terrorists' capacity to exploit it—significantly expands the scope of counterterrorism, impedes efforts to increase friction for terrorists, and also creates unprecedented parameters to be taken into consideration, in several, perhaps seemingly unrelated, areas of government activity.

In other words, the reality of globalized media critically influences the terrorist threat by increasing the "inequality of arms" [7] between terrorism and counterterrorism. Such concerns can never overshadow the innumerable benefits of global communication networks in most areas of human activity, but, as long as counter-terrorism is considered a priority, these concerns do need to be taken into consideration. They have a bearing on any decisions which may have a direct or indirect impact.

The aspects mentioned in this chapter demonstrate some of the multiple dimensions of the issue. Policies, which may be very successful in achieving their primary objectives, may also have important repercussions for terrorism or counterterrorism. Decision-makers cannot disregard potential secondary effects of their actions, even if they do not see themselves or their decisions as directly involved with counterterrorism.

The current communication environment, by increasing the level of asymmetry regarding the terrorist threat and altering its character, has created a new reality to which the terrorist movement seems to be adapting successfully. Although terrorists cannot be allowed to determine governmental policies, the new dynamics are, nevertheless, influencing various elements of the threat. Therefore, our understanding of and approach to counterterrorism need to reflect such changes. The new asymmetry in relation to terrorism needs to be taken seriously into consideration by decision-makers. Either the necessary policy adjustments need to be initiated, or, at the very least, better-educated decisions must be made, which address the actual threat environment.

References

[1] Morley, David, 2000. *Home Territories.* Taylor & Francis
[2] Morley, David, and Kevin Robbins, 1995. *Spaces of Identity.* Taylor & Francis.
[3] Passy, Florence, 2002. "Social Networks Matter. But How?" *Social Movement Analysis: The Network Perspective.* Eds. Mario Diani and Doug McAdam, 2002. Oxford University Press.
[4] Arquilla, John, David Ronfeldt, and Michele Zanini, 2000. "Information-Age Terrorism". *Current History* 99:179-185.
[5] Said, Edward W., 1997. *Covering Islam.* New York: Vintage Books.
[6] International Crisis Group, 2005. "Understanding Islamism". *Middle East/North Africa Report* 37.
[7] Akerboom, E.S.M., 2003. "Counter-terrorism in the Netherlands". *Tijdschrift voor de Politie.*

The Seductive Web: Terrorism, Media and Psychological Effects

Hypermedia Seduction for Terrorist Recruiting
B. Ganor et al. (Eds.)
IOS Press, 2007

Countering Web-based Islamist Narratives: Conceptualizing an Information War and a Counter-propaganda Campaign

Dr. Katharina VON KNOP
Adjunct Professor, Program on Terrorism and Security Studies George C. Marshall Center, Germany, K.von-Knop@web.de

Abstract. Terrorist groups use the Internet to drive every aspect of their business: psychological warfare, data mining, fund-raising, recruitment and mobilization, planning and coordination, and Internet indoctrination. Radicalization of individuals is a dynamic group process often activated or boosted by radical Islamic websites. Terrorism and anti-terrorism are based on narratives, and whichever story is able to persuade the majority of people will win. I argue that the West is losing the war on terrorism because our story is not persuasive enough, and because we do not use the gateway of mass persuasion—the Internet—effectively enough. Fighting radical terrorist groups will take more than firepower; the battle of ideas has to be fought and won in the digital battlefield. The terrorists' infrastructure and their belief system has to be attacked in the air, via the Web and on the ground. The goal of this chapter is to provide an analysis of possible tactics tailored for countering Islamist narratives on the Internet, and to emphasize elements of strategy for countering those narratives. The first part of the tactical analysis will discuss methods of neutralization of terrorist websites and chat-rooms. The second part of the tactical analysis deals with the Web as a counter-propaganda tool. The basis of the findings has been a broad analysis of political Internet campaigns.

Keywords. Terrorism, counterterrorism, Internet, Islamist narratives, PsyOps, information war, counter-propaganda.

Introduction

After a year marked by riots over cartoon portrayals of Muhammad, a major terrorist attack in London and two further attempted ones in that city, as well as continuing wars in Iraq and Afghanistan, most Muslims and Westerners are convinced that relations between them are generally bad these days. Many in the West see Muslims as fanatical, violent, and lacking in tolerance. And Muslims in the Middle East and Asia generally see Westerners as selfish, immoral and greedy, as well as violent and fanatical [1]. This reciprocal mistrust has to be changed if the West wants to counter terrorism effectively in the long run.

What kind of individual is seduced, captivated and persuaded to follow the radical ideology of Al-Qaida and its affiliates? Two-thirds of people belonging to Al-Qaida whom Marc Sageman has studied came from the upper or middle-class. Al-Qaida is very much a middle-class movement. Only 13 percent of Marc Sageman's sample were Madrassa educated, which means that 87 percent were not. Sixty-two percent had been to college [2]. These are the best and the brightest members of their societies [3].

Furthermore, 70 percent had joined the jihad in a country other than the one in which they grew up. When we look at the people who were involved in the attacks and attempts in Europe, it turns out that the majority of them were born in the country in which they carried out an attack or attempted to do so.

An adherence to the ideology of Al-Qaida is a phenomenon of diaspora. Sixty-eight percent of Marc Sageman's sample had pre-existing friendship bonds; another 20 percent had kinship bonds. That means a group of men had collectively decided to join the jihad. This was true in Hamburg, Madrid (March 2004), Casablanca (May 2003), and London (July 2005) [3].

These people came from good families; they had no trauma in their backgrounds and they had enjoyed a higher education [4]. Based on Sageman's research, there is no immediately identifiable common profile for the individuals, but it is possible to identify two groups of people. The first group was born in a non-Western country and came to Europe to get a university degree. These people became homesick and started to look around for people in their age-group with whom to build up friendships. They went to mosques, and there they met other men, and they became friends.

The second group consists of men and women, second-or third generation economic immigrants who are intent on becoming upwardly mobile. These individuals no longer live within the culture of their parents or grandparents, and they are not able to adapt to the culture of the countries they live in. They are searching for an identity and a reason for living. They discover the radical ideology of Al-Qaida as a lifestyle which gives them a clear guideline for their lives. This is also the reason why 10 percent of the people who were directly involved in attacks or attempted attacks are Christian converts to Islam. Groups of friends drifted collectively towards a certain religion and, eventually, towards radical Islam.

So joining the jihad basically is a bottom-up, self-nominating activity. The social bonds come before any ideological commitment. These people were not recruited in the classical sense; they applied to Al-Qaida. The radicalization is a group-specific collective phenomenon. As a group, they can do collectively what they do not have the courage to do individually. The point is that the social bonds that existed among the men before they joined any such groups eventually came to provide the infrastructure for their operational activity.

Thus, the Al-Qaida terrorism we have observed inside Europe is mostly self-generated and self-organized from the bottom up; it is a local initiative with considerable flexibility. This diaspora phenomenon has its roots in the individual's lack of embeddedness in the society within which he or she lives.

The Internet plays a highly important role in the process of this collective radicalization. Firstly, the Web spreads the radical Islamist message; secondly, it allows the user anonymity; and thirdly, the Internet is capable of creating a virtual community. Given the estimated more than 4,500 radical Islamist websites which currently exist, visitors to those sites can soon develop the illusion that the whole world shares these beliefs. The Internet has become an invisible hand in the organization of terrorism and specifically terrorist cells.

I am arguing that the movement Al-Qaida would not exist today in the shape which confronts us, without the Web. According to Prof. Gabriel Weimann, the number of terrorist websites has shot up from 12 in 1997 to more than 4,500 today [5].

As stated above, the radicalization of individuals is a dynamic group process, and in this case the process is often activated or boosted by radical Islamic websites.

The next argument which this chapter presents is that terrorism and anti-terrorism are based on storytelling, and whichever story is able to persuade the majority of people will win out. Certainly terrorism and counter- and anti-terrorism are the most brutal ways of telling stories, but in this sense they are weapons of mass persuasion, when used with skill. Considering the growing number of attacks and attempted attacks by Al-Qaida, and the growing radicalization of Muslims inside Europe, the West is losing.

At this point I would like to suggest the West is losing the war on terrorism precisely because our story is not persuasive enough. As Western military and intelligence agencies fight radical terrorist groups, it will take more than firepower to win the battle of ideas being fought in the digital battlefield. The terrorist infrastructure and belief system has be targeted in the air, via the Web and on the ground. These target areas are interdependent. When we are able to disturb one, there will be an impact on the others too.

The goal of this chapter is to provide an analysis of possible tactics tailored for countering Islamist narratives on the Internet, and to emphasize elements of strategies for countering those narratives. The concept of countering the narratives and thus the ideology of Al-Qaida comprises Political Warfare (PolWar). PolWar strives to create an alternative social contract and dependency infrastructure, and to induce their popular adoption. In terms of terrorism, this is commonly achieved through efforts of agitation, subversion, rioting, propaganda, psychological warfare operations, disinformation, diversionary diplomacy, economic manipulation and attacks [6].

Information warfare, including InfoOps (information operations) and PsyOps (psychological operations), can be understood as a part of PolWar. The "Dark Web"[1] can be understood as a war zone. The concept of fighting a war on the Web is called information warfare. This refers to the offensive and defensive use of information and information systems to deny, exploit, corrupt, or destroy an adversary's information, information-based processes, information systems, and computer-based networks, while protecting one's own. In the current context, such actions are designed to achieve advantages over the movement of Al-Qaida.

1. Al-Qaida's Web-based Operations

The underlying assumption of this study is that Al-Qaida and its affiliates have two fronts of warfare: the terrorist attack and the Internet. The Web is nowadays the essential platform for Al-Qaida and its affiliates to run their war against the West. Their use of the Web has expanded beyond routine communication and propaganda operations to training, organizing logistics for their campaign, and developing their strategic intelligence and virtual communities.

Weimann [5] highlights the innovative ways in which terrorist groups use the Internet to drive every aspect of their business: psychological warfare, data mining, fund-raising, recruitment and mobilization, planning and coordination, and Internet indoctrination. We examine each of these briefly in turn.

---

1 The term "Dark Web" will be used to describe terrorist websites collectively.

Psychological Warfare: The Internet is used to deliver threats and disseminate multimedia content designed to create fear and panic, as seen in Iraq. Nearly all insurgent groups in Iraq have media teams that post statements and create videos and Web broadcasts.
Data Mining: A sophisticated array of open source technologies is used, including search engines and website analytics, to collect intelligence on enemies and potential recruitment and funding targets.
Fund-Raising: Terrorist groups leverage Internet user demographics and online front groups to execute aggressive funding drives, collecting vast amounts of money through online payment systems that are difficult to track.
Recruitment and Mobilization: The Internet and advanced technology provide powerful tools for recruiting and mobilizing group members, through integrated communications.
Planning and Coordination: Terrorist groups take advantage of new technologies such as encryption, voice-over-IP and secure messaging systems to improve the ease, speed and cost of their communications. This enables the sharing of information such as training videos and manuals, enhancing their planning efforts and agility in an ever-changing environment.
Internet Indoctrination: The Internet was an integral part of the group's indoctrination and operations which we have seen in recent history. The 17 terrorism suspects in Canada, as well as the group of the London 2005 attack and the group responsible for the London 2006 attempt, made extensive use of the Internet. Authorities believe the Internet-savvy suspects did much of their communicating online, where they also developed their radical ideology.

Al-Qaida websites have increased in number, technical sophistication, content, and media richness. Terrorist sites are also extremely dynamic. They suddenly emerge, frequently modify their formats, and then swiftly disappear or change their URLs [7]. Al-Qaida's site locations and contents change almost daily. Furthermore, because of their politically controversial nature and the depiction of glaring details in some video clips, they are often hacked by hackers or shut down by their Internet Service Providers. The Dark Web provides rich and diverse terrorism-related data such as training manuals, forum postings, images, audio records, video clips, and fundraising campaigns.

2. Seductiveness of Islamist Narratives

To develop a strategy and to identify appropriate and constitutional tactics to counter Islamist narratives, it is necessary to gain a deeper understanding of the role these narratives play in the seduction and persuasion of target audiences. Using a working "theory of story", we can gather insights into why stories are so important for structuring human thought. The most salient point here is that stories are often rich in metaphors and analogies; metaphors, in turn, affect our most basic attitudes toward the world [8].

These narratives address their audience on tactical and strategic levels, and as such they can be qualified as information operations or psychological operations. Thus, counter-measures must also be divided into tactical and strategic information operations or psychological operations. Narratives and InfoOps share the same goal, namely to seduce, captivate, persuade and influence the audience.

The concept underlying this article about hypermedia seduction is two-fold. Firstly, there is the metaphorical orchestration of specific actions or responses in a user's behavior. These actions, or particular elements thereof, facilitate the onset of desired behavioral change. The achievement of desirable behavioral change relates to the idea of the system as the seducer, where change is believed to occur due to the system's ability to sustain its appeal [9]. So the system itself becomes the seducer. As mentioned earlier, the persons involved in the latest terrorist attacks carried out under the umbrella of Al-Qaida in Europe followed a group dynamic behavior of radicalization, which was initiated and/or boosted by the Internet.

Secondly, the act of seduction defines the particular path or course of a relationship, as formed as the consequence of interaction between user and system. The process of seduction is believed to enhance the user's perception of the nature of the communication. This, we believe, allows the subject to be swept outside and beyond his or her usual experience, resulting in the formation of a more meaningful relationship [10]. The anonymity of the Web allows the user to search, obtain and become a part of those narratives which provide him or her with the previously missing sense of social identity.

To be seduced by such narratives, the user must be in a condition of emotional neediness. To be persuasive, the narratives have to suggest goal fulfillment, namely the achievement of a sense of identity. To provide that identity, personal motivation must become influenced by a higher motivation, which in turn determines the goal formation. Users may quickly develop the illusion that the whole world shares the same beliefs.

Once inside this online community, users gain acceptance, attention and approval. Additionally, the real world community, namely the users' friends, provide feedback loops. This confirmation from the real world is important to create a more stable sense of identity, which in turn ensures that the users themselves are transformed into a human guidance system that is programmed for a one-way trip.

The term "narratives" is used in this chapter as it enables us to broaden our perspectives to identify appropriate countermeasures, tailored to address the seductiveness of Islamist narratives channeled through the Web.

Narratives play multiple critical roles during a life circle of a terrorist organization. Terrorist organizations follow typical life circle concepts, and undergo times of acceleration and times of deceleration. It is possible to identify a genesis point, a point of consolidation and a point of organizational and ideological transformation. An understanding of this process is essential, if we wish eventually to influence terrorist movements to die out or to morph into peaceful non-state actors. The following section examines the life cycle of such movements.

During the genesis or gestation stage, stories seduce the audience and influence their actions by

1. providing incentives for recruitment, particularly by providing "justice" frameworks which serve to mobilize discontent,
2. helping to justify the need for an organization to the community in which it will be embedded, and to first-round stakeholders,
3. reinforcing pre-existing identities friendly to the nascent organization,
4. creating necessary identities where none exist,

5. setting the stage for further growth of the organization,
6. solidifying founding members into leadership roles,
7. defining the possible space of actions as the organization blossoms,
8. reinforcing role-specific obligations so as to ensure group members continue to accomplish their functions,
9. providing "fire walls" against attempts to discredit foundational myths,
10. articulating ideological niches for the organization, and
11. making salient to organization members the environmental conditions conducive to organizational growth [11].

During Al-Qaida's transformation from a terrorist organization to a movement, its narratives seduced and persuaded the audience by
1. smoothing the transition to new organizational forms,
2. helping to ease the organization into a different set of stakeholder relationships,
3. providing the foundation for the revivified identities which could be used in whatever new form the organization adopted, and
4. serving to demobilize those portions of the organization which had served their purpose or were no longer needed.

3. Political Warfare: Running a Mass Persuasion Campaign

"There are but two powers in the world, the sword and the mind. In the long run the sword is always beaten by the mind." Napoleon Bonaparte

The Internet plays a massive supporting role for terrorist groups, by facilitating their communications and streamlining their operations. The speed and power of the Internet to shape perceptions and influence opinions makes it a vital tool for such organizations. Donald Rumsfeld has stated that "Today we are fighting the first war in the era of e-mail, blogs, BlackBerries, instant messaging, digital cameras, the Internet... The US government still functions as a five-and-dime store in an eBay world." [12]

We now live in a complex, multimedia environment where individuals, groups and populations are targeted by sophisticated online psychological operations. The good news is that the technologies and techniques that are being used to spread the ideology of terrorism can also be used to combat it.

3. 1 Tactics to Counter Islamist Narratives: Guiding Questions

To identify appropriate tactics to counter Islamist narratives, five questions have to be answered.

1. What are the tactical interests and objectives of the countries countering Al-Qaida?
The tactical interests and objectives of countries countering Al-Qaida are to prevent terrorist attacks, to neutralize the key players and members, and to stop the spread of Al-Qaida ideology.

2. What are the threats to those interests and objectives?
There is the threat that not only the people belonging to, or who are in favor of, the movement of Al-Qaida will be targeted. There is also the chance that Muslim support will not be gained, or will be gained and then lost again. Finally, there is a threat in terms of the possible reduction of democratic values, such as freedom of speech and of the media. Correlating with this threat, there exists the threat that police states may be created.

3. What are the appropriate tactical responses to those threats?
Tactical responses can occur on two levels: firstly, neutralization of terrorist websites and chat-rooms; and secondly, using the Web as a counter-propaganda tool.

Assessment is needed regarding which measures would work best in accordance with democratic values, and also which would gain the support of Muslim citizens. International and inter-agency coordination is essential. For instance, it would not be fruitful should security institution of Country A observe a specific website to gather intelligence, while at the same time this website is hacked by an individual or non-governmental institution in Country B.

Keeping the fluent international character of websites in mind, an international legal framework needs to be created, and decisions should be made regarding which institutions would be responsible for which measures. It should be identified which national or international governmental institutions have the best capabilities for which measures. Various efforts by private groups or individuals who observe and take down Islamist websites should be included in daily governmental work.

NATO, OSZE, and EU, together with Europol, Eurojust, Interpol, the Security Alliance, and the Club of Bern, plus all national intelligence agencies, have to work together in countering Islamist narratives. We are facing the same threat, and there is no need to duplicate functions. Already existing capabilities must be adjusted and improved.

4. What principles should guide the tactics used by countries countering Al-Qaida?
The principles which should guide our tactics in countering the ideology of Al-Qaida should be based on how we define ourselves. When we want to counter the narratives and ideology of Al-Qaida, we have to be clear about what and who we are. Without a strong identity we cannot persuade others to follow us instead of following Al-Qaida. The countries of the European Union, in particular, need to offer their Muslim citizens an identity that allows them to live as proud Muslim citizens within those respective countries.

5. What are the consequences of the chosen tactical responses?
A clear cost-benefit analysis about every tactical measure conducted to counter Al-Qaida's ideology must be drawn up.

3.2 The First Six Steps of a Web-based PolWar

Before any tactical or strategic measures can be carried out, an analysis on the context and content of Islamist websites is essential. This process of "learning about the enemy" should be ongoing. Non-governmental research projects like Internet Haganah (conducted by Aaron Weisburd) and PRISM (conducted by Reuven Paz) are already doing very fruitful work in this field.

It is possible to identify six steps which must be taken into consideration before any tactical or strategic campaigns are decided on:

1. *Identify terrorist groups or individuals.*

2. *Identify an initial set of websites created by these terrorist groups.*

3. *Expand this website list through link analysis.*

4. *Collect contents and context of the identified terrorist websites.*

An automatic Web crawler is applied to collect the contents of these sites. All types of contents from terrorist websites, including textual files (e.g. HTML files, plain text files), multimedia files (e.g. images, audio/video files), and archive files (e.g. ZIP files, RAR files) should be collected [13].

There are two general approaches to collecting domain-specific Web documents:

- manual selection and
- automatic Web crawling.

The manual approach is often used when the relevance and quality of information from websites is of the utmost importance. However, this approach is labor-intensive and time-consuming, and often leads to inconclusive results. The automatic Web crawling technique is an efficient way to collect large amounts of web pages. This can be done using retrieval systems such as Convera RetrievalWare.

Given the sheer volume of data, technological tools are essential for an effective and efficient intelligence agency. Intelligence agencies need to use analysis and decision-making support tools, foreign language tools, sample-analyses instruments and foresighted modeling tools. Data collection needs to be done automatically, and whichever system is utilized, it should be able to analyze and sort the data, and raise the alarm if necessary.

One major concern in using Web crawlers is that off-topic documents are often introduced into the collection, due to the limitations of Web crawling technologies.

The web information we are interested in is often not in English. Cross-lingual information retrieval (CLIR) can help break language barriers by allowing users to retrieve documents in foreign languages, via queries in their native languages [14]. Most reported CLIR approaches translate queries into the document languages and then perform monolingual retrievals [15]. This method helps experts to explore global Dark Web information without first having to learn foreign languages, and reduces the need for human translators working in the domain of terrorism research.

5. *Observing chat-rooms:*

Observation is essential to learn about the target groups, to gather intelligence, to identify key players, to prevent attacks, and to neutralize the people who are spreading radical ideology.

6. *Wiretapping:*

   This measure will enhance public safety, and ensure that the surveillance needs of law enforcement agencies continue to be met as Internet-based communications technologies proliferate.

3.3 Information Warfare

Websites have become an essential weapon for Al-Qaida and its affiliates. As such, the Web itself can be understood as a war zone. Following this understanding, terrorist websites have to be treated as weapons of war, and those who run them have to be treated as enemies and legitimate targets for military action.

The concept of information warfare is used when referring to fighting an enemy on the Web, but the same term implies a focus on military measures, and that would exclude many fruitful approaches. To avoid this misunderstanding, the use of the term "information warfare" in this chapter must be clarified. It is used here with reference to the offensive and defensive use of information and information systems to deny, exploit, corrupt, or destroy an adversary's information, information-based processes, information systems, and computer-based networks, while protecting one's own.

Such actions have to be designed to achieve advantages over the movement of Al-Qaida. Information warfare encompasses numerous forms of attacks, and not every technique is an appropriate tool for attacking a non-governmental movement like Al-Qaida. Many actual attacks involve combinations of vulnerabilities. Examples of vulnerabilities include stack overflow attacks (where you pass an over-long parameter to a program that carelessly executes part of it) and password guessing; both of these have been used by the Internet worm.

A common strategy is to get an account on any machine on a target network, then install a password sniffer to get an account on the target machine, then use a stack overflow to upgrade to a root account. The exact vulnerabilities in use change from one year to the next, as bugs in old software get fixed and new software releases a new crop of bugs. Still, there are some patterns, and some old favorites that keep coming back in new guises. An overview of these follows.

Extreme Hacking
In effect, hacking is being progressively deskilled, while defense is becoming unmanageably complex. The most important attack is the stack overwriting attack, and the second most important is password guessing. At this point it is useful to list and briefly to explain the most commonly known possibilities for attack.

1. A stack overflow attack on the BIND program, used by many Unix and Linux hosts for DNS, giving immediate account access.
2. Vulnerable CGI programs on Web servers, often supplied by the vendor as sample programs and not removed. CGI program flaws are the most common means of taking over and defacing Web servers.
3. A stack overflow attack on the remote procedure call (RPC) mechanism, used by many Unix and Linux hosts to support local networking, and which allows intruders immediate account access.

4. A bug in Microsoft's Internet Information Server (IIS) (Web server) software, which allows immediate access to an administrator account on the server.
5. A bug in sendmail. One of the recent flaws can be used to instruct the victim's machine to mail its password file to the attacker, who can then try to crack it.
6. A stack overflow attack on Sun's Solaris operating system, which allows intruders immediate root access.
7. Attacks on NFS and their equivalents on Windows NT and Macintosh operating systems. These mechanisms are used to share files on a local network.
8. Guesses of usernames and passwords, especially where the root or administrator password is weak, or where a system is shipped with default passwords that people don't bother to change.
9. The IMAP and POP protocols, which allow remote access to e-mail but are often misconfigured to allow intruder access.
10. Weak authentication in the SNMP protocol, used by network administrators to manage all types of network-connected devices. SNMP uses a default password of "public".

Note that none of the above attacks is stopped by encryption, and not all of them by firewalls. For example, vulnerable Web servers can be kept away from back-end business systems by putting them outside the firewall, but they will still be open to attack. If the firewall runs on top of an operating system with a vulnerability, then it can simply be taken over. Although some of these attacks may have been fixed, the underlying pattern is fairly constant. Most of the exploits make use of program bugs, of which the majority are stack overflow vulnerabilities. The exploitation of protocol vulnerabilities (such as NFS) vies with weak passwords for second place.

11. Spam and address forgery: The most that is commonly done is a look-up of the hostname against an IP address using DNS. So someone who can forge IP addresses can abuse the facilities. The most common example is mail forgery by spammers.
12. Attacks on Local Networks: Let's suppose we have a machine attached to the target person's LAN, and we want to take over an account in someone else's name to commit a fraud. Given physical access to the network, we can install packet sniffer software to harvest passwords, get the root password, and create a suitable account.
13. Gaining control over a chat room: Another common way of bringing down a host is known as smurfing. This exploits the Internet Control Message Protocol (ICMP), which enables users to send an echo packet to a remote host to check whether it is alive. The problem arises with broadcast addresses that are shared by a number of hosts. Some implementations of Internet protocols respond to pings to both the broadcast address and their local address (the idea was to test a LAN to see what is alive). So the protocol allowed both sorts of behavior in routers. A collection of hosts at a broadcast address that responds in this way is called a smurf amplifier. The attacker constructs a packet with the source address forged to be that of the victim, and sends it to a number of smurf amplifiers. The machines there will each respond (if alive) by sending a packet to the target, and this can swamp the target with more packets than it can cope with. Smurfing is typically used by someone who wants to take over an Internet relay chat (IRC) server, so they can assume control of the chatroom. The innovation was to automatically

harness a large number of "innocent" machines on the network to attack the victim.

14. Distributed Denial-of-service Attacks: A more recent development along the same lines is the distributed denial of service (DDoS) attack. Rather than just exploiting a common misconfiguration, as in smurfing, an attacker subverts a large number of machines over a period of time, and installs custom attack software in them. At a predetermined time, or at a given signal, these machines all start to bombard the target site with messages. The subversion may be automated using methods similar to those in the Morris worm. So far, DDoS attacks have been launched at a number of high-profile Web sites.
15. Spoofing attacks: We can combine some of the preceding ideas into spoofing attacks that work at long range (that is, from outside the local network or domain). Say that Charlie knows that Osama and Ayman are hosts on the target LAN, and wants to masquerade as Osama to Ayman. He can take Osama down with a service denial attack of some kind, and then initiate a new connection with Ayman. A simple way of guessing the sequence number of Ayman, which worked for a long time, was for Charlie to make a real connection to Osama shortly beforehand, and use the fact that the value of Ayman changed in a predictable way between one connection and the next. Modern stacks use random number generators and other techniques to avoid this predictability, but random number generators are often less random than expected—a source of large numbers of security failures. If sequence number guessing is feasible, then Charlie will be able to send messages to Ayman, which Ayman will believe come from Osama (though Charlie won't be able to read Ayman's replies to him). In some cases, Charlie won't even have to attack Osama, just arrange things so that she discards Ayman's replies to him as unexpected junk. This is quite a complex attack, but no matter; there are scripts available on the web that do it.
16. Routing attacks: Routing attacks come in a variety of flavors. The basic attack involves Charlie telling Osama and Ayman that a convenient route between their sites passes through his. Source level routing was originally introduced into TCP to help get around bad routers. The underlying assumptions—that "hosts are honest" and that the best return path is the best source route—no longer hold, and the only short-term solution is to block source routing. However, it continues to be used for network diagnosis.
17. Identity theft: In the United States, about half a million people are the victims of this kind of fraud each year. If criminals can steal someone's identity, counter-terrorists can do the same. Fraudulently making a video or audio tape as Osama bin Laden and passing it to al-Jazeera, or faking articles and videos to be placed on Al-Qaida's websites,, should not be that difficult.
18. Passwords: In a typical system, human users must authenticate themselves to a client. These authentication passwords can be gathered by password-snooping attacks, but also by more subtle routing attacks. The following are some common password attacks:

    Targeted attack on one account: This entails guessing a particular user's password. People who are allowed to select their own password or PIN will often choose the same one for a number of systems, so it's easy to remember. This method of password attack includes "shoulder surfing" by physical observation of the target person when he or she uses a public computer.

Attempt to penetrate any account on a system: To obtain a logon as any user of the system. This might be done so as to steal the service directly.
Attempt to penetrate any account on any system. We want an account on any system in a given domain; it doesn't matter which one. For example, terrorist motives are to get a place to hide Information or pirated software, or a platform from which attacks can be launched anonymously on other systems.
Service denial attack: The attacker may wish to prevent the legitimate user from using the system. This might be targeted on a particular account, or system-wide.
Attacks on Password Entry: Password entry is often poorly protected.

19. Intrusions: An intrusion occurs when an attack is able to successfully exploit a vulnerability.

Many other forms of attacks exist, such as viruses, worms, Trojan horses and so on. The above list merely serves to give the reader an idea of the technical possibilities for running an InfoWar against the ideology of Al-Qaida.

3.4 The Challenges

One of the main obstacles is the nature of the Internet itself. It is relatively open and unregulated, and highly interconnected. While the Web has evolved into a global platform for people to use in disseminating and sharing ideas, terrorists and extremists are also utilizing the Web for their relocation, propaganda, recruitment, and communication purposes.

Relatively few countries have effective written laws to recognize as crimes the words and images typically found on militants' websites. Internet technology is expanding exponentially from year to year, and the legal framework has yet to catch up in many countries. Often, depending on the country, it is not illegal to post anything at all on the Internet. Sometimes, countries do have decency rules and laws, but these do not extend to laws against incitement to riot or incitement to violence on the Internet.

Even in the United States, where the recently enacted Patriot Act gives law enforcement officers expanded powers to combat terrorism, convicting the authors of suspected terrorist sites has proven difficult. This Act (which was passed in the wake of the 11 September 2001 terrorist attacks) makes it illegal to advise or assist terrorists, for example via an Internet website. Yet in a closely watched case in Idaho, a Saudi Arabian student attending school there was recently acquitted of charges that a website he was running was raising funds for terrorist groups. The student, Sami Omar Al-Hussayen, argued that he was merely hosting the information, and was not himself the author, and that the site was protected by the free speech provisions of the US Constitution. The main charges were dismissed for lack of evidence identifying him specifically as a terrorist.

One of the main steps taken by law enforcement agencies is the identification of the web server of a particular website; this is usually the source which law enforcement agencies apprehend so as to undo terrorist websites. But the task is not all that easy. While each server and personal computer on the Internet has a unique address, locating an offending website is not as simple as, for example, tracing a telephone call.

There are several problems preventing the effective and efficient discovery of Dark Web intelligence. The first problem is mainly associated with information overload. The amount of data available on the Web is often overwhelming and unmanageable to terrorism researchers and experts. Also, Dark Web data are scattered across many terrorist websites, making it hard for terrorism experts to obtain a comprehensive picture. One of the tricks terrorist groups use is to link a website from server to server, or to "bounce" it, as the practice is known. When somebody goes to an Al-Qaida sympathizer site, the user or the site is bounced to two or three other sites. It is thus difficult to pinpoint exactly who is putting up the site. Even when a host computer or user is found, the problems have just begun. This tactic, in effect, conceals the identity of the original server and the site's author.

Identifying codifications can be very challenging for computer systems. Another important problem is that terrorist data posted on the Web are not consistent. Websites may suddenly emerge, frequently modify their formats, and then swiftly disappear, or, in many cases, seem to disappear by changing their URLs but retaining much of the same content [16].

As already mentioned, another challenge is that intelligence agency of Country A may be observing a specific website, while at the same time the site will be hacked by a non-governmental institution or intelligence agency of Country B.

4. Web-based Counter-Propaganda

The tools discussed in this part of the chapter are psychological operations and propaganda. The term "propaganda" has negative connotations, and in the current context it is replaced with the term "clarification". Thus psychological operations are a part of clarification.

The basis of the findings discussed here has been a broad analysis of political Internet campaigns. *Moveon.org* (http://www.moveon.org/) is a prominent example of a web-based political campaign. The reason political Internet campaigns are so successful is that websites make it easy for volunteers to sign up and distribute information, providing a list of e-mail address contacts for the campaign to utilize for mobilization purposes [17].

The purpose of political Internet campaigns can be compared with the purpose of the online campaign conducted by Al-Qaida and its affiliates. Both attempt to seduce the online user by engaging him or her in a sensory experience, influencing his or her needs, suggesting the fulfillment of a goal, and inspiring or guiding the user to make a choice by providing a higher-level motivation. Captivation occurs on the fulfillment of the goal, and the function at this point is to facilitate the formation of a relational bond between the user and the system [18]. The role of persuasion is to promote recognition of the gratification achievable through the attainment of higher goals.

In comparison, websites of democratic political parties want to persuade the citizen to become a voter who punches the ballot at the right place and donates money to the organization. The mission of Al-Qaida and its affiliates is also to persuade the user to follow the organization and to make donations. Both campaigns want to seduce, captivate and persuade the audience by using the Web. For this reason, it is worth analyzing the highly professional market of political Internet campaigns so as to identify methods which are able to seduce, captivate and persuade the target group.

In the following section, methods of political Internet campaigns that have proven successful will be identified, and it will be assumed that these methods can be adapted to address the people who are interested in the ideology of Al-Qaida. At this stage of research, however, the findings have not been tested.

Observers of the growth of political campaigning on the Internet have uncritically accepted the premise that utilizing the Internet as a campaign medium and strategy have revolutionized the practice of campaigning. Our opponents, the terrorist groups, have rapidly learned to use the Web as their platform for political campaigning. Understanding the underlying logic behind such campaigns, as well and understanding the uniqueness of the Internet as a vehicle of political communication, reveals this trend.

Lost in the rush to embrace the Internet as a campaign medium is what campaigns are all about, namely a series of tasks involving resource mobilization. As one author who has studied the use of technology in campaigning puts it, candidates must mobilize the predisposed, demobilize hostile voters, convince the undecided, and convert the initially hostile. They must do so by designing persuasive messages, communicating these messages, monitoring the responses, and facilitating the desired behavior [19]. Political communication serves as the vehicle in attempting to accomplish these tasks, and is a process that campaigns must maintain some control over; otherwise, they are likely to lose their contests.

However, if campaigning is done correctly, considerable benefits can result [20]. Campaigning via the Internet provides a great opportunity to control the process. Internet campaigning can also replicate feelings of personal contact [21]. A smartly-constructed site can convey a positive image for a candidate; in addition, anticipated and/or commonly-asked questions can be reproduced as website material, pre-empting the need for the candidate or his surrogates to answer such questions personally.

For national, state, and larger local constituency contests, websites can stand in for personal contact, creating an opportunity heretofore unattainable for these types of campaigns. The asynchronous aspect of the Internet, namely allowing prospective voters to access the carefully selected material and painstakingly crafted imagery of a high-quality site, is a blessing for the larger campaign, in particular.

Before all of this can happen, however, government agencies need to understand two things; firstly, the psychographic profiles of those susceptible to recruitment, and secondly, the messages that affect them. They also need to understand how these individuals are influenced: what channels are meaningful to them, whom they listen to, the effect of peer networks, and so on, in order to learn how to reach them most effectively.

Scholars at the *Institute for the Internet and Democracy* at George Washington University found in their research that 69 percent of individuals who accessed the website of a political party or candidate from November 26 to December 31 2003 were "influentials". This means that these individuals comprise the ten percent of the US population who engage in two-step-flow[2], or tell their neighbors what to buy, what politicians to support, where to go for vacation [22].

---

[2] Two-step-flow theory: The two-step flow of communication hypothesis was first introduced by Paul Lazarsfeld, Bernard Berelson, and Hazel Gaudet in *The People's Choice,* a 1944 study focused on the process of decision-making during a Presidential election campaign. Katz and Lazarsfeld developed the two-step flow theory of mass communication. This theory asserts that information from the media moves in two distinct stages. First, individuals (opinion leaders) who pay close attention to the mass media and its messages receive the information. These opinion leaders then pass on their own interpretations in addition to the actual media content. The term "personal influence" was coined to refer to the process intervening between the media's direct message and the audience's ultimate reaction to that message. Opinion leaders are quite influential in getting people to change their attitudes and behaviours, and are quite similar to those they influence. The two-step flow theory has improved our understanding of how the mass media influence decision making. The theory has refined our ability to predict the influence of media messages on audience

Influentials are engaged in one or more of the following ways: making a political contribution to a candidate or political party; reading political e-mail, and forwarding it on; visiting a political web log; participating in a political chat room; or visiting a news site for political information. It is probably accurate enough to assume that the influentials amongst Islamist groups and societies similarly visit radical websites, and seduce and persuade their friends to share the given ideas and views. If this is the case, it is even more important for the West in general to develop an online strategy to address these opinion leaders.

To run such a strategy, a political Internet campaign against terrorism must use tactics which have proven successful and which can be transferred. Certainly these methods should be adapted one by one, since they must be tailored to capture the attention of the opinion leaders, and must be able to seduce, captivate and persuade these leaders. The following list provides proven methods for political Internet campaigns which have the ability to be adapted to counterterrorist efforts.

4.1 Methods for an Online Campaign

4.1.1Content Methods

- Fix target and channel; use existing channels. The medium may or may not be the message, but the Internet does act as a considerable leverage point, and is becoming a well-defined entry-point to the media cycle. Once in the cycle, stories feed on themselves, and propagate through the more "conventional" media outlets;
- Target pressure points. Demographics on the Internet work highly in favor of targeted messages, providing numerous specialty forums with near-ideal spreads in income and age factors;
- Stress micro at the micro level, stress macro at all levels. The Internet is an international mechanism that can be used to manage local or topical messages, and with the same stroke of the pen, have wide distribution;
- Test messages and iteratively design them. While careful controls to limit distribution of test messages need to be used, a "natural selection" takes place that tends to kill messages that are non-viable, while propagating viable ones;
- Be flexible; run the operation in place. Newcomers cannot expect to manage propaganda efforts on the Internet, but once established inside of certain communities, operators can manage quite well;
- Know the context. The Internet is well structured to assimilate newcomers into the rules and nomenclature, providing a continuity of context that is quite striking;
- Set the tone properly. The Internet is designed to move information and it does so quite well; it also acts as a valuable forum to release information that is beneficial to the operation and acts informatively;
- Timing, duration, and repetition of message are critical. The Internet has an extremely fast cycle of turn-over, but also has a way of rehashing topics and messages continually;

behaviour, and it has helped to explain why certain media campaigns may have failed to alter audience attitudes and behavior.

- Keep the content simple and emotional. This requires skill in constructing the content for the Internet, which tends to apply logic more than most media; a direct message still cuts through the noise;
- At the same time, websites must be rich in content and frequently updated to continue the seduction of the users;
- Use no new issues; exploit existing ones. Hijacking old topics and putting a new "spin" on things is another favorite "indoor sport" of the Internet;
- Aggregate the message. Starting with the basic concept of any message, and evolving it over time works well on the Internet, but only if other rules are not violated (such as those to do with trust and context);
- One of the problems of the Internet, from a PsyOps perspective, is that many cyber communities are already highly skeptical because of the education and experience of individuals and institutions using the net;
- Disinformation tends to be harder to manage, but somehow always manages to find willing minds. Deception, the minor religion of the intelligence community and Internet alike, finds an abundance of opportunities in the net, particularly via the use of back-end active measures to damage perceptions of data or channels;
- Evoke group identifications. If managed in context, the Internet is a structured yet highly fractured social community, and evocation of group identities is greatly situation-dependent;
- The Muslim community itself plays a critical role in helping to insulate its members from extremist ideology on the Internet;
- Local Muslim leaders or grassroots organizations have to be won over to take their "fight" to the Internet;
- Establish trust. Voluntary communities on the Internet are structured with de facto reputation markets, and past performance is a major factor in how a message is received and interpreted;
- Request volunteer recruitment effort. Use rewards of gifts (books etc.) and request that visitors and e-captains (web influentials) send messages on to friends, neighbors etc.;
- Organize presentations of weekly activities for Web activists;
- Advise e-captains to raise concerns and issues with their local talk radio host;
- Target e-captains individually with messages;
- Build a community among essential target groups;
- Create a media letter-writing campaign;
- Post names of local neighbors on the website, who are able to encourage others who might wish to become active;
- Choose alternative Muslim opinion leaders to express their views on these websites;
- Don't misstate facts. Present alternative interpretations; this kind of activity is already the mainstay of the Internet;
- Websites addressing women and children should be also created;
- They key to success is personalization and creativity;
- Websites should not be obviously pro-Western;
- Create a superior Web presence;
- A single website doesn't constitute a Web strategy;
- Create a good email presence;
- Place advertisements, pop-ups and links on radical websites;
- Conduct online discussions (meetups, forums, talks);

- Learn about your visitors.

4.1.2 Functional Methods

- Websites have to be easy to navigate;
- Websites have to be interactive;
- Choose a good name for the web address;
- Update friendly blog URL's;
- Advise visitors on how to download information and send it out to these and other blogs.
- Place Trojan horses on radical websites to collect email-addresses;
- Send spams with your material to these email addresses.

4.2 Videos and Games

A single effectively placed video clip can find its way into the hands of several thousand opinion leaders who, in turn, can reach millions. That makes it possible to make a difference. When it comes to influencing opinions, video packs a greater emotional punch than text or photographs. The power of video to communicate ideas to large audiences has not been lost on politicians, corporations or the clergy. *YouTube* (http://www.youtube.com, for example, is the largest of the video-sharing sites, and it attracts more than 16 million viewers per month and presents 100 million videos a day.

Sammy Studios, a startup video game manufacturer faced with a limited marketing budget, no previous fan base and no previous games, wanted its first game to be in the Top Ten games of the year. How did Sammy achieve this goal? By creating a "dot.org" community website that was managed by an individual host who chatted on boards, personally answered e-mails and posted news notes. The website was designed to have a one-to-one feeling, and provided three distinctive kinds of content to members:

1. Ongoing flow of exclusive previews to tease the visitor;
2. Community features such as message boards, polls, chats, and pictures;
3. Rewards: young males are a competitive lot, and were awarded redeemable points for recruiting members.

The result was that several thousand hard-core fans caused the sale of millions of *Sammy Studies* games during the launch.

Another example of the power of video clips occurred when the British government needed to educate young people on the dangers of sexually transmitted diseases (STDs). Traditional warning messages were being ignored as the holiday party season approached. The government's response was to hire a media firm to produce an interactive, entertaining viral game. The "12 Days of STD" was targeted at human resources professionals who could forward it to employees within their organizations.

The result? From an initial mailing to 34,000 human resources professionals, the government achieved a pass-on rate of 4.5, an estimated total of 445,000 viewings. The following year the viral campaign went even further, as the game reached more than two million people and generated significant press and online coverage.

5. Towards a Strategy for Countering Islamist Narratives

Before a web-based clarification campaign can be started, an internationally accepted strategy should be developed. The concept of the war or battle of ideas has been mentioned is several governmental strategy papers, such as the *National Strategy of Combating Terrorism* of the United States (September 2006), or the *Strategy of Countering International Terrorism* of the United Kingdom (July 2006). The Strategy of the United Kingdom is in this context the most precise one, and it also opens the door for the integration of Muslims into the wider society.

All strategies that focus on terrorism tend not to address the specific issue of the ideology of Al-Qaida and its affiliates sufficiently, and they overlook the Internet as one of the most important channels of this ideology. But "a really big hole" in the US strategy, as one counterterrorism official said, "is that we focus on the terrorists and very little on how they are created. If you looked at all the resources of the US government, we spent 85, 90 percent on current terrorists, not on how people are radicalized." [23]

To be effective, a "countering Islamist narrative" (CIN) strategy has to be conducted in the air, on the Web and on the ground. Such a strategy should also be embedded in international institutions. NATO, UN, OSZE, EU and Interpol can take responsibility for specific parts of the strategy based on their capabilities. These institutions should work as interlocking not as interblocking institutions. Furthermore, every country should have a department whose purpose is to counter terrorist narratives.

For the development of such a strategy, the following questions need to be answered.

1. What are the strategic interests and objectives of the countries countering the narratives of Al-Qaida?
2. What are the threats to those interests and objectives?
3. What are the appropriate strategic responses to those threats?
4. What principles should guide the development of Western policy and strategy?
5. What are the consequences of the chosen strategic responses?

In short, what should be the grand strategy confronting the narratives of Al-Qaida?

The answers to these questions are not completely different from the answers to questions which examine the tactical level. But the difficulty in developing a strategy indicates our inability to find a common definition of terrorism. We are not even able to define our enemy.

Additional questions could be:

- What would we like our targeted audiences to see, and what impact do we wish to have?
- Do we want them to "like" us?
- Do we want them to question and doubt the information they get from their own governments, as we did with Radio Free Europe during the Cold War?
- Do we wish them similarly to cease supporting militant jihadists in their midst?

The basic problem here is axiomatic in all military and political strategies: one can never factor out the enemy [24]. You never can be sure how your enemy will behave.

An effective CIN strategy must be multi-faceted. This chapter argues that a grand counterterrorism strategy that actually produces the results we desire will depend on our having a subtle, psychologically-informed understanding of narratives and terrorism. A counterterrorism strategy would benefit from a comprehensive consideration of the stories terrorists tell. Understanding the narratives which influence the genesis, growth, and transformation of terrorist organizations will enable us to better fashion a strategy for undermining the efficacy of those narratives, so as to deter, disrupt and defeat terrorist groups.

Such a "counter-narrative strategy" would have multiple components, with layered asynchronous effects. While effective counter-stories will be difficult to coordinate and will involve multiple agents of action, their formulation is a necessary part of any comprehensive counter-terrorism effort. In practice, effective counter-narrative strategy will require understanding the components and content of the story being told, so that we can make accurate predictions. In other words, we need a sophisticated understanding of "strategic rhetoric".

Strategic communication is a vital component of national security. In a crisis, it must be transformed with a strength of purpose that matches the commitment to diplomacy, defense, intelligence, law enforcement and homeland security. The current problem is that national strategic communication concentrates on nation states not on global networks. Strategic communication requires a sophisticated method that maps perceptions and influences networks, identifies policy priorities, formulates objectives, focuses on "doable tasks", develops themes and messages, employs relevant channels, leverages new strategic and tactical dynamics, and monitors success.

The components of strategic communication are public diplomacy, public affairs, PsyOps and InfoOps. These are briefly discussed below.

Public diplomacy involves the exchange of people and ideas. The aim is to build lasting relationships and receptivity to a nation's culture, values, and policies, to influence attitudes and mobilize the public in ways that support policies and interests. Time horizons in this context refer to decades and news cycles.

Public affairs refers to communication activities that are intended primarily to inform and influence the media and the citizens. Of relevance is the way events and people are depicted in the media.

PsyOps refers here to *strategic* psychological operations, which advance broad or long-term objectives. Global in nature, they may be directed toward large audiences or at key communicators. *Tactical* psychological operations are more limited, and are used by commanders to secure immediate and near-term goals. In this environment, these force-enhancing activities serve as a means to lower the morale and efficiency of enemy forces.

Finally, military **InfoOps** must be coordinated and energized.

This approach has to build on in-depth knowledge of radical Islamism and factors that motivate radical or terrorist behavior. To be effective, a communication strategy has to engage in a respectful dialog of ideas with Muslims, which would begin with listening and assume decades of sustained effort. Such a strategy would influence attitudes and behavior, and would enable political leaders and policymakers to make informed decisions on changes in strategy policies, messages, and choices among instruments of statecraft. To be persuasive, strategic communication must be credible

A CIN strategy must focus on the destruction of myths, and should implement a shift in the metaphors that are being used. Myth creation involves the weaving together of the narrative elements of a story with facts about past and present situations, so as to create an emotionally compelling background. It is precisely such a background that very often directly influences the susceptibility of a population to manipulation by "myth mongers." The fanatical devotion shown by Al-Qaida operatives stems in large part not from any rational deliberative process, but rather from the success Osama bin Laden and others have had in fashioning a coherent and appealing foundational myth. Thus confronting Al-Qaida's narratives must be a critical mission requirement.

Some concepts relating to an effective communication strategy will now be discussed.

Credibility: We have to listen to the Arab narrative before we can counter the Islamist myth. Muslim participants have repeatedly stressed that the key is not only the content and context of the message, but also the credibility of the messenger. In order for a CIN strategy to be effective, the West and the targeted Muslim countries must regain their credibility in the Muslim world through policy actions.

Terminology: A CIN strategy has to address the terminology used in the press and by the politicians. The focus must be shifted away from discussing religion ("moderate Islam vs. Islamism"), and should rather be on encouraging the reform of secular and religious education. Terminology plays a significant role in this process. We need to understand better the nuances of Islam and Arab culture, or we will inadvertently validate Islamist ideology by talking about it in religious terms. For instance, should the terms "jihad" and "jihadist" be replaced with terms such as "war" and "warriors"? The term Islamism should be replaced with the term anti-Islam. Countering anti-Americanism, anti-Israelism and anti-Islam should be one of the primary goals.

Traditions: We should look for solutions from within Muslim traditions, because solutions derived from Western traditions will by definition be rejected as illegitimate. Finding alternative images for the ones being used by the followers of Al-Qaida is just one necessity for a web-based clarification campaign. It is also necessary to create a situation in which the distinction between Islam and anti-Islam becomes obvious.

Partners: In order to be truly effective in de-legitimizing terrorism, discussion about religion needs to come from within Muslim religious leadership, not from the West. The goal is to disseminate the message that the West is not at war with Islam, that this is not a religious or cultural war.

Identity: Western societies have to decide if they want Muslims to be a part of their national identity.

Act local, think global: We must have a long-term perspective on how to deal with the Middle East and the Muslims inside Europe. This long-term perspective starts with the offer that state institutions can provide education, medical treatment and social welfare. Furthermore, universities in Cairo, Beirut, and Istanbul, as well as schools and NGOs, could provide useful venues in which to open up the "modernization" debate. The same debate can be continued via Islam education in public schools throughout Europe, articles in journals, and the efforts of individual intellectuals. Such debate would allow all parties to hear different views and engage in a discussion that can truly educate, without appearing from the start to be a narrowly construed instrumental propaganda effort.

Students need more exposure to Muslim scholars who have grappled with modern problems and understand how to respond to such problems in ways that are compatible with sacred Islamic texts. Furthermore, higher education can emphasize the critical study of classical Muslim texts; it can also provide more exposure to the diversity of opinion that will broaden students' worldviews, and help them to recognize Westerners as human beings who are equal to themselves. Again, these educational reforms need to come from within the Muslim world, although the West can support it with its resources.

An effective CIN strategy must be multi-faceted, addressing all these aspects. This is a long-term undertaking that needs to center on education and development. Most importantly, CIN cannot be divorced from policy. In the final analysis, the effectiveness of our message will rest on our credibility and willingness.

6. Conclusion

The goal of this chapter was to provide an analysis of possible tactics tailored for countering Islamist narratives on the Internet, and to emphasize elements of strategy for countering those narratives. The first part of the tactical analysis discussed methods of neutralizing terrorist websites and chat-rooms. The second part of the tactical analysis surveyed the Web as a counter-propaganda tool.

The conclusion now presented is that effective CIN policies can counter the Al-Qaida narrative by focusing on the movement's center of gravity. This would entail interrupting the process of educating new terrorists by demonstrating that the extremists' story is not valid. In a multistage radicalization process, a network of participants communicates with and shares information with the individual, which means that the credibility of the information source is critical.

It must be understood, for example, that a government spokesperson on the 10 p.m. news would carry less credibility for the new terrorist than would the creator of his or her favorite blog. This is a particular challenge when dealing with people who have become radicalized. In addition, the vehicles of messages which challenge the tenets of the group are strictly censored. Simultaneously, the individual gets bombarded by cause-related information such as videos, lectures and literature.

The more an individual becomes seduced and persuaded by radical ideology, the more his or her critical thinking decreases, and it becomes harder to access the mindset of this person. Without access to alternative information sources, it is more difficult and costly to reach these individuals. However, if counterterrorism agencies are willing to understand this process, it is possible to identify those stages where the target is still open to alternative ideas and to counterterrorist opinions.

References

[1] The PEW Global Attitudes Project The great Divide: How Westerners and Muslims View Each Other. Europe's Muslims More Moderate, http://pewglobal.org/reports/display.php?ReportID=253.

[2] Marc Sageman: Alternative Views of the Terrorist Threat, in *International Studies Review*, December 2005, p. 676.

[3] Marc Sageman: Alternative Views of the Terrorist Threat, in *International Studies Review*, December 2005, p. 676.

[4] Ibid. p. 677.

[5] Gabriel Weimann: *Terror on the Internet: The New Arena, the New Challenges,* Potomac Books, 2006.

[6] Michael Wilson "Considering the Internet as an Intelligence tool," *Decision Support Systems*, Inc. , http://www.metatempo.com/NetIntelligence.pdf

[7] Classic works here include George Lakoff and Mark Johnson's *Metaphors We Live.* See also Gabriel Weimann: www.terror.net: "How Modern Terrorism Uses the Internet", United States Institute of Peace, *Special Report 116*, March 2004.

[8] Emran Qureshi & Michal A. Sells: *The New Crusades: Constructing the Muslim Enemy*, Columbia University Press, 2004.

[9] Chanel Mbakwe & Daniel Cunliffe: "Conceptualizing the Process of Hypermedia Seduction", Paper presented at the International Meeting of Sience and Technology of Design. 25-26. September 2003, Lisabon, Portugal, p. 2.

[10] Ibid.

[11] William Casebeer & James A. Russel: Storytelling Towards a Comprehensive Counter Narrative Strategy, in: *Strategic Insights*, Volume IV, Issue 3, March 2005, p. 6.

[12] Donald Rumsfeld: "New Realities in the Media Age", speech at the Council on Foreign Relations, 17.02.2006, http://www.cfr.org/publication/9900/new_realities_in_the_media_age.html.

[13] See: Yilu Zhou, Jialun Qin, Quanpi Lai Reid, Hsinchun Chen: "Buliding Knowledge System for Researching Terrorist Groups on the Web", Proceedings of the Americas Conference on Information Systems, Omaha, NE, USA, August 2005, p.4.

[14] Ibid. p.5.

[15] L. Ballesteros & B. Croft: "Dictionary Methods for Cross-Lingual Information Retrieval", in *Proceedings of the DEXA Conference on Database and Expert System Application,* Zürich, Switzerland, 1996, pp. 791-801.

[16] Gabriel Weimann: *Terror on the Internet: The New Arena, the New Challenges,* Potomac Books, 2006.

[17] Deb Price: "Candidates Hit E-Campaign Trial: They Tap the Powers of the Net to Reach Voters, their Pockets", in *Detroit News*, 26.11.1999, p. 1.

[18] Chanel Mbakwe & Daniel Cunliffe: "Conceptualizing the Process of Hypermedia Seduction", Paper presented at the International Meeting of Sience and Technology of Design. 25-26. September 2003, Lisabon, Portugal, p. 3.

[19] Stephen E. Frantzich: *Political Parties on the Technological Age*, New York, 1989.

[20] Judith S. Trent & Robert V. Friedenberg: *Political Campaign Communication: Principles and Practises,* 3rd ed. Wesport, 1995.

[21] Gary Selnow: *Electronic Whistle-Stops: the Impact of the Internet on American Politics,* Westport, 1998.

[22] Montague Kern: Web and Mass Media Campaigns by Political Candidates, MoveOn.org and the Democratic Party in the 2003-04, Presidential Primaries, Struggling for Votes and Unity in a Year of Web Innovation. School of Communication, Information and Library Studies Rutgers University April 19,2004, p.2.

[23] Karen deYoung: Spy Agencies Iraq War Hurting U.S. Terror Fight, in: *Washington Post*, 24.09.2006, http://www.washingtonpost.com/wp-dyn/content/article/2006/09/23/AR2006092301130.html

[24] Harry G. Summers, *On Strategy A Critical Analysis of the Vietnam War*, Dell Publisher, 1984, p. 71.

Hypermedia Seduction for Terrorist Recruiting
B. Ganor et al. (Eds.)
IOS Press, 2007

Design and Identity – Visual Culture and Identity Politics

Karen Lisa G. SALAMON©
Anthropologist, Ph.D, Associate Professor,
Danmarks Designskole/ Denmark's Design School
Copenhagen – Denmark
kls@dkds.dk

Abstract. The Danish "Cartoon Crisis" of 2005 to 2006 is presented in this chapter as a case study in the current culturalization of politics. This culturalization yields fertile ground for terrorist recruiting, particularly via the Internet and other mass media. Building on contested claims of authenticity, a hypermediated culture war is echoing the escalation of militant conflict. A number of issues are addressed in this chapter. In terms of *identity politics*, both on the side of the newspaper that published the cartoons and on the side of the protesting Muslims, the cartoons were used as a pretext for capturing public discourse and the public imagination. They were also used to present particular, strategic versions of authentic identity formations and distinctions between "our side" and "their side". The issue of the *aesthetization and culturalization of politics and popular culture* is also addressed. The escalation of the crisis shows to what extent all parties involved regarded the electronic and other mass media as predominant political machines. Claims to fame happen via successful claims to a particular identity, or via aesthetically resonant acts. Mass-mediated cultural representations are regarded as serious political material, which contributes to the further aesthetization of politics. Another issue is that *viral marketing, commercialized narratives and strategic branding all influence grass-root activism and militancy*. All parties in the cartoon conflict either created or modified particular strategic narratives. They were inspired in this by some of the branding technologies that have appeared along with the growth of the Internet and hypermedia. Finally, the question of the *role of social distinctions and claims to authenticity* is addressed. All sides of the conflict appealed to their constituencies via particular articulations of authenticity, in relation to who could speak on behalf of whom or what. Of these various themes, this chapter particularly addresses that of authenticity, which seems the most irreconcilable and potentially violent. A conclusion is that authenticity claims, their production and mediation via mass media, and their perception by users, must be further decoded if we are to deconstruct the ideological basis and cultural "machines" of fundamentalist militancy today.

Keywords. Identity politics, design, authenticity, militant fundamentalism, cultural war, aesthetization of politics, anthropology, strategic branding, Danish "Cartoon Crisis".

Introduction

My particular contribution to the discussion of hypermedia seduction and terrorist recruiting is focused on the concept of authenticity. I examine the role of authenticity in privileging particular constructions of cultural identity and forms of cultural production, and thus subsequently also its role in privileging certain forms of discourse and certain social groups. I am interested in better understanding the ways in which claims for authenticity are used to produce certain cultural and political agendas, both in mainstream democratic society, and amongst the radical, militant groups that try to overthrow this society.

I will take the liberty of discussing the topic of identity and design, irrespective of the kind of media that present them. I am no expert on hypermedia, and for the purpose of my particular discussion, it seems that the difference between linear and non-linear media is not particularly significant. Much of the visual material that appeals to mass audiences via the Internet essentially functions similarly to other mass media, but happens to be distributed via hypermedia-technology.

Of course, I do acknowledge that the hypermedia play a special role. This is due not only to their almost global reach, high speed, instant but flexible timing and seductively multi-layered grid, but also to certain socially and aesthetically appealing qualities and their structure of accessibility[1] [1]. The hypermedia facilitate a sense of social presence and group belonging, as has been documented by a number of studies [2] and recent news reports.[2] They also make possible a multi-media communication that is in itself affectively and poetically appealing. In addition, they make possible a sense of virtual community for users, whilst guaranteeing relative anonymity and the freedom of individual interpretation of consequences for practice[3] [4].

I will touch on some of these issues later, but will not address them in depth, as my focus of this presentation lies elsewhere. My concern here is with the contemporary interplay of identity issues and design, which happens to be mediated today via the technologies of the hypermedia. This phenomenon involves an array of issues such as

- identity politics,
- an increased aesthetization and culturalization of politics and popular culture,
- the influence of viral marketing, commercialized narratives and strategic branding on grass root activism and militancy,
- the role of authenticity claims and social distinctions in contemporary cultural production.

I would like to expand on these issues in this chapter. I do realize that these are rather abstract and very comprehensive issues that cannot be addressed in either detail

---

[1] In collaboration with Dr. Ida Engholm, I have elsewhere discussed the role of aesthetic and socially symbolic norms for web design, and the strategic use of creating brand identities and distinctions by breaking norms in interface design. [1]

[2] The Danish press recently reported that youngsters challenge each other to dangerously daring activities via Internet clubs. For example, a 14-year old boy was caught challenging other youths on popular chat-sites to climb tall buildings. Other young people have arranged "fighting-dates" via chat-sites, subsequently meeting in person to fight each other in physically destructive ways. The young people are described as wanting to "test the limits", and the Internet challenges are called "a fashion-craze" [3]

[3] These issues are discussed further in relation to cyber-mediated religions by Højsgaard and Warburg [4]. See also [2]

or depth here. However, I would like to introduce them as viable issues to include in the analysis of web-seduction and the related growth of militancy.

My own anthropological research is concerned with the question of subjectivity in terms of authenticity and identity, and how this is linked to formations of cultural production. Such cultural products include the design industry and forms of governance. I have particularly looked at the role of popular religion in strategies of governance through cultural production, for example, the use of New Age concepts and models for corporate human resources management and branding [5]. What I would like to do in this chapter is briefly to state my argument in relation to the interplay of design and identity issues.

In February 2006, the Secretary-General of the United Nations commented on a political crisis that was triggered by twelve drawings having been published in a daily newspaper in my country, Denmark [6]. It is not very often that Denmark makes it to UN headlines, and possibly even less common that drawings and design end up on the world diplomatic agenda. As an academic and social anthropologist working in Denmark's School of Design, I obviously found this event interesting in its own right, and all the more so in the present context of a seminar on "Hypermedia Seduction for Terrorist Recruiting".

The "Cartoon Crisis" is illustrative of several urgent contemporary political issues. These range from antagonisms between Islamist political ambitions and Western forms of democratic governance, to the role of visual communication in popular culture. The development of the crisis, which began in the latter part of 2005, involved issues of visual culture, the contemporary role of mass media in the global system, and a very concrete political battle over the popular imagination.[4] I shall not go into a deeper discussion of the Cartoon Crisis as such, but mention it here in relation to my main interest.

What I am particularly concerned with are the ways in which visual culture comes to serve as the medium for a certain aspect of identity politics, namely battles over authenticity.[5] I would like to suggest that a significant part of the Cartoon Crisis centered around issues of authenticity, on both the Islamic and cartoonist sides. Before examining the significance of battles over authenticity—which also affect design—I will briefly outline the Cartoon Crisis and the authenticity issues that it entailed.

---

[4] It may also be noted that Denmark is part of the allied forces currently in Iraq, and thus may also be associated with broader—even global—issues of security and imperialism.

[5] In this context I will not discuss the definition of either authenticity or of identity politics as theoretical concepts per se, or as philosophical terms. My approach to the concept of authenticity will partly become clarified through my discussion. I do not work from a particular philosophical understanding of "the authentic" as a natural or ontological category, but rather study the ways in which different cultural discourses claim "authenticity", and how they construct the term, and its repercussions in particular contexts. I use the term "identity politics" in its popular sense, as political strategies that argue from the concerns (and usually also exclusivist speech positions) of particular social groups identified on the basis of gender, religious denomination, race, ethnicity, or sexual orientation. The term has often been associated with suppressed minorities, such as Afro-Americans or Queers, and thus has acquired a somewhat "progressive" tone related also to the term "political correctness". I use the term in a broader sense, where "identity-politics" covers a particular social strategy, irrespective of its inherent political vice or virtue.

1. "The Most Serious Diplomatic Crisis Since World War II"

Firstly, I offer a brief summary of the events that led to almost global negative reactions from Muslim communities, and an international diplomatic crisis for Denmark that is regarded as the most serious since World War II. I will mention just a few of the occurrences that seem relevant to our discussion, rather than providing an extensive description of the development of the crisis.

On September 17 2005, the Danish liberal daily newspaper *Politiken* ran an article discussing a difficulty encountered by a writer. The writer had been unable to find an illustrator for his Danish-language children's book on the Qur'an and the life of the Prophet Muhammad. According to the newspaper, three artists had declined the writer's proposal, before a fourth had finally agreed to assist—but anonymously. The alleged refusal by the artists was seen as evidence of self-censorship, and led to much debate. During this time, a popular comedian declared that he would, hypothetically, dare to urinate on a Bible on television, but would not dare to do the same to the Qur'an, thus implying that he would not risk reprisal.

On September 30 2005, the newspaper *Jyllands-Posten* published an article entitled "The face of Muhammad" [7]. The article appeared alongside twelve cartoons, which either depicted Muhammad or commented on Islamic taboos or on Danish Islamophobia [8]. The cultural editor supplemented the drawings by stating the following:

> Some Muslims reject the modern, secular society. They demand a special position, insisting on special consideration of their own religious feelings. It is incompatible with contemporary democracy and freedom of speech, where you must be ready to put up with insults, mockery and ridicule. It is certainly not always attractive and nice to look at, and it does not mean that religious feelings should be made fun of at any price, but that is of minor importance in the present context. [...] We are on our way to a slippery slope where no one can tell how the self-censorship will end. That is why *Morgenavisen Jyllands-Posten* has invited members of the Danish editorial cartoonists union to draw Muhammad as they see him. [...] [7].

The newspaper explained that it had invited about forty different artists to give their interpretations of Muhammad, but only twelve caricaturists had chosen to respond, each submitting a single drawing. More than five months later, the same editor would feel compelled to further explain his motives to the *Washington Post.* He stated: "The cartoonists treated Islam the same way they treat Christianity, Buddhism, Hinduism and other religions. And by treating Muslims in Denmark as equals they made a point: We are integrating you into the Danish tradition of satire because you are part of our society, not strangers. The cartoons are including, rather than excluding, Muslims" [9, 10, 11, 12, 13, 14].[6]

About a month after the publication of the drawings, eleven Islamic ambassadors asked for a meeting with Danish Prime Minister Anders Fogh Rasmussen, in order to

[6] The Cartoon Crisis has been analysed and strongly debated in a number of Danish language books. My quotes are from English translations of the original sources on Wikipedia, September 2006.

discuss what they perceived as an "on-going smearing campaign in Danish public circles and media against Islam and Muslims".[7]

In a letter, the ambassadors mentioned not only the issue of the Muhammad cartoons, but also a recent indictment against a neo-Fascist radio station, as well as discriminatory utterances by politicians, including one of the ministers. It turned out that the ambassadors had received petitions from Danish imams on these issues.

The Danish government replied with a letter only: "The freedom of expression has a wide scope and the Danish government has no means of influencing the press. However, Danish legislation prohibits acts or expressions of blasphemous or discriminatory nature. The offended party may bring such acts or expressions to court, and it is for the courts to decide in individual cases." [9]

The Egyptian Minister of Foreign Affairs wrote several letters to the Prime Minister of Denmark, EU's coordinator of foreign policy, the general secretary of OSCE and to the secretary general of the UN, explaining that they only wanted "an official Danish statement underlining the need for and the obligation of respecting all religions and desisting from offending their devotees to prevent an escalation which would have serious and far-reaching consequences".[8]

The crisis went on to include a trip to Muslim countries in the Middle East by a delegation of Danish imams, who lobbied for international reactions against Denmark. They brought along the cartoons, but also other visual material that turned out to be forgeries. In March 2006, a boycott against Danish products began in most Islamic countries. In Denmark itself, anti-Islamic sentiments grew, to the extent where several percent of voters shifted political allegiance, according to opinion polls [15].

Muslim organizations in Denmark sued the newspaper, and the imams and various Muslim groups demanded an excuse, whilst right wing politicians demanded that the imams rather excuse themselves and show loyalty to Denmark, in which country they had attained citizenship. Since then and until the time of writing (November 2006), the boycott of Danish produce has continued to various degrees in a number of Muslim countries and communities worldwide. These boycotts have been carried out in an uneven and apparently diplomatically and media impressionable manner.

Furthermore, the cartoonists still have to live "underground" (personal communication with cartoonists' acquaintances, September & October 2006); Danish companies have lost well more than a quarter of a billion Danish Crowns in income; right wing extremists from the British National Party have allegedly distributed 500,000 handbills in Britain showing one of the drawings; and Iranians have renamed Danish pastry "Mohammad-roses", and have issued a cartoon competition for the best anti-Semitic drawings [8].

---

[7] Letter of 12 October 2005 from the ambassadors of Turkey, Iran, Egypt, Algeria, Libya, Saudi Arabia, Pakistan, Indonesia, Bosnia & Herzegovina, Morocco and the head of the Palestinian General Delegation to Denmark, to the Danish Prime Minister Anders Fogh Rasmussen , www.rogerbuch.dk/jpabrev.pdf (accessed 27 February 2007).

[8] Quoted in Danish *"Alligevel havde vi [...] forventet en officiel dansk erklæring, der ville understrege behovet for og endda forpligtelsen til at respektere alle religioner og afstå fra at krænke deres tilhængere, for at forhindre en eskalering, der kunne få alvorlige og vidtrækkende konsekvenser"* in news item "Egypten gav Fogh mulighed for forsoning", in the daily newspaper *Politiken*, 22. February 2006. Copenhagen, Denmark.

2. Four Themes

Why is the Cartoon Crisis story of interest within the context of hypermedia seduction and terrorist recruiting? I find it relevant for several reasons. Earlier in this chapter I mentioned a number of issues related to the interplay of identity constructions and cultural productions such as design. Without going into an extended argument at this point, I would like briefly to point out where I believe the Cartoon Crisis corresponds with these issues:

Identity politics: Both on the side of the newspaper that published the cartoons and on the side of the Muslims (including Danish Muslims, the Islamic ambassadors, and the tens of thousands of Muslims who demonstrated or boycotted worldwide), the cartoons are being used as a pretext for capturing public discourse and the public imagination. They are also being used to present a particular, strategic version of the distinctions between "our side" and "their side". The role of the actual cartoons is relatively minor, but they have come to represent iconic cultural positions on both sides of the conflict. On the side of the cartoonists and the newspaper, the cartoons represent the democratic freedom of expression; they may also represent the negation of autocratic, patriarchal and medieval traditions that might otherwise infringe upon liberal Danish culture[9] [16]. On the side of the imams and other Muslim voices, the cartoons represent disrespect for Islam and for Muslims living in Europe, as well as a Christian Crusade attack against the central values of Islam.

Aesthetization and culturalization of politics and popular culture: The escalation of the crisis shows to what extent all parties involved in the conflict accept the premise that cartoon drawings in a regional Danish newspaper are serious political material. I doubt that this would have been the case some twenty or thirty years ago. Even the liberation of pornography in Denmark (which was the first country in the world to liberate porn, as far as I know) did not seem to draw anywhere near this amount of serious political attention. The electronic mass media, not least of all television, might play their role in the aesthetization of politics, but there are, of course, also other factors. Cultural theorists have debated these issues, both with respect to various forms of modern totalitarian political agitation and with respect to post-modern visual culture.

Viral marketing, commercialized narratives and strategic branding influence grass-root activism and militancy: All parties, from the newspaper that published the cartoons, to the imams who lobbied against the newspaper, to the ambassadors who made an event out of the case, to the media agencies that broadcast the case worldwide, and finally the Danish political parties—which have created new narratives of national loyalty relating to the cartoon crisis—all of these parties seem to be inspired by some of the branding technologies that have appeared along with the birth of the Internet and hypermedia.

---

[9] The governing Danish Minister of Culture, MP for the Conservatives Brian Mikkelsen, expressed himself thus at the annual meeting of the Conservative Party in September 2005, five days prior to the publication of the cartoons: "A medieval Muslim culture will never become as valid here in this country as the Danish culture, which as it were, has grown out of that piece of old ground which is situated between Skagen and Gedser and between Dueodde and Blaavandshuk". (My unauthorised translation from Danish of: "*En middelalderlig muslimsk kultur bliver aldrig lige så gyldig herhjemme som den danske kultur, der nu engang er groet frem på det stykke gamle jord, der ligger mellem Skagen og Gedser og mellem Dueodde og Blåvandshuk*") [16].

News about the cartoon could not have spread this quickly, effectively, or with such visual appeal, without the Internet and hypermedia.

The role of authenticity claims and social distinctions in contemporary cultural production: All sides in the cartoon conflict seem to appeal via particular articulations of authenticity, in relation to who can speak on behalf of whom or what. Who would be the authentic spokesperson for the Danish Muslims? Who would authentically represent original Islamic views regarding visual representations of the Prophet? Who has a true understanding of the culture related to freedom of expression? How could an authentic Danish company "sell out" by trying to recapture Muslim market-shares, through giving an excuse on behalf of the newspaper—in which it has no formal share whatsoever? Is a Danish-born Muslim, who demands censorship on religious caricatures, authentically "Danish"? And so on and on.

Of these four themes, the issue of authenticity seems the most irreconcilable and potentially violent. In a discussion of conspiracy theory, social theorist Mark Featherstone has argued that "theories of absolute difference may inadvertently create an atmosphere of conspirational thinking" [17]. The inability to recognize the authenticity of other positions might be related to this conclusion. The cartoonists might be seen as guilty of what Featherstone calls "performing exclusionary violence at the level of text/discourse" [ibid. p. 34].

However, the Muslim interest groups involved in the conflict may be seen as equally representative of a monolithic ideology, which tries to silence critique. Such an ideology claims the existence of a natural order of reality, whilst disregarding any oppositional stance "by arguing that critical thinking is constitutive of an 'un-natural' construction" [ibid. p. 35]. It is worth noting that the word "inauthentic" could have been used in this quote in place of the word "un-natural".

The present context of the war in Iraq, with Denmark actively involved on the allied side, in addition to the post-9/11 cultural climate, further aggravate these tendencies towards exclusionary and conspirational thinking.

3. The Return of the Authentic?

In Denmark, over the past few years—even before the Cartoon Crisis—we have been debating the role of design in society. In the Danish media, design has often been described as "national heritage", typically with a reference to the Danish Modern style. In one of the largest daily newspapers (*Politiken,* where the debate of self-censorship in relation to images of the Prophet began), original handicrafts were recently contrasted with industrial design.

Architect and design-researcher Ahnfeldt-Mollerup was quoted as follows: "I think people are looking for something that they perceive as more authentic and more rooted than in the 1990s, when people were enormously concerned with brand items. Something that can contribute to give them a special identity. This is true for all fields, such as clothing, interior design, kitchen goods, anything."[10] [18]

---

[10] My unauthorised translation into English of the following original Danish quote: *"Jeg tror, folk søger noget, som de oplever som mere autentisk og mere forankret end i forhold til 1990'erne, hvor man var*

There is nothing new in linking design consumption with identity issues. As Gabriel and Lang state, "[t]he consensus of otherwise irreconcilable perspectives appears to be that in late capitalism, consumption is the area where personal and group identities are fought over, contested, precariously put together and licked into shape. [...T]he Western consumer readily transfigures into an identity-seeker" [19].

Identity is reflected in, or rather produced through, consumption, in this case the consumption of design and new media. However, the above quote, which refers to the renewed consumption of "original handicrafts" in design, suggests more than this. It addresses the phenomenon of middle-class Europeans searching beyond mass-produced brands and industrial design to find a unique, meaningful sense of identity and authenticity.

One might speculate that hypermedia seduction and terrorism are far removed from European, middle-class design consumption. I would suggest otherwise. Several of the perpetrators of recent terrorist acts against civil society have turned out to be educated, middle-class consumers, who have turned to extremist movements due to what they perceive as a religious search. We might classify this search as a quest for identity, and possibly also for authenticity. This is my hypothesis here.

In a Danish newspaper feature in the intellectual weekly *Weekendavisen*, during April 2006 (thus published during the Cartoon Crisis), a reporter visited a family of second-generation Turkish immigrants to Denmark. The grandfather of the family had immigrated to Denmark as migrant laborer in the 1960s, taking along the father (now the owner of a bicycle shop), who was then a teenager. The mother of the family had come to Denmark in order to marry the father. The two daughters, one of whom was a primary school teacher and the other a high school student, were both still unmarried at the time of the interview.

The family was interviewed by the journalist, and explained how they had gradually become more and more religious, and how they trusted the Internet for religious advice. The journalist also interviewed a young imam and facilitator of "Denmark's United Cybermuslims" [20], which was amongst the favorite mailing lists used by the two young women. The list, which addresses its readers as "brothers and sisters", is connected to the site http://www.islam.dk/. The young Danish imam and list-facilitator explained that "The imams just service the parental generation, and the parents have completely mixed their religion with the traditions of their homeland [...] it is from the Net that the young people can receive the most reliable knowledge of Islam". [Ibid.]

At the time of the interview, the two young Danish Muslim women were regularly consulting his site, along with such sites as islamcity.com, islamfortoday.com, and sunnipath.com. More recently the young women have checked these sites to find out whether it is *halal* to pluck one's eyebrows. Inspired by the mailing lists, they have stopped listening to music and started wearing veils as head-coverings, and more recently also a full *jilbab*-dress. One of the young women said that she now either recites the Qur'an, or rehearses dates from Danish history, instead of listening to music, when she exercises on her running machine. She has also convinced her mother to begin praying five times a day, whereas her father is more reluctant: "He is proud of us, because it is nice to have some daughters who are strong believers. It makes him

---

enormt optaet af mærkevarer. Og noget som er med til at give dem en speciel identitet. På alle områder, indenfor beklædning, indretning, køkkenting, alt."

happy that we know so much and can guide him, even if we are younger than he. But sometimes it is also difficult for him. Then he might say that he can't stand listening to us anymore" [21].

A Danish scholar of cyber-religion, MT Höjsgaard, suggests that we are witnessing a Muslim youth rebellion, comparable to the Lutheran reformation of the Roman-Catholic church:

> Like Luther, the young Muslims want to return to the original starting point of their faith. They want to read the old scriptures and liberate themselves from the traditional interpretations, and here the Net makes it possible to search for those interpretations that suit you precisely [...] The Net lends more power to the individual human being, but in some strange way what happens is that many [use the Net to] search for that which is authoritarian, reactionary and fundamentalist [21].

In other words, both the young imam and the scholar interpret the increased religiosity and newfound fundamentalism of young Danish Muslims as a search for something rather more monolithic and authoritative, but also as a trend of youthful rebellion. Phrased in my terms, we seem to be witnessing yet another form of "consumption of authenticity", which at the same time is actually a "production of authenticity". Young, educated, middle-class Danish Muslims surf the Internet and consult hypermedia in order to organize their lives in terms of identity-political strategies.

Authenticity seems to resurface as an issue during certain epochs. In my anthropological analysis of contemporary management thinking, I have also touched upon this phenomenon, mainly as it is expressed in the search for totality or ideological holism [22]. Various critical cultural studies have, in various ways, attempted to deconstruct or attack the concepts of authenticity and essence. Theodor Adorno analyzed what he called a "jargon of authenticity" in relation to German ideology [23]. And post-modernists as well as critics of Western colonial thinking have in different ways attempted to describe and deconstruct the figure of authenticity. They might do so, for example, by deconstructing the subject, by addressing the transcendental status of language and meaning, or by questioning representation and replacing essence with simulacra [24, 25].

The current presentation does not attempt to synthesize these efforts, or to present an independent theory of authenticity. It merely tries to sketch out the role of authenticity as a contemporary design issue. It does also suggest that analyzing the widespread contemporary search for authenticity might be one of the keys to understanding and countering the seductive appeal of the new fundamentalism and militancy, which is mediated by well-designed hypermedia.

4. Authenticities, Ownership and Social Prestige

I would like to suggest that the concern with authenticity is central to the appeal of well-designed media for other reasons too. Authenticity is at play in branding, patenting, claims for originality and authorship. It is thus fundamental to the assessment of identity, value and ownership rights. It is also intricately linked with the claim to social prestige.

Authenticity has to do with assessing what a thing or message really *is*, and thus is a matter of identity and ontology. As mentioned above, ownership of authentic design can be a way of producing identity by consumption. The consumption of authenticity is generally characterized by being expensive or in other ways exclusive. Often it is a way of displaying either great wealth (i.e. the ability to buy original works of art or unique services) or socially privileged identity and insider-knowledge (i.e. belonging to the "pure race" or having the ability to discern when a quote from the Qur'an is authentic). By buying an expensive and authentic piece of art, an economically rich consumer can hope to procure some of the social and cultural authenticity that symbolically flows from the artwork, but which he might lack due to a personal background that is viewed as non-authentic by his social surroundings.

Similarly a young Muslim Dane, brought up in an assimilated, increasingly secularized, culturally rather ignorant, middle-class family, might gain a sense of authentic Muslim identity and "brotherhood" (or any sense of identity and social belonging at all) by "consuming" websites that claim to represent authentic Islamic knowledge.

The association between authenticity, privileged forms of identity and economic wealth cuts both ways. The "authenticity formula" also facilitates access to economic wealth for those who can claim authentic identity and knowledge. By joining the right social networks and having the right kind of knowledge and appearance—or "blood and race"—one might gain access to economic opportunities or other forms of social privilege. In the extreme version, one might gain access to Paradise and absolute, unquestioned identity and authenticity by literally sacrificing oneself for the cause.

This phenomenon is what the French anthropologist Pierre Bourdieu explained as the exchange of various forms of capital. For example, there could be an exchange between cultural, social and economic capital, with trends and preferences often being expressed in the distinctive taste-regimes of different social groups [26]. Most of us are so familiar with this social and cultural mechanism that it has become invisible to us, and only becomes evident when presented in more exotic or extreme versions. Such extreme versions include the identity politics and authentification battles that take place over the definition of native or aboriginal art and design, in such places as Australia and North America.

The issue of race and blood is linked to cultural production, design and economic exchange, all of which are also recognizable in the hypermediated fundamentalist and conspirationalist, seduction that is our main topic here. On this important point, I quote MF Brown at some length:

> ...The Indian Arts and Crafts Act of 1990, which strengthened similar legislation passed in the 1930s. The broad intent of the law is simple: to guarantee the authenticity of Native American art by prohibiting the sale of products falsely claimed to be made by American Indians or Alaska Natives. The law was designed to protect native artists who find their livelihood threatened by counterfeit goods, some produced on near-industrial scale in Mexico, Pakistan, India, Thailand, and the Philippines. One might think that certifying works as authentically Native American would be a straightforward business, but it is not. The law defines as

> Indians and Alaska Natives only those officially registered as members of a recognized tribe or community. One result is that people who are clearly Indians, but who lack the requisite tribal membership, are prevented from identifying their work as Native American. [...And in Alaska] Artists are issued [...an authentificating] tag that identifies a work of art as an "Authentic Native Handicraft from Alaska". Artists are issued the tags after submitting an application that includes verification of Alaska Native identity (blood quantum of one-fourth or greater)[...] [27].

Although the topic of Native American handicraft might sound rather remote from that of hypermedia and terrorist recruiting, it may be clear by now that battles over authenticity do play a significant role in both of these contexts. Furthermore, the above scenario illustrates how cultural production, in the form of design, is used in national identity politics and in various local battles over race and identity *as access to privilege*. Branding, authenticity and identity are closely linked, and it has become possible today more than ever before to "consume" the identity of other people. One may even consume their authenticity, as the Native American example shows. In such instances, design becomes a tool for defining legal property and (cultural) privilege, and even personal identity.

Thus, both governmental agencies and minority groups may attempt to construct structures of privilege, by trying to regulate the access to claims for authenticity. This practice finds expression in strategies of consumption.

Militant Islamist, Christian Patriot and Neo-Nazi groups further elaborate upon such forms of identity politics. These groups generally support their claims to legitimacy by insisting that they are built on ancient, perennial, exclusivist and authentic foundations. They tend to use naturalist and essentialist metaphors and images, typically relating to anchors of human identity such as blood, race, family (motherhood, brotherhood), death and sacrifice.[11]

But in spite of their naturalized, particularist and essentialist claims, such groups also reflect a modern, fleeting, mass-cultural and near-global trend. This is detectable in the following characteristics:

- their use of hypermedia and other mass-media
- their constructions of metaphors
- their play on (post-) modern formats of identity and absolute difference
- their conspiracy theories and uses of pseudo-scientific discourse, and
- their rationalized obsession with particular forms of symbolic order.[12]

---

[11] Examples of this are the calls for martyrdom issued by Islamists, eg. Sheikh Abdulla Azzam's claim that "History does not write its lines except with blood." [28]; or video-speeches by young suicide-bombers, such as the one by Mu'min Rajab Rajab Abu Hafs, prepared before his suicide attack on December 7, 2004 and published on *Hamas*' website in February 2006: "My dear mother, you who have cared for me, today I sacrifice my life to be your intercessor [on Judgment Day]" [29].

[12] The modernity of fundamentalism and similar forms of discourse has been argued by several scholars of culture and society [30, 31]. I will not argue this point further here, but rather take this analytical standpoint as a preconditional claim for my further arguments.

These fundamentalists claim a privileged access to original, eternal Truth—or to "pure blood", or the "highest race", or a "secret knowledge" about others' conspiracies, and so on. Subsequently, they assert an authentic "birthright" to execute their ideology universally. But although their claims reach back into pre-modernity and deep into the inner essence of nature, the ontological grounds on which they try to conquer the world belong to modern, capitalist, individualist consumer culture and its legal and economic foundations, as much as to a privileged source in a golden past or a higher racial order.

5. The Design of Authenticity

The interplay of identity issues and design obviously also appears in formats other than the hypermedia and various contemporary mass media. It is worth noting that conspiracy theories and destructive violence have been mediated via mass media such as books, newspapers, folksongs, and graffiti or cartoons for centuries.

However, the Danish Cartoon Crisis has served as a case study for the development of an important argument. This argument throws light on the role of visual media and other forms of design in identity politics and in certain (re)constructions of new identity formations, and thus also platforms for action—some of which might lead to terrorist activities.

We are up against massive social traditions, political trends and cultural formations here. However, it is of the utmost importance to work against the further proliferation of conspiracy theories and of militant fundamentalism. One way of countering such phenomena is via analyses of the means by which they create images of authenticity for themselves. From our own experiences, we know that television commercials do not really work unless we sense some level of authenticity in them, either in terms of the product they sell, in terms of their "cultural capital" (which we might recognize as "good taste"), or because they somehow manage to seduce us by speaking to our needs for identification.

Looking at the Danish material that I have briefly presented here, I would suggest that we might consider hypermedia seduction in a similar light. We need to address the ways in which authenticity is produced and perceived by users, who dream of becoming "authentic" themselves.

References

[1] K.L.G. Salamon & I. Engholm, "Blue is the colour of banking. A design-analytical perspective on norms of web design and the strategic use of creating distinctions by breaking norms." Websites and values - genres and styles as socio-cultural indicators. *Futureground.* Design Research Society. International Conference 17 -21 November 2004, Design Research Society, 2004, p. 13

[2] L.L. Dawson & D.E. Cowan, (Eds). *Religion online. Finding faith on the internet*, Routledge, New York 2004.

[3] H. Mols, "Internetklubber bag livsfarlig leg," *MetroXpress*, Copenhagen, 2006, p. 2.

[4] M. T. Højsgaard and M. Warburg (Editors), *Religion and cyberspace*, Routledge, (imprint of Taylor & Francis Books Ltd), London, 2005.

[5] K. L. G. Salamon, "Faith brought to work. A spiritual movement in business management, Anthropology in Action." *Journal for Applied Anthropology in Policy and Practice* **7**(3) 2000. pp. 24-29

[6] K. Annan, "Statement by the secretary-general at the opening session of the second meeting of the high-level group for the alliance of civilizations", Office of the Spokesman for the Secretary-General. Doha, Qatar, 26 February 2006, 2006.

[7] F. Rose, " Muhammeds ansigt," *Jyllands-Posten*, 2005.

[8] A. Spiegelman, "Karikaturmord," *Politiken*, Copenhagen, 2006, pp. 1 & 6-7.

[9] Wikipedia, "Jyllands-posten muhammad cartoons controversy," 2006.

[10] P. B. Thomsen, "Muhammedkrisen - hvad skete der, hvad har vi lært?", *People's Press*, Copenhagen, 2006

[11] K. Rothstein & M. Rothstein, "Bomben i turbanen", *Tiderne Skifter Forlag A/S,* Copenhagen, 2006.

[12] L. E. Andersen, *Den tabte uskyld*, Syddansk Universitetsforlag, 2006.

[13] D. Trads, "Islam i flammer - danskerne og det muslimske oprør", *Høst og Søns Forlag A/S*, 2006.

[14] T. Seidenfaden & R. E. Larsen, "Karikaturkrisen", *Gyldendals Forlag*, 2006.

[15] R. DR-Nyheder, "Muhammed-krisen dag for dag," DR Nyheder/ Tema, 2006.

[16] B. Mikkelsen, "Kulturminister brian mikkelsens tale", *Konservative Landsråd,* 2005.

[17] M. Featherstone, "The obscure politics of conspiracy theory," *The age of anxiety: Conspiracy theory and the human sciences,* J. Parish and M. Parker (Eds), Blackwell, Oxford, 2001, pp. 31-45.

[18] L. Ø. Andersen, "Det kunstfærdige håndværk," POLITIKEN Copenhagen, 2004, pp. 24-25.

[19] Y. Gabriel and T. Lang, The unmanageable consumer. Contemporary consumption and its fragmentations., SAGE, London, 1995.

[20] F. Alev, "Danmarks forenede cybermuslimer," *Islam.dk*, 2006.

[21] J. Mygind, "Danmarks forenede cybermuslimer," *Weekendavisen,* Copenhagen, 2006, p. 2.

[22] K.L.G. Salamon, "No borders in business: The management discourse of organisational holism," *Cultural capitalism: Politics after new labour*, T. Bewes and J. Gilbert (Eds), Lawrence & Wishart, London, 2000, pp. 134-157.

[23] T. W. Adorno, *The jargon of authenticity*, Northwestern University Press, Evanston, IL, 1973.

[24] J. Baudrillard, *Simulations, Semiotext(e)*, New York, 1983.

[25] J. Derrida, "The end of the book and the beginning of writing," *From modernism to postmodernism*, L. Cahoone (Editor), Blackwell, Oxford, 1996.

[26] P. Bourdieu, "La distinction." *Critique sociale du jugement,* Les editions de minuit, Paris, 1979.

[27] M. F. Brown, *Who owns native culture?* , Harvard University Press, 2003.

[28] S. A. Azzam, "Martyrs: The building blocks of nations," *Religioscope.* Archives., 2001, p. 4 pp.

[29] M. m. R. R. A. Hafs, "Suicide terrorists farewell video. Prepared before suicide attack on December 7, 2004," http://www.pmw.org.il/, 2004.

[30]. Z. Bauman, *Modernity and the holocaust*, Cornell University Press, Ithaca, N.Y., 1989.

[31] P. Heelas, The new age movement, Blackwell Publishers, Oxford, 1996.

Further references:

Beaudouin, V. & Velkovska, J. "The cyberians : An empirical study of sociality in a virtual community," *Proceedings of esprit i3 workshop on ethnognraphic studies in real and virtual environments: Inhabited information spaces and connected communities*, B. K. (Ed.), Edinburgh, 1999, pp. 102-112.

Mols, H. "Internetklubber bag livsfarlig leg," *MetroXpress*, Copenhagen, 2006, p. 2.

Salamon, K.L.G. "Going global from the inside out – spiritual globalism in the workplace" *New Age and globalization*, Mikael Rothstein (Ed.), Renner Publications & Aarhus University Press & Headington, Aarhus/ Oxford, 2001, pp. 150-172.

---, "Prophets of a cultural capitalism: An ethnography of romantic spiritualism in business management", *FOLK Journal of the Danish Ethnographic Society* **44** (2003), 89-115.

---, "Magic business times. Spirituelle entwicklung am arbeitsplatz," *Norm der abweichung*, M. v. Osten (Ed.), Edition Voldemeer /Springer Verlag, Zürich, 2003, pp. 39-55.

---, "Le management par la transformation personnelle – les nouveaux modes de relation au soi comme moyens de pouvoir et/ou de connaissance," *Gouvernement, organisation et gestion: L'héritage de michel foucault*, A. Hatchuel, É. Pezet, K. Starkey and O. Lenay (Eds.), Les Presses de l'Université Laval, Quebec, 2005, pp. 235-252

---, "Possessed by enterprise. Values and value-creation in mandrake management," *Magic, culture and the new economy,* O. Löfgren and R. Willim (Eds.), Berg, Oxford, 2005, pp. 47-55.

Hypermedia Seduction for Terrorist Recruiting
B. Ganor et al. (Eds.)
IOS Press, 2007

Turning the Tables: Harnessing Media Means to Counter Radicalization

Michael E.G. CHANDLER[1]

International Consultant on Terrorism, Security and Related Matters
Member of the International Advisory Councils to the International Institute for Counter-Terrorism (ICT), Herzliya, Israel and the International Centre for Political Violence and Terrorism Research (IDSS), Singapore

Abstract. The theme of this publication is "Hypermedia Seduction for Terrorist Recruiting", which concentrates on the role of the Internet—specifically, the use, abuse or misuse of the Internet by terrorist groups, their supporters and sympathizers. However, television broadcasting plays a complementary role. This chapter looks at how the broadcast media are exploited, and how—without the freedom of the press being compromised—such media could fulfil a more positive and supportive function in countering the radicalization which currently subverts Muslim youths. This subversion of youth sometimes reaches the point where youths are prepared to go as far carrying out suicide attacks against the incorrectly-perceived enemies of Islam.

Keywords. Television, broadcasting, United Nations, world community, radicalization, media, responsibility, Islamist, extremism.

Introduction

The attacks in London on 7 July 2005 presented modern-day Europe with one of the worst of all terrorist threats, namely that from suicide bombers. To add to the pain and horror of that day, these were not Islamist terrorists from North Africa or the Middle East; these were British subjects, three of whom had been born and raised in the United Kingdom, albeit the sons of immigrant parents of Pakistani origin.

Then, on 10 August 2006, the authorities in London announced that the Security Services and the police had foiled an alleged plot in which improvised explosive devices (IEDs) were to have been detonated on trans-Atlantic flights. The IEDs were to have been disguised as liquids and smuggled on board five to ten flights which were to depart from UK airports. If these allegations prove to be true, then this intended terrorist atrocity would have involved at least another ten to twenty young British Muslims committed to supporting the would-be suicide bombers.

---

[1] Michael Chandler was Chairman of the former UN Security Council Monitoring group dealing with sanctions against Al-Qaida and the Taliban. He is the author, along with Dr Rohan Gunaratna, of *Countering Terrorism – Can we meet the threat of Global Violence?* [1]

As if this was not enough confirmation of the way in which this particular threat is evolving, on 9 November 2006, the Head of MI5, Dame Eliza Manningham-Buller, gave yet further chilling details. In a speech to the Mile End Group in London she stated that the Security Service was aware of thirty suspected plots, and 200 groups or networks involving about 1,600 individuals in the UK connected with Islamist extremist terrorism [2].

The prime reason being put forward for Muslims turning against their fellow citizens, and being prepared to undertake mass murder in London's morning rush-hour and over the Atlantic or US cities is radicalization. Although Al-Qaida's first attempted attack in the UK took place in the year 2000, the current radicalization stems from the movement's reaction to more recent events:

- the foreign policies of Mr Blair, who stands alongside Mr Bush and the US in his so-called "war on terror";
- the invasions of Afghanistan and Iraq;
- the purported "occupation" of Muslim lands, which in this context is perceived as a direct attack on Islam (or at least that is how those responsible for spreading the "rhetoric of radicalization" present their case).

One of the most interesting and important aspects of this Islamist indoctrination process is the means by which the rhetoric is spread. The use of the Internet by terrorist groups to further their causes has been recognized for some years. During that time, the phenomenon has been studied in detail by small groups of individuals who have sensed and understood the enormity of the problem. It is only more recently that a much greater understanding and more widespread study and analysis of the subject has become more commonplace in a number of circles. These include academia, intelligence communities, security and law-enforcement services, armed forces, and other counterterrorism operatives.

The exploitation by terrorist groups of modern global communications systems and facilities has not only kept pace with advances in technology, but in some instances appears to have been ahead of the forces of law and order and their supporting elements. Notwithstanding these developments—and despite a natural tendency on the part of many to see the various ways in which the Internet has been and is still being exploited to the terrorists' advantage—it is important to keep these advances in perspective. There are still many parts of the Muslim world where there are no internet cafés, nor even mobile or cell-phone coverage, but where there is television coverage.

Such television coverage is not necessarily via satellite. It may be confined to one national or government-sponsored channel. But those who live in these remote areas tend to gravitate to the places where news (and possibly the means to watch football matches) is available. Consequently, it is important to take into account the impact of the other electronic media, in tandem with the Internet, when analyzing the effect on terrorist recruiting and radicalization processes. It is essential to see how one media format complements the other, and how they feed on one another.

Despite the use by a variety of terrorist groups of the television—an everyday part of contemporary life, this chapter will focus only on those aspects of the subject which are related to Islamist extremism, or to terrorist groups associated with this strain of fundamentalism.

1. Television and Terrorism

Television is used extensively, not only by political leaders and government officials in their attempts to keep the public abreast of the terrorism threat with which they are faced, but also by the terrorists themselves. Whereas politicians and government members will make every effort to portray the terrorists in a criminal context, the terrorists exploit the medium in a variety of different ways.

In this age of global satellite television, terrorists can rely on prime time, worldwide coverage—free of charge—of any attacks or other events which they may stage. Some terrorists have learnt to exploit statements made by enemy politicians and their spokespeople to their own advantage. Continual references to the "war on terror", "the global war on terror" and an "axis of evil" have achieved little. If anything, these phrases have merely produced widespread skepticism, and not only amongst Muslims. Such political statements, which are intended both for home consumption and as a message to terrorists and their supporters and sympathizers, have sometimes become double-edged swords. Punchy, hard-hitting sound-bites like these have a tendency to be turned on their heads by the rhetoric of the terrorist leadership.

Sending a message that countries in the West intend to be tough on terrorism is crucial. However, it is also important to be sure that one can deliver effectively on such promises, if credibility is not to be lost; concurrently, the door must be kept ajar for an effective "hearts and minds" campaign. Such campaigns need to be structured to woo the support of the majority of the Muslim population away from the extremists, towards moderation and support for a more peaceful coexistence between differing cultures.

To achieve this, it is important to "accentuate the positive"—a theme which will be repeated throughout this discourse. Straightforward countering of terrorist rhetoric will often play into the hands of the terrorists, and may been seen by target audiences as pro-Western, anti-Islamic propaganda, even though it is directed only at Islamist extremists.

Over the years, the "hearts and minds" approach has proved to be a key ingredient in the successful conclusion of counter-insurgency campaigns; one ignores this important facet at one's own peril. The same approach is equally important and applicable in the implementation of counterterrorism strategies. Knowing and understanding how to tackle the "hearts and minds"—not so much of the terrorists themselves, but of those to whom they look for support, the communities they try to indoctrinate with their rhetoric of hate, and those groups of individuals from which they draw new recruits—is even more important in asymmetric warfare.

In a counterterrorism environment, the "hearts and minds" approach does not end with the ways in which the various services and agencies conduct their "kinetic" operations and deal with the target communities. It has to go much deeper, into the clever and subtle use of the media. In so doing it must counter terrorist abuse of the Internet, and get the right message broadcast over public media, in particular television. In developing such a media strategy, it is of prime importance to keep in mind the ways in which terrorists have developed their usage of the Internet, and thus the possibilities for complementing and undermining this use and abuse of hypermedia.

Some terrorist groups have their own television channels, such as the Lebanese Hezbollah with its *Arabiya TV*. This was used to great advantage by Hezbollah during its summer 2006 confrontation with Israel. Here again we see the corollary effect between the group's own TV channel and the websites associated with it. Other groups, like Al-Qaida, make extensive use of existing channels such as the Qatar-based *Al-Jazeera.* Although *Al-Jazeera* existed well before 9/11, it was the events of that day that catapulted the channel into the international limelight.

The kind of notoriety *Al-Jazeera* enjoys raises an important question, namely the role of the media with regard to terrorism. Should the media be free to promote terrorism, either wittingly or unwittingly; or should it assist only in countering the effects of terrorism—without, of course, compromising the freedom of the press?

It appears that *Al-Jazeera* holds exclusive rights to screen audio and visual coverage of pronouncements made by Usama bin Laden and his deputy Ayman al-Zawahiri. Doubtless, if *Al-Jazeera* was to stop acting as the mouthpiece for the remnants of the original Al-Qaida leadership, another broadcaster would be happy to scoop up the baton. In any event, Al-Qaida publicity would still get broadcast over the World Wide Web. In fairness to *Al-Jazeera,* on some occasions that is exactly what they—along with numerous other outlets—report; namely, a repetition of what has already been sent out on an Al-Qaida website.

Evan Kohlmann, in a program published on MSNBC television, provided a simple explanation of how video tapes featuring bin Laden and his deputy al-Zawahiri are recorded, "polished" for presentation, and then distributed [3]. Kohlman's article included a simple interactive presentation of the process. Reference to the article can also be found on the *Counterterrorism Blog* [4], a US-based medium that provides a timely, comprehensive and informative view of global terrorism; this includes news, analysis and broad commentary.

The exploitation by terrorists of the World Wide Web notwithstanding, it appears that on those occasions when *Al-Jazeera* is the exclusive broadcaster of Al-Qaida's messages, the channel promotes terrorism and the actions proscribed by the terrorist leadership. It knowingly broadcasts anti-Western and anti-Israeli propaganda on behalf of Al-Qaida. Together with broadcasting such hate rhetoric, it also calls on Muslims to take up arms, and generally acts as the mouthpiece of Al-Qaida.

These facts all bring into question the responsibility of the media. Freedom of speech notwithstanding, this particular role that is being played by *Al-Jazeera* is stretching the bounds of tolerance. It is one thing for a terrorist organization to have its own television channel, which can be dealt with accordingly via electronic means if it oversteps the mark of internationally acceptable standards of broadcasting. But for an international channel to be seen to be promoting terrorism is a different "ball-game" altogether. The issue of how the international community or elements of it—and the media is a very important element—play their different roles, in terms of responding to the threat from trans-national Islamist terrorism, highlights the question of media responsibility.

2. Media Responsibility

A more positive approach to problems such as those surrounding *Al-Jazeera* would be a clear, firm collaborative effort on the part of all international media to deny Al-Qaida or any other terrorist groups the airtime they seek. Hate-filled and bigoted extremists, who bring their own religion into disrepute, need to be seen for what they are and ignored. If states and key public individuals are genuine in their outright condemnation of terrorism *per se,* they need to—metaphorically speaking—"put their money where their mouth is" and ban terrorist propaganda from the airwaves.

This in itself would start sending a clear signal from the international community that terrorism will not be tolerated. Denying terrorist groups this "free" means of spreading their propaganda, whilst not necessarily silencing them, would at least send the correct signal to them and their supporters. That message would state clearly that "you don't feature in the modern world". Inciting hatred and calling upon people from across the Muslim world to take up arms against those who do not conform to their dogma, namely a politicized Islam, cannot be considered as acceptable forms of either freedom of speech or freedom of expression. It has to be outlawed.

On the subject of media responsibility, and in the context of Islamist propaganda, it should be not be too much to expect—nor should it be unrealistic to demand—that coverage of events be balanced and accurate. Tragically, this is not always the case. An imbalance was demonstrated, for example, in the coverage of the July/August 2006 conflict in Lebanon between Israel and Lebanese Hezbollah, as well as during the aftermath of that conflict. The "doctoring" of images by a Reuter's photographer, in such a way as to portray the results of Israeli air strikes as being far more devastating than they in fact were, is now well documented.

Reporting by other, erstwhile respected international television news channels was also deemed by many to be biased in favour of Hezbollah. In addition to the incidents involving tampering with photographic and television images, there were occasions when casualty figures were grossly exaggerated. This misreporting was inevitably biased to influence international opinion against Israel.

The subject of embedded reporters raises further questions. There will always be journalists who hanker after a "whiff of cordite", who will want to be right in the front line, where the action is, because that is the nature of their work. However, if, in the interests of transparency, the practice of embedded reporters is to be allowed—as it was during the initial stages of the invasion of Iraq by Coalition forces in 2003—then there are certain standards that need to be observed. Such standards should be upheld in the interests of consistency in reporting the facts, as a strong antidote to extremist propaganda, and in keeping with the development of a sound, effective "hearts and minds" campaign.

In the heat of the moment, it may be easy to forget how the specter of Muslims dying at the hands of "unbelievers" or "infidels" is likely to be viewed in Muslim countries, Islamic states and even amongst Muslim diaspora in Western countries. No matter how righteous the cause, a less "gung-ho" and more sensitive approach might be better; the battle for the hearts and minds of those people whose support is being sought can all too easily be lost. It is crucial, both in short- and long-term counterterrorism strategies, that the "hearts and minds" campaign remains a high priority in the de-radicalization process.

3. Removing the fuel from the fire

The situation of conflict between Israel and the Palestinians has been continuing for far too long, as has the unresolved dispute in the Kashmir region. The resolution of these two conflicts, in particular the former, would remove from television networks some of the images which have strengthened anti-Western sentiments on the part of Muslims in many parts of the world. This will be no easy task. But as we saw during July/August

2006 in Lebanon (and as just mentioned), the propaganda harvest that terrorists can reap from these situations is significant, and easily plays into the hands of the Islamist extremists worldwide.

4. Revaluing the Currency of the International Community

The conflict in Lebanon highlighted yet another deficiency in the de-radicalization process. Clearly, a comprehensive and coordinated approach to that upsurge in cross-border violence was missing within the international community. The speed with which the United Nations Security Council responded was lamentable. So, too, was that of the Arab League.

It is one thing for the latter organization to have gathered in emergency session and condemned the Israeli response, and then to have called upon the UN to respond as a matter of expediency. But in terms of anyone condemning the firing of barrages of highly lethal rockets into Israeli towns and cities on a daily basis, the silence was deafening. Once again, these responses emphasized the international community's lack of "currency".

It is this kind of credibility that requires urgent revaluation. Renewed credibility would improve the stature of the United Nations in its key role of ensuring peace and security in the world. If the "world body" was seen to act with firmness, to be resolute in spite of significant national interests, and do so in a fair and unbiased way, this would also help to defuse the extremists' efforts. The positive publicity thus generated would naturally counter the radicals' rhetoric.

5. Accentuate the Positive

Despite the somewhat negative tone of parts of this chapter, certain events and situations indicate that there are positive aspects to the question of radicalization. Unfortunately, good news does not sell prime time television. Tragically, broadcasters seem to be of the opinion that viewers are more ghoulishly attracted to "bangs, bullets, bodies and blood." Consequently, when there is good news, it invariably comes in the form of a documentary, the majority of which tend to be shown during off-peak times, or on days when large numbers of potential viewers are away from their sets (e.g. at weekends).

Nonetheless, if the producers of television news were to stop bombarding their audiences with blood, body parts and blown-up vehicles, the majority of viewers might actually appreciate a more positive information flow. This does not mean that there should be a misrepresentation of the situation on the ground. Coverage should always be factual. However, much of the good that is being done or has been achieved in recent years—in myriad conflict situations—gets ignored in favour of the failures. The question is: how can such a significant change in the media be achieved?

6. International Collaboration

On 8 September 2006, the UN General Assembly (GA) adopted a new Counter Terrorism Strategy [5]. Although the GA still has not been able to agree on a universal definition of terrorism, this Strategy did contain one clause that is relevant to the current discussion. It deals with countering terrorism on the Internet, and reads as follows: "...to explore ways and means to (a) coordinate efforts at the international and regional level to counter terrorism in all its forms and manifestations on the Internet, and (b) use the Internet as a tool for countering the spread of terrorism..."

Despite the adoption of this Strategy not being binding on member states, the above clause does provide a framework within which states can establish the means to cooperate and even collaborate with each other, against the use, abuse and misuse of the Internet by terrorist groups and their sympathizers.

A similar initiative, in terms of bringing the appropriate government departments together regarding the management of television channels, might identify a further way forward in denying the terrorists the airtime they currently enjoy. This could be done without compromising the freedom of the press. Unfortunately, this idea was not included in the above Strategy, but it could nonetheless be developed regionally.

7. Conclusions

The role played by television in "hypermedia seduction for terrorist recruiting" remains significant. Despite dramatic technological advances in global communications over the past decade, television is still the only means of news available to many people. As such, it is an invaluable propaganda tool for terrorists, and this is not restricted to underdeveloped countries in the "global South".

Consequently, the potential of television needs to be harnessed in the fight against Islamist extremist terrorism, to the extent that it becomes an integral component of national and regional counterterrorism strategies. Television broadcasts are a key ingredient of any successful "hearts and minds" campaign, and programs need to be designed in a clever way so as to counter the current trends in radicalization.

References

[1] Chandler, M. & Gunaratna, R. *Countering Terrorism—Can we meet the threat of Global Violence?* Reaktion Books Ltd, London, November 2006.

[2] http://www.timesonline.co.uk/article/0,,2-2447690.html "International Terrorist Threat to the UK", Eliza Manningham-Buller, Director-General MI-5, November 9, 2006

[3] http://msnbc.msn.com/id/13842265/ , "American greases al-Qaida media machine", 14 July 2006

[4] http://www.counterterrorismblog.org/

[5] http://daccessdds.un.org/doc/UNDOC/GEN/N05/504/88/PDF/N0550488.pdf?OpenElement

Author Index